AF892955

Springer Tracts in Mechanical Engineering

Series Editors

Seung-Bok Choi, College of Engineering, Inha University, Incheon, Korea (Republic of)

Haibin Duan, Beijing University of Aeronautics and Astronautics, Beijing, China

Yili Fu, Harbin Institute of Technology, Harbin, China

Carlos Guardiola, CMT-Motores Termicos, Polytechnic University of Valencia, Valencia, Spain

Jian-Qiao Sun, University of California, Merced, CA, USA

Young W. Kwon, Naval Postgraduate School, Monterey, CA, USA

Francisco Cavas-Martínez, Departamento de Estructuras, Universidad Politécnica de Cartagena, Cartagena, Murcia, Spain

Fakher Chaari, National School of Engineers of Sfax, Sfax, Tunisia

Francesca di Mare, Institute of Energy Technology, Ruhr-Universität Bochum, Bochum, Nordrhein-Westfalen, Germany

Hamid Reza Karimi, Department of Mechanical Engineering, Politecnico di Milano, Milan, Italy

Springer Tracts in Mechanical Engineering (STME) publishes the latest developments in Mechanical Engineering - quickly, informally and with high quality. The intent is to cover all the main branches of mechanical engineering, both theoretical and applied, including:

- Engineering Design
- Machinery and Machine Elements
- Mechanical Structures and Stress Analysis
- Automotive Engineering
- Engine Technology
- Aerospace Technology and Astronautics
- Nanotechnology and Microengineering
- Control, Robotics, Mechatronics
- MEMS
- Theoretical and Applied Mechanics
- Dynamical Systems, Control
- Fluids Mechanics
- Engineering Thermodynamics, Heat and Mass Transfer
- Manufacturing
- Precision Engineering, Instrumentation, Measurement
- Materials Engineering
- Tribology and Surface Technology

Within the scope of the series are monographs, professional books or graduate textbooks, edited volumes as well as outstanding PhD theses and books purposely devoted to support education in mechanical engineering at graduate and post-graduate levels.

Indexed by SCOPUS, zbMATH, SCImago.

Please check our Lecture Notes in Mechanical Engineering at https://link.springer.com/bookseries/11236 if you are interested in conference proceedings.

To submit a proposal or for further inquiries, please contact the Springer Editor **in your region**:

Ms. Ella Zhang (China)
Email: ella.zhang@springernature.com
Priya Vyas (India)
Email: priya.vyas@springer.com
Dr. Leontina Di Cecco (All other countries)
Email: leontina.dicecco@springer.com

All books published in the series are submitted for consideration in Web of Science.

Fang Chen · Jacques Terken

Automotive Interaction Design

From Theory to Practice

机械工业出版社
CHINA MACHINE PRESS

Springer

Fang Chen
Changsha Automotive Innovation Research Institute (CAIRI)
Changsha, China

Jacques Terken
Department of Industrial Design
Eindhoven University of Technology
Eindhoven, The Netherlands

ISSN 2195-9862 ISSN 2195-9870 (electronic)
Springer Tracts in Mechanical Engineering
ISBN 978-981-19-3447-6 ISBN 978-981-19-3448-3 (eBook)
https://doi.org/10.1007/978-981-19-3448-3

Jointly published with China Machine Press, Beijing, China
The print edition is not for sale in China Mainland. Customers from China Mainland please order the print book from: China Machine Press.
ISBN of the Co-Publisher's edition: 978-7-111-68718-4

© China Machine Press 2023
This work is subject to copyright. All rights are solely and exclusively licensed by the Publisher, whether the whole or part of the material is concerned, specifically the rights of reprinting, reuse of illustrations, recitation, broadcasting, reproduction on microfilms or in any other physical way, and transmission or information storage and retrieval, electronic adaptation, computer software, or by similar or dissimilar methodology now known or hereafter developed.
The use of general descriptive names, registered names, trademarks, service marks, etc. in this publication does not imply, even in the absence of a specific statement, that such names are exempt from the relevant protective laws and regulations and therefore free for general use.
The publishers, the authors, and the editors are safe to assume that the advice and information in this book are believed to be true and accurate at the date of publication. Neither the publishers nor the authors or the editors give a warranty, expressed or implied, with respect to the material contained herein or for any errors or omissions that may have been made. The publishers remain neutral with regard to jurisdictional claims in published maps and institutional affiliations.

This Springer imprint is published by the registered company Springer Nature Singapore Pte Ltd.
The registered company address is: 152 Beach Road, #21-01/04 Gateway East, Singapore 189721, Singapore

Preface

This book is a revised version of an earlier book that we wrote in Chinese for the Chinese market and which was published in 2021. We have studied similar areas, and we have been friends for many years. When we attended a conference in China in 2019 together, Fang Chen, who started working as a consultant for Huawei in 2016 mentioned that, while the Chinese automotive market is booming, the field of automotive HMI design is still lagging behind, and people working in that area face considerable challenges. They face the lack of understanding from other people involved and often lack basic knowledge, systematic education or the support of basic research. We concluded that it would be useful to have a book to support practitioners in the area of automotive HMI design and that such a book could also be used as part of training programmes for future automotive HMI designers. The next day we had the outline of the book. We discussed the outline together, made changes to the details and also divided the work.

Originally, we decided to write the book for people working in automotive interaction design in China. We wanted the book to provide them with enough basic knowledge, enough theory and methods to give them the tools to do what they need to do. Once the book was available, the idea arose that the book might also be useful to readers who are not familiar with Chinese, so we decided to produce an English version. In doing so, some parts that were specific to the Chinese market were removed, and topics that were not included in the Chinese version but still were considered relevant were added.

The title of this book integrates three different elements. It is a book about the automotive Human–Machine Interface. While the term User Interface is more common in the general domain of Interaction Design, we use the term Human–Machine Interface (HMI) because that term is more common in the domain of automotive engineering. So, it is a book about how people interact with and control the functions that are available in vehicles. In the second place, it is a book about design. That means it is not primarily a book providing an introduction into scientific theories about the interaction of people with technology. Although the body of scientific knowledge is highly relevant to the design of the HMI and needs to be part of any book about automotive HMI design, the design methodology is equally important. In the third place, we take a human-centred approach. In the engineering domain, technical considerations, market share and consumer value are often dominant, resulting in featurism. However, in our opinion, the design process should take into consideration how people actually use their vehicle and the functions available in the vehicle and should be organised such that this human-centred focus is effective right from the start of the design process.

The book consists of five parts. Part I sets the stage and introduces the field of Interaction Design. The main disciplines contributing to Interaction Design are listed, and main terms are introduced. Part II summarises the body of knowledge that is relevant to automotive HMI design. This consists mainly of knowledge about how people perceive and understand information and use this information to decide and act in the world. However, there is more to driving than information processing, and therefore, Part II also considers theoretical insights in the area of user experience. That is, it considers driving behaviour and its psychological foundations. Finally, it summarises more recent work about how driver behaviour may be influenced and changed. Part III summarises theoretical frameworks in the area of interaction design and prevailing interaction techniques that are applied in the design of user interfaces. Part IV contains an introduction into the methodology of designing automotive user interfaces. It introduces how the design process is structured and describes methods for the different stages of the design process. Also, it goes into platforms for evaluation of design concepts, in particular driving simulators. Finally, Part V discusses developments in the area of automated driving and the associated Human Factors issues.

The book is meant primarily for people who work in the area of automotive HMI design. While we acknowledge that there are many people working in this area who have solid training in the field of cognitive psychology and research methodology, who may consider parts of this book rather superficial, our experience is that few people span all fields relevant to interaction design, in particular, both the fields of research and design. Furthermore, while there are books about interaction design and about applied cognitive psychology/cognitive engineering, none of these books focuses on the automotive domain. Our aim is to bring the different fields together and present a book that focuses on the automotive domain.

This book is *not* a thorough introduction into the different sub-fields of knowledge relevant to automotive interaction design. Within the scope of a book like this, may topics that are relevant to design and design research can only be introduced, and for in-depth treatments, the readers need to consult the literature and courses in university. Furthermore, while driving is a social activity, when summarising psychological theories, we do not go into social psychology. Also, it is *not* a book about topics such as electric vehicles, car sharing or insurance of automated driving systems, or a book about societal issues as sustainability. Likewise, we do not go into areas like transport modelling and traffic infrastructure. Yet, we hope that this book is a helpful introduction to the main fields of expertise in the domain of automotive interaction design.

Changsha, China
Eindhoven, The Netherlands
February 2022

Fang Chen
Jacques Terken

Contents

Part III Interaction Design/UI Design

Part IV Design Process and Research Methodology

Part I
Interaction Design for Automotive

Chapter 1
Introduction

Abstract As an introduction to the book, in the first chapter, trends in automotive innovation such as advanced driver assistance systems and automation are considered, and it is argued that such developments create challenges for designers to create systems and associated user interfaces that customers understand and appreciate. It is concluded that a human-centred design approach is needed to satisfy the needs of the customers.

The car has been used as a means of transport for over 100 years, with the driver manipulating the steering wheel and pedals to make the car go the way s/he wants it to, a process that relies heavily on individual driving skills. Rapid advances in sensor and computer technology have led automotive design engineers to explore ways in which new technologies can assist in driving tasks and even automate them, with the first attempts to do so beginning in the 1980s when CMU (Carnegie Mellon University, USA) played a pioneering role. Today, assistance driving technology has come a long way and has matured to the extent that most cars have one or more assisted driving systems as standard, such as (Adaptive) Cruise Control ((A)CC), Lane Keeping Assist (LKA), Collision Warning, etc., Furthermore, many companies are offering initial automated driving software, and are doing on-road and real-world driving tests of higher levels of automated driving.

The main motivation for working with automated driving is the safety of traffic. In the United States, the number of traffic deaths in 2015 was 55,000. In Europe, 25,000 people died in traffic accidents in 2017. In China, more than 63,000 people were killed in traffic accidents in 2017. Analysis of the causes of accidents shows that the vast majority of accidents are caused by human error (drunk driving, speeding violations, bad judgement, distractions are the most important causes). It is therefore widely believed that accidents caused by human error could be eliminated if drivers were better assisted, or even taken out of the driving task altogether, allowing people to get out of the human–machine-road driving loop (out of the loop), thus significantly reducing the number of accidents and fatalities. Other motivations and other arguments for the development of autonomous vehicle research include convenience (automation enables people to engage in other activities while in a car), comfort (automation can be adjusted to occupant preferences), connectivity (automation and

© China Machine Press 2023

F. Chen and J. Terken, *Automotive Interaction Design*, Springer Tracts in Mechanical Engineering, https://doi.org/10.1007/978-981-19-3448-3_1

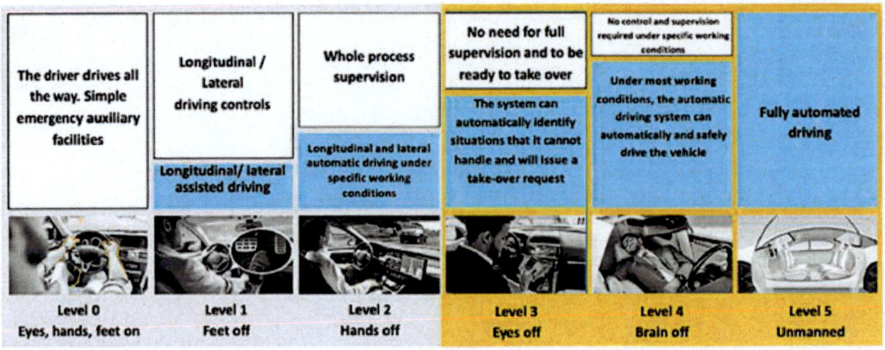

Fig. 1.1 Six-level classification of automated driving systems proposed by SAE International. The white boxes represent the driver's task, the blue boxes represent the system's task

connectivity to the internet allow better adaptation to traffic conditions, various road environments), sustainability (automated systems will drive more smoothly, avoid unnecessary acceleration and deceleration, and reduce energy consumption) and mobility-for-all (driving automation will make auto-mobility accessible for people who are traditionally unable or not allowed to drive, such as intoxicated people, the elderly, young people and the visually impaired).

Automating car driving is a complex process that cannot be achieved overnight. The control part of the vehicle may be relatively easy, and currently self-driving vehicles can perform well under controlled conditions and in confined areas, but the dynamic nature of the real world (traffic conditions, infrastructure, other road users' behaviour and weather conditions which can change unpredictably at any time) increases the complexity of the driving task by an order of magnitude. In order to standardise the discussion, the Society of Automotive Engineers (SAE) has proposed (and subsequently revised) a six-level automation classification in their standard report J3016 (SAE 2021), which has been adopted worldwide and replaced other classifications such as those proposed by BASt (Germany) and NHTSA (USA). Figure 1.1 shows a simple interpretation of this classification from a human factors perspective.

- Level 0 (for classification completeness), called No Driving Automation, is fully manual driving. Driver support features such as Automatic Emergency Braking, Blind Spot Warning and Lane Departure Warning may be available, but the driver is responsible for performing all dynamic driving tasks.
- At level 1, called Driver Assistance, at most one automated driving feature is available, typically adaptive cruise control or lane centring, to support the driver's task, but the driver is driving.
- At level L2, called Partial Driving Automation, (at least) two automated driving features are available, most commonly adaptive cruise control (longitudinal control) and automated lane centring (lateral control). This level of automation is characterised by the fact that, although the vehicle can drive automatically

in relatively simple traffic conditions, for example on motorways under normal conditions without any unexpected events, the driver needs to be aware that s/he is the driver, even if the feet are off the pedal and s/he is not steering. The driver needs to monitor the vehicle's behaviour and the traffic situation continuously to identify situations which the system cannot handle and take over the actual driving again.
- Level L3, called Conditional Driving Automation, indicates that the system itself can handle the task of driving, but only under specific conditions, such as normal conditions on the motorway. However, unlike level 2, where the driver needs to identify which situations the system cannot handle, at level 3 the automated driving system itself can identify situations it cannot handle, in which case the system asks the driver to regain control of the driving (allowing the driver a certain response time). Thus, when the vehicle is in L3 automated mode, the driver does not need to monitor the vehicle's behaviour and traffic conditions and can engage in non-driving related activities, until the system issues a Take-over Request. Regaining control typically requires the driver to perform an action such as pressing two buttons (to prevent accidental interruption of the automated driving system by accidentally hitting a pedal).
- Level L4, called High Driving Automation, is similar to Level 3, with the difference that the system is able to handle more road conditions and therefore more scenarios in which it can drive automatically. Also, if the driver does not respond to the system's request to regain control (for example, if the driver has a heart attack), the vehicle's automated system is able to execute a risk mitigation strategy, e.g., driving the vehicle safely to the emergency lane, bringing it to a halt and automatically notifying the emergency services.
- At level L5, called Full Driving automation, the vehicle can handle all kinds of road conditions that may be encountered while driving. At this level of automated driving, primary controls such as pedals and steering wheel may no longer be available. The driver is not required to monitor the driving, nor is there any situation where the driver is required to take over control, so there is no need for the "driver" to be qualified to drive, and the vehicle can even be driven without a driver.

The development of Internet of Things (IoT) technology makes a further contribution to assisted and automated driving by connecting vehicles to other vehicles (V2V) and to the infrastructure (V2I), bringing many benefits to traffic management and individual vehicle driving. For example, it is possible to collect large amounts of data on traffic flows via the IoT, which can be used to provide navigation guidance for individual vehicles, enable traffic managers to dynamically open and close lanes when required, and provide decision makers with information on traffic and traffic density. V2V connections allow vehicles to be notified of upcoming anomalies on the road ahead and in vehicles, and to coordinate dynamic control of intersections. V2V also plays a major role in following vehicles.

The development of these technologies creates a vast design space and at the same time raises two important questions: (1) Which are the technologies that meet the needs of society and individual users: Which assistance systems should be developed, what should their functions be, and how should automated driving systems be developed? (2) How will we design and present these technologies so that users understand the technology, understand how to use it, and do not begin to misuse or abandon it? How can we design it so that users get the most value out of it?

This requires a human-centred design approach, which ensures that the needs, preferences, abilities and limitations of users are taken into account in the design process, with the aim that the outcome of the product innovation process, through interaction design, will better meet the needs of users. In this context, the term 'user' also includes customers, drivers and sometimes other road users. When developers start to apply relevant and especially innovative technologies, failure to consider the needs of the user can lead to unintended or even counterproductive results. A concrete example is Tesla's Autopilot (automation level 2: a combination of adaptive cruise control and automatic lane keeping). As it is a Level 2 system, the user is required to hold the steering wheel at all times and monitor the vehicle's behaviour, so that s/he can take over driving control in the event of an exceptional situation. However, people often misunderstand this system and think that they can stop monitoring and engage in other activities. The authors' interviews with Tesla drivers in China and the Netherlands show that most drivers who observe that the system works as expected, actually treat it as a level 3 system and start engaging in non-driving related activities (texting, playing games on their smartphones). Another example is that if the autopilot does not meet their current needs (for example, slowing down when passing an intersection or quickly rushing through a yellow light, or displaying braking behaviour due to the unexpected presence of pedestrians), people may consider the system poorly designed and turn off the autopilot. A third example is that people may not understand how the system works. A specific case in point is adaptive cruise control, which is available in two versions: full speed range and a conventional version (which only operates at speeds above 30 km/h). An acquaintance of ours drove a car equipped with ACC and, because he found it so convenient, he asked for a car with ACC when his own car had to go to the dealer for maintenance. Not being aware of the existence of two different forms of ACC, he assumed that ACC would always work across the whole speed range, since that was what he was used to. However, when encountering a traffic jam that drove at a speed below 30 km/h, he discovered that the rental car's ACC worked only at speeds over 30 km/h … Should we expect this person to read the exhaustive manual before starting to use the system? Probably not. Another example is the acceleration performance of ACC systems, which is usually considered to be slow. Another acquaintance of ours solved this problem as follows. He found out that the amount of acceleration depends on the difference between the actual speed and the target speed: the bigger the difference, the higher the acceleration. So, what he does is setting the cruising speed on the motorway to 180 km/h. As he says, his car doesn't actually speed up to a speed of 180 km/h, since there are always cars ahead of him travelling at 120 km/h. However, at one occasion he had to get off the highway ramp; as there was no other traffic at the exit, he found that this method posed a significant risk … These cases show that when people use the new technology in question, they may use it in a way that is very different from what the designers expected. Or they may not even be aware that a particular assistance system is available, so that the targeted benefits are not realised. A recent Dutch study showed that many people were unaware that their vehicle was offered with an assisted driving system.

Other examples of how automated driving technologies may not meet people's abilities or preferences are discussed later in the book (in particular, Chap. 16). As the functionality offered by assisted/automated driving will become more complex, inferring from the above examples, we may assume that there will be an increasing difference between what designers believe will meet user preferences and abilities on the one hand and how people actually understand and use the technology on the other. Due to the rapid development of technology, it seems difficult to obtain information from and about users that helps to give guidance to the design process.

The book contains 16 chapters, divided over five parts. Part I introduced the challenges and the field of Interaction Design. Part II covers the basics of cognitive psychology required for the design of automotive interactions. Multimodal interaction technologies such as speech, graphics and haptics are discussed, as well as the application of intelligent technologies in the creation of predictive interfaces. Part III covers theories related to automotive interaction design. Part IV covers different design processes, especially human-centred design processes, and various design methods and testing methods. To enhance the study of vehicle interaction design, Chaps. 14 and 15 are devoted to issues of driving simulators and the basics of experimental methodology. Finally, in Part V we discuss human factors issues at different levels of autonomous driving, exploring both the challenges that arise at different levels of automation and the challenges that arise as autonomous vehicles will interact with other road users, and categorise various driving scenarios and possible events in order to design how to provide intelligent assistance to drivers.

The main audience for this book is those working in the automotive industry who are involved in automotive design. It can also be used as a textbook for courses taken by students in automotive colleges. It contains some of the basic concepts and theories of interaction design. The examples used and the theoretical and interaction design methods chosen are essential for automotive design. This is a basic textbook for interaction design. For those who need an in-depth understanding of interaction design in cars, especially when it comes to the design of a specific system, further references to other literature and studies are required.

Reference

SAE (2021) Taxonomy and definitions for terms related to driving automation systems for on-road motor vehicles. J3016_202104. https://www.sae.org/standards/content/j3016_202104/

Chapter 2
Introduction to Interaction Design

Abstract In this chapter, the field of interaction design is introduced, and the different disciplines involved in interaction design are considered. Main concepts such as usability and user experience are defined, and design objectives for usability and user experience are considered.

Automotive design has been around for a hundred years. Throughout the decades, the greatest focus has been on the technical aspects of the car. Interaction design in automobiles, on the other hand, has not been raised to a high level. This is because, compared to the rest of the industry, automotive systems were too simple and therefore the challenge for interaction design was not very strong. But with the advent of assisted driving systems and the development from partially to fully automated driving, the technology has become more platform-based, creating new and unprecedented challenges, and proper interaction design has gained importance. In the last two decades in particular, interaction design in cars has attracted increasing attention.

It is always a challenge to get the interaction design right in the car. Generally speaking, we divide in-car systems into two parts. One part is directly related to driving and the other part is not. The part that is not directly related to driving includes various infotainment systems, etc. In between there is the in-car navigation system, and various functional systems associated with geographic location, which are not related to the driving operation itself, but to the car as a means of transport, since modern people's driving has become increasingly dependent on electronic navigation systems. In the future of driving, more and more infotainment systems will be introduced into the car. The personalised, intelligent design of the car is the way forward and a challenge for the future.

In relation to automated driving, the different levels of automation, in terms of the technical performance of the car, are divided into six levels (Fig. 1.1), but for interaction design, we are really only concerned with two states: the car is driving itself or a human is driving. From L0 to L2, the car is driven by a human, L3 is a car that can drive partially automatically, but in many road conditions, the car is still driven mainly by a human, and at L4, the car is basically able to drive automatically in nearly all road conditions, but a human can take over the driving. At L5, the car is basically driverless and can perform all driving tasks. This shows that at different

© China Machine Press 2023

F. Chen and J. Terken, *Automotive Interaction Design*, Springer Tracts in Mechanical Engineering, https://doi.org/10.1007/978-981-19-3448-3_2

levels of automation, the role played by humans in the whole driving process is different and the interaction design will face different problems.

2.1 Misconceptions and Challenges of Interaction Design

Interaction design is often misunderstood as just the design of the user interface, a part of art design, and therefore many people think that the focus of interaction design is aesthetics, to make people's five senses feel pleasant and comfortable. Accordingly, many colleges put interaction design in art schools, and many companies branch out into interaction design according to people's five senses: hearing, vision, smell, body sensation (vibration), or confuse interaction design with interaction technology, thinking that interaction design is all about voice technology, gesture input technology and so on. In fact, interaction design is a highly integrated and applied discipline based on in-depth understanding not only of the technologies involved, but also of human cognitive psychology and human physiology, social science and aesthetics.

Interaction designers need to have in-depth knowledge of the needs of people, the possibilities of technology and the context in which it may be applied. By having a good understanding of people and a good knowledge of technology, the designer applies this knowledge to deliver good design outcomes. Therefore, this book takes a look at several aspects, including cognitive psychology, some basic physiology, methodologies for studying human needs, design methodologies and methods for evaluating the design of different systems, standards and so on.

Interaction design in the car is a rapidly evolving discipline, and the challenges and problems it faces seem to be increasing. Nowadays, people deal with many products that require interaction on a daily basis, such as our mobile phones, our computers, kitchen appliances, televisions, air conditioners and many more. There are countless products that require interaction between humans and machines. If the user's needs, preferences, abilities and limitations are systematically studied and the insights applied in the design process, these products should be easy, smooth and comfortable to use. However, there are many products on the market that people feel do not work well and don't know how to use. These products often do not take into account the human factor, the operational characteristics of people when using the product. To make the product work well, to make it more convenient and simple, to make the user feel competent and happy, is the aim of the interaction design engineer. Of course, not every product is designed to be easy to use. High-end, expensive cameras are not meant be as easy to use as a simple camera. But even in those cases good design can help the user to master the challenges without getting frustrated, and the resulting feeling can be a sense of achievement.

With the development of electronics and networking technology and the improvement of intelligent technology, more and more functions are being introduced into cars and human–machine interaction is becoming more and more complex. Interaction design in cars requires more than just simple usability, it also requires safety. This safety is not only for the driver and passengers, but also for other road users. Therefore, designing the interaction with in-car systems introduces additional challenges compared to the design of general consumer electronics.

It is only in the last decades that so much attention has been paid to in-car interaction design in the world. One of the authors remembers buying a nice German car some years ago and during the two years she owned it, she didn't know how to turn on the dome light and gave up after making various attempts. Then she went to a friend's house who lived in Paris. They found out that his car was the same as hers. He turned on the dome light to check the map on his way to drive her to the airport (GPS was not common at that time), which excited her and she asked him how he did it. He proudly told her that he put his hand on the lower left side near the front door shaft, there was a key that was used to control the various lights, pull out this rotary lever and the dome light was on! He told that a lot of people didn't know this trick. Then, to prove that she wasn't the stupidest person, she did some research and found that almost no one, professor or Ph.D., who drove the car had managed to find out how to turn on the dome light, and the funny thing was that everyone's manuals were in the front bucket of the car and no one bothered to look them up, because no one would spend half an hour looking up the manual for the small matter of controlling the dome light. To a designer, it may seem like a good idea to have all the light controls in one place, but for a user, there's no way to imagine that the dome light switch would be there, let alone that the swivel can be pulled outwards.

Don't think this is a story from the past. Nowadays, in-car systems are becoming more and more complex and many older drivers are at a loss for what to do with many of them. Erik Hollnagel, who is a renowned expert in the field of automotive HMI, once told the following story from his own personal experience. He was once driving home on a French motorway when his newly purchased car suddenly popped up with a warning message "There is a problem with your car's engine!" He looked at it and, thinking it was a serious problem, pulled into the nearest petrol station at a motorway rest area. He went to ask the staff at the station how to deal with the problem and the staff, who looked like an experienced man, said, "If it was a mechanical car, I could help you find the problem and fix the fault, but now it's electronic and I can't figure it out". At the professor's request, he followed him to the car. Then he said "Start the car and see what happens". When the professor started the car again, the warning had disappeared. He drove the rest of the way with a lot of trepidation.

The two stories here are both related to in-car interaction design. To design a product that interacts with people, you need to consider the following questions: (1) Who will use the product? (2) How will these users use it? (3) In what environment or context will they use it? The design of the user interface, the way information is entered and presented, needs to match the users' behaviours. For interaction designers, we often need to ask ourselves an important question: How can my design optimise the user's interaction with the product in its particular environment, so that the product or system supports the user's activities in a useful, effective, usable and enjoyable way (Sharp et al. 2019, p. 9)? Because users often operate systems in their instinctive, taken-for-granted ways, the process of using a system is likely to be very different from what the designer imagined, or hoped for, so every designer needs to consider the characteristics of the user. What are they good at? What are their weaknesses? How can we make them better at what they want through our design?

2.2 Definition of Interaction Design

What is interaction design? Sharp et al. (2019, p. 9) give the following definition.

> Interaction design is about designing interactive products to support the way people communicate and interact in their everyday and working lives
>
> Interaction design is about creating spaces for users to communicate and dialogue with the system, enhancing their experience of using the product in their work and everyday lives, while increasing productivity, pleasure and satisfaction.

In the automotive domain, Interaction Design is often carried out in the department of Human–machine Interaction, and concerned with the design of the Human–Machine Interface (HMI). Interaction design is a very comprehensive discipline encompassing many fields, three of which are often confused with the concept of interaction design: ergonomics, human factors (HF), and human–computer interaction (HCI). It should be noted that interaction design encompasses all these disciplines, which have a longer history than interaction design and are the basis of interaction design. But interaction design is not the same as any of these disciplines. It also encompasses general psychology, cognitive psychology, engineering, computer science, software engineering, sociology, anthropology, ubiquitous computing, human physiology, acoustics, aesthetics, biomechanics and many more. In design, it includes graphic design, product design, art design, industrial design, information engineering, even film and television design, media, marketing and more. No single person is capable of understanding so many areas, so interaction design is usually done by a team, especially in the case of cars. The interaction design team also needs to be made up of people from a variety of backgrounds rather than a single one (Fig. 2.1).

There is no single interaction design team in the world that encompasses all of these disciplines, which means that the composition of team members' knowledge backgrounds will vary from product to product. In many cases, ad hoc teams are formed to complete a particular project. The advantage of this is that people from different backgrounds working together can often produce more creative design ideas for innovative products. Of course, when people from different backgrounds work together, there is also a period of adjustment. Automated driving systems are complex systems that involve many different technologies, such as mechanics and electronics. The design of these systems brings together many different disciplines and specialisms, each with its own design approach. Within the engineering domain, we can distinguish between the disciplines of mechanical engineering and software engineering (computer science and electrical engineering). Both disciplines have developed their own methods to guide developers in delivering technically sound products. However, the fundamental principle of interaction design is that the process

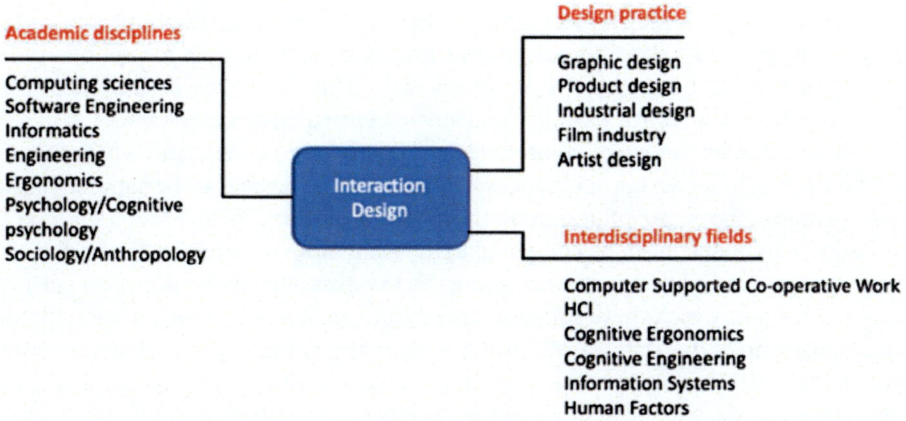

Fig. 2.1 Interaction design requires an integrated team

of developing technically sound products needs to take into account the consumers' point of view and their needs or, more generally, the human point of view, in order to achieve a result that is satisfactory to both the individual user and society. In Part IV of the book, we examine the methodologies of the technical disciplines, summarise the motivations and characteristics of human-centred design, and consider how human-centred design approaches can be combined or harmonised with technical design approaches.

2.3 Human-Centred Design

There is a fundamental difference between human-centred design and technology-centred design. Until the 1980s, for consumer products, there was a relatively direct relationship between the form of the product and its function. Most products contained a number of buttons or similar controls and it was a relatively simple matter to figure out how to use the available functions by manipulating the controls. More complex systems were usually intended for professionals or for very specialist amateurs (as in the case of photo cameras), and their use usually required more or less explicit training and practice. In the case of complex systems aimed at the general public (e.g., vehicles), the law required people to obtain a licence to use the system, which involved also explicit training. However, in the late 1980s, personal computers became available to non-professionals and personal computers began to enter the homes of the general public. At the same time, developments in electrical engineering made it possible to extend the functionality of devices such as car radios. For these new devices, the relationship between form and function was no longer simple and clear. Often there were few controls, and a single control would provide control over many functions. On the other hand, often people were not willing to spend much time learning and training in operation. This has led developers to understand that they must strive to design user interfaces that are easy to use, otherwise these products and features will be abandoned. This led to the development of a conceptual framework for usability and a user-centred design approach.

When it comes to human-centred design, it is good to understand what makes it different from other design approaches. Human-centred design, as the name suggests, puts people at the centre of design. In this context, the term ‘people’ refers primarily to users. This concept can be compared to the technology-centred design of the past. Technology-centred design is more about the functions that technology can perform and the form in which these functions need to be expressed in such a way that the user understands the state in which the technology is being used. Human-centred design considers the needs of people first, and then looks for technologies that can meet those needs, and the interaction between people and systems takes into account their perceptions, habits, abilities and limitations and real needs.

The advantages of adopting human-centred design are manifold: (1) increased productivity; (2) increased usability and user experience; (3) reduced training and after-sales service costs; (4) reduced work stress and discomfort; (5) increased market competitiveness; and (6) benefits for the sustainability of the product. The theory and methodology of human-centred design is at the heart of our book and will be the subject of Chaps. 8–13.

Before we go any further, we need to distinguish a few basic concepts.

UX: User experience
UI: User interface
IxD: Interaction design

These concepts are often mixed or confused. In fact, there is a clear distinction between them. Interaction Design refers to the design work of a designer who uses different theories and techniques to design a product according to the needs of the user, and who presents the product to the user as a User Interface. The user interacts with the system through this interface in order to complete the tasks that s/he wants or needs to complete. The User Experience is the feeling that the user gets from operating the interface and completing the task. Figure 2.2 illustrates these relationships.

2.4 Usability

There are several levels of interaction design for the car, as shown in Fig. 2.3. Firstly, there is functionality. This concerns the different functions that the in-car interaction system offers to help the user perform the tasks s/he needs to do, such as driving the car, listening to music, communicating with the outside world, helping to check road conditions and so on. Each design therefore serves one or more functional objectives. The second level is safety. Safety is always at the forefront of car design. When cars become more automated, safety will be entirely taken care of by automated systems, and safety may no longer be a primary concern in the interaction design process of the user interface. But other aspects of safety, such as information security, will still be present. The third layer is usability. The fourth layer is user experience.

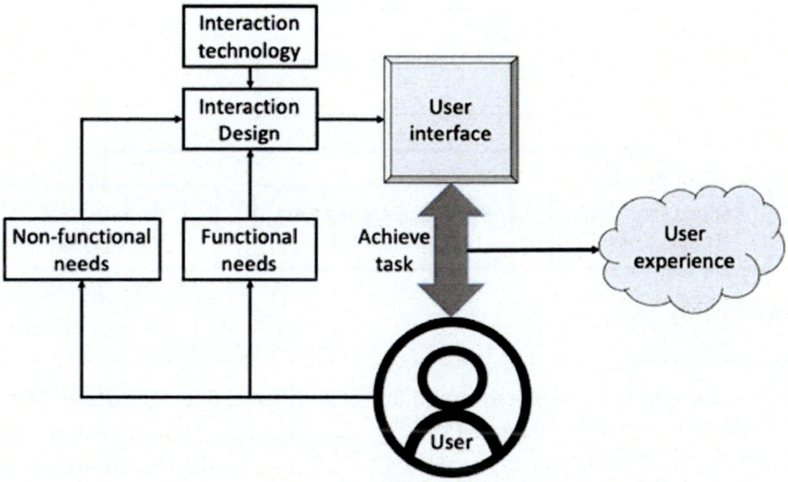

Fig. 2.2 Interaction design, the relationship between user interface and user experience

Fig. 2.3 Hierarchy of car interaction design

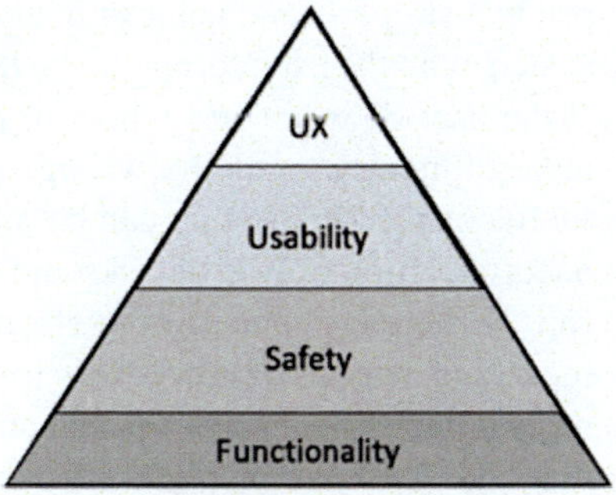

ISO standard 9241–11 defines usability as.

> The extent to which a specific user can use a product to achieve a specific goal with effectiveness, efficiency and satisfaction in a specific usage environment.

Figure 2.4 expresses the concept of usability. Several things need to be noted about this definition. Firstly, it was devised as part of an attempt to make usability measurable, i.e., to replace the need for experts or professionals to make global judgements about whether a given product is easy to use, by a method to measure a product's usability in terms of actual use. Secondly, it involves a specified user, a specified target and a specified context of use. In other words, a product's usability is only valid for people with certain characteristics, for whom a product is or is not easy to use when they use it under certain conditions and for certain purposes. If these conditions do not apply, then complaints about poor usability are irrelevant. The relevant characteristics of the population are usually related to issues such as background knowledge and communication skills (language proficiency). For instance, if a system is designed for a professional audience, and a layperson finds it difficult

Fig. 2.4 Usability goals

to use, the complaint has no bearing on the system's usability. Similarly, if a system is intended for normal people and a blind person says s/he cannot get it to work, this comment has no bearing on the system's usability. Thirdly, the definition contains the terms of effectiveness, efficiency and satisfaction. These concepts need further elaboration.

Effectiveness is a question of whether a user is able to complete the task that the product supports and achieve his/her goals. For example, in the case of a navigation system, can the user complete the task of entering the intended destination/itinerary? If the user does not know how to proceed during the process of entering the destination/itinerary, or if the wrong destination is specified, the navigation system is ineffective. Effectiveness can be measured for instance by asking a number of target users to complete a set of tasks and count the number of tasks that are solved successfully. Efficiency concerns the effort required to complete the task. Efficiency can be measured in terms of time ("how long it takes the user to enter a destination/itinerary") or, in other cases, in terms of the amount of mental effort it takes to perform a particular task (for example, how long the user needs to concentrate to perform the task), or, in yet other contexts, the number of people required to complete a particular task. Both effectiveness and efficiency can be measured objectively, without having to rely on subjective judgement. For example, to improve effectiveness, a test might be arranged with 50 participants who perform a set of tasks, and we can calculate the percentage of tasks successfully completed. For efficiency, an improved version of the user interface could be tested again with the same participants and the time taken to complete the tasks could be calculated and compared with the time it took with the previous version of the interface. In addition, during such tests, errors that occur are usually analysed to identify usability bottlenecks that need to be addressed. The distinction between effectiveness and efficiency also implies a distinction between fatal errors, i.e., errors that prevent the user from completing the task successfully, and non-fatal errors, i.e., errors that do not prevent the user to complete the task successfully but slow down the interaction. For effectiveness, only the fatal errors are relevant, while efficiency is affected by non-fatal errors. Furthermore, it should be noted that effectiveness is usually the primary concern for novice users and infrequent use, while efficiency is the primary concern for experienced and frequent users, for whom effectiveness is no longer an issue because they already know how to complete the tasks successfully. For companies, efficiency may be also a concern in the case of novice and infrequent users. For instance, an inefficient user interface at a toll booth at the highway may result in unnecessary queues and a request to install more toll booths.

The definition of usability does not only rely on objective metrics, but also includes a subjective component, namely satisfaction with use. Satisfaction is usually measured by asking test participants to fill in a questionnaire or by conducting interviews. In addition, questionnaires for measuring satisfaction are often combined with questionnaires for usefulness and usability, such as the SUS questionnaire (Sect. 13.5). This allows developers or testers to compare objective data about usability with the subjective opinions of users.

Other definitions of usability include additional elements. For example, Nielsen (1994) suggests including aspects such as learnability and memorability. Learnability relates to ease of use at first contact, and memorability relates to the knowledge of interactions with the system that people recall after a period of non-use. In other words, while the ISO definition does not explicitly mention that people's knowledge of their interactions with the system may change over time, Nielsen's definition explicitly recognises the relevance of changes in people's knowledge to repeated use. However, as the ISO definition has become the accepted standard, we will use the standard definition of ISO 9241-11.

2.5 User Experience

In the latter half of the 1990s, product and system developers began to recognise that interacting with products and systems is not only about understanding how to use the system, but also about the full range of human emotions. The emotional aspects of interaction with the product/system, previous experiences, and expectations became the focus of the user experience component. ISO standard 9241-210 defines user experience as "the personal perceptions and responses that result from the use or intended use of a product, system or service".

Hassenzahl's (Hassenzahl and Tractinsky 2006) framework of user experience includes usefulness, usability and enjoyment. Firstly, a product or system should be useful to people, where 'usefulness' can be understood as 'allowing people to engage in activities that are meaningful to them'. Meaningful activities are considered to be those that relate to basic human needs, such as the famous Maslow's Hierarchy of Needs (Maslow 2013). Maslow's theory divides human needs into five categories: Physiological needs, Safety needs, Social needs, Esteem needs and Self-actualisation needs, in ascending order. At the self-actualisation level, there is a process of learning and growth, aesthetics and self-transcendence. Secondly, the product should be easy to use. Thirdly, the interaction with the product or system should also be emotionally satisfying or pleasurable, or at least not emotionally intrusive (frustrating, annoying, etc.). A product or system can provide a good user experience if it provides meaningful functionality, is easy to use and provides an emotionally satisfying experience.

It is clear from the ISO definition that user experience not only depends on what happens during actual use, but that the user's expectations also play an important role. A product or system with the same target characteristics may produce a completely different experience depending on the level of expectations of the user. Expectations may be generated by advertisements, experience with previous products from the same company, comments from the public opinion community, personal communication between friends/relatives, etc. Therefore, it is important that advertisements do not overstate the benefits that the product or system offers; otherwise, users may be disappointed with the actual experience.

So far, the definition of user experience is rather vague, and it seems that many factors are involved in user experience, for example, memory also contributes to user experience. If a person's interaction with a product or system changes over time, we might ask how an instantaneous interaction can determine the overall user experience. Is the overall experience the average of all momentary experiences, or is there a more complex relationship? The peak-end rule, proposed by Kahnemann (Kahneman et al. 1993) goes some way to answering this question. The peak-end rule states that the memory of an experience is largely determined by the peak experience (positive or negative) and the experience of the final moments of the actual interaction. For example, your experience of using your in-car system for a whole year is no match for the mental imprint left by a traffic accident. Therefore, excessive negative experiences should be avoided as much as possible, as they can seriously affect the overall user experience, and one bad experience can replace 100 good ones. In addition, designers should pay particular attention to the final part of the interaction, as this can also be an important factor in determining the overall user experience.

The desired goal of the user experience varies somewhat from product to product, but in general terms, it consists of the following descriptors.

1. Goals to be achieved

Comfortable	Helpful	Interesting	Delightful	Driven
Exciting	Superior	Challenging	Surprising	Cognitively stimulating
Motivating	Engaging	Supportive of creativity	Emotionally fulfilling	Entertaining

2. Outcomes to be avoided

Boring	Disappointing	Discouraging	Overly pampering	Sinful
Making you feel stupid	Fed up with	Annoying	Childish	Fancy

From these terms, we can see that a good user experience refers to the positive feeling that a user has after using a product. This feeling varies from person to person. As different people have different cultural backgrounds, different levels of education,

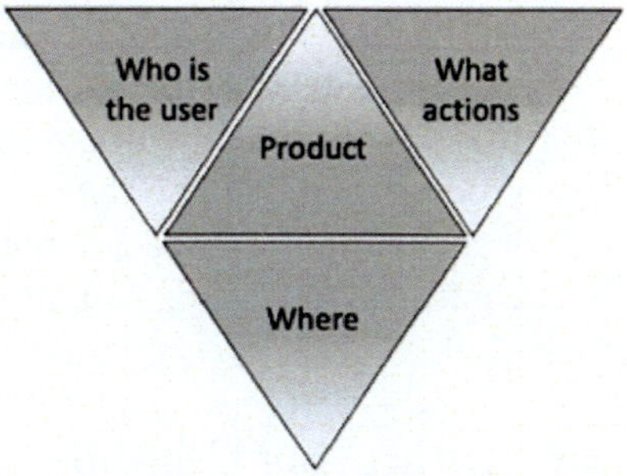

Fig. 2.5 Factors to consider in interaction design

different past experiences and needs, different expectations and so on, all of them will have different user experiences when using the same product. It is clear from this that the user experience will change with the knowledge, understanding and continuous use of the product. Therefore, we would say that there is no way to design the user experience, but rather that design should aim to establish a good experience through the product. We will discuss more about user experience in Chap. 6.

This being the case, let us turn back to what interaction design is. As can be seen in Fig. 2.2, interaction design is the process of designing the user interface with the user's needs in mind, that is, the user interface is created through interaction design, and the user experience is shaped by the user's interaction with the user interface. In other words, interaction design is the process of helping a technical product talk to its users, or as Norman says in his book "Design Psychology": "Design is really an act of communication, and designers need to have a deep understanding of the people they are communicating with." (Norman 2013).

So, what are the design considerations? Figure 2.5 provides an illustration. If we take the middle triangle to represent a product, then the immediate factors to consider in interaction design are: Who is the user? Where does the user interact with the product? What actions will the user use to interact with the product?

The German industrial designer and academic Dieter Ramsay has identified ten principles of design that are equally important for interaction design.[1] Any good design will have the following characteristics.

1. It is innovative. It does not only give us a deeper understanding of the product, but also allows for a different experience than before.
2. It is useful and long-lasting. Any product needs to have a practical value for its users. Good design stands up to changing trends, because it covers universal topics that are always relevant to you. Well-designed products play an important role in how we perceive things and how we perceive ourselves.
3. It is aesthetic, comprehensible, simple, honest and thorough down to the last detail. A product needs to be pure, leaving only the necessary functionality, understated but sophisticated and attractive. Even if only 10% of the details are

[1] https://www.interaction-design.org/literature/article/dieter-rams-10-timeless-commandments-for-good-design.

not perfect, it will detract from the overall feeling. At the same time, the user can see straight away how to use it without having to learn further.

4. It is unobtrusive, as little design as possible, and environmentally friendly. A product should not be "over-designed", with too many things that we don't use, it should not foster social motivators such as status, wealth and power, and not contribute to wasting natural resources.

2.6 Improving Usability

Human cognitive abilities and limitations are discussed in detail in Chap. 3. Since it is often difficult to translate these scientific insights into convenient design tools, theorists such as Nielsen, Norman and Schneidermann have provided usability principles or rules to guide design. In The Design of Everyday Things (Norman 1988) Norman identifies three principles that must be met in order to achieve good usability: mapping, affordance and constraints.

- Mapping is about the relationship between the form of an element in an interface and its function. In order to achieve a particular goal, it should be clear for the user how to operate the elements in the interface. For example, if three light switches are adjacent to each other, there should be a clear mapping from the physical arrangement of switches to that of the lights, so that it is clear which switch controls which light.
- Affordance is about the physical characteristics of a control. The physical shape of the control suggests how to operate it. For example, a button is shaped so that people, when they see it, know that it is for pressing, while a knob elicits a rotating movement. The handle of a car door is a cue for where to put one's hand when opening the door.
- Constraints refer to limitations on how people can or are expected to interact with a system. Physical constraints limit the physical interaction between people and objects. For example, a typical light switch can only be placed in two different states, 'on' or 'off'. Because it is always in one of the two states, it can only perform one action. Cultural constraints are learned practices that influence people's perceptions of how they interact with the system. For example, "green" usually means "pass/go on", while "red" usually means "stop". Using such constraints in the interaction automatically elicits the proper behaviour. It should be noted that cultural constraints may acquire characteristics similar to affordance, and hence prompt for action. For example, because icons on touchscreens are not physical elements, there is no inherent affordance for pressing, but even so, because people have learned this association, the icons are felt to be pressable elements.

It may be argued that these principles, at the outset of use, help the user to generate assumptions about how to interact with the system based on the information provided by the interface. If the principles are violated, the user will find that s/he has to learn and remember the way to interact with the system. This additional learning requires effort and reduces the ease of use of the system. Furthermore, the ability to generate correct assumptions also depends on the user's background knowledge. Someone who has no experience with touchscreens may not think that the corresponding function on the touchscreen can be activated by touching an icon. For someone with experience with touchscreens, it is simple enough for that person to figure out which icon needs to be clicked to activate a particular application. Ease of use is therefore very much dependent on the background knowledge of the user. Usability and ease of use will therefore vary from person to person.

In addition to these general principles, theorists have proposed various rules or guidelines for interaction design (for example, Schneiderman's eight golden rules[2] and Nielsen's ten design heuristics[3]). These aspects are explained in more detail later. Some of the most salient points are.

- Visibility: The system should always make the user aware of what is happening through appropriate feedback within a reasonable time. For example, it should be clear which mode the system is in to avoid mode confusion. Similarly, it should be clear that if a response cannot be displayed immediately, the system needs to show that it is performing a particular action and how long it will take to complete that action.
- User control and freedom: the user should feel in control of the interaction and be able to predict the changes that will occur in the system after each action, rather than being passively asked to perform various actions.
- Consistency and standards: the use of different terms to describe the same thing, the same situation or action should be avoided. The operation of the system also requires adherence to a number of recognised conventions, various design guidelines and standards.
- Flexibility and efficiency of use: shortcuts ("accelerators") should be provided so that skilled users can speed up interactions and thus interact more efficiently, without complicating the interface for novice users. In other words, skilled users should have the possibility to set up shortcuts. As mentioned before, novice users are interested in effectiveness, whereas for experienced users, the principle of effectiveness is usually satisfied and efficiency becomes more important.

It is worth mentioning here that these heuristics or rules were proposed in the 1990s, when the development of electronics had greatly increased the functionality of systems and the design needed to enable the user to handle the large number of functions available. With the advent of intelligent systems, systems do not only react to the user's actions, but also perform certain actions automatically according to the user's behaviour and context. Nevertheless, the heuristics and rules mentioned above are equally valid for ensuring good usability and no research has yet been conducted to explicitly oppose these regulations.

---

2 https://www.interaction-design.org/literature/article/shneiderman-s-eight-golden-rules-will-help-you-design-better-interfaces.

3 https://www.nngroup.com/articles/ten-usability-heuristics/.

Usability can be achieved by applying the design principles and heuristics described above and by conducting continuous testing throughout the design process. As the design principles and heuristics are still abstract, designers need more specific guidelines and criteria, including interface design and methods for evaluating how the interface design is judged in terms of achieving the goals of the system. Guidelines and criteria will be discussed in later chapters.

It should be mentioned that pleasantness may be influenced by cultural differences. In particular, aesthetic preferences may differ between cultures. What is considered beautiful in Europe (e.g., minimalist design) may not be considered attractive in China. However, as Chinese people become more accepting of Western culture and as globalisation progresses, they are becoming more appreciative of the simple, beautiful design style of Scandinavia.

Finally, it should be noted that, while the usefulness, ease of use and pleasantness of a product all affect the user experience, the contribution of ease of use is asymmetrical: while a badly used product may negatively affect the user experience, good usability does not necessarily have an equally positive impact. Instead, good usability is often taken for granted. Nonetheless, usability should still be valued so that poor usability is avoided, but design for a good user experience undoubtedly also requires attention to the aesthetics, resulting in a pleasing experience.

2.7 Know Your Users

Understanding your user is an important part of interaction design. Understanding the user includes understanding his life, his work, his education, his environment, his relationships and so on, so that you can design a product that meets his needs. Understanding the end user, understanding a group of users, is not the same as understanding the individual user. Each individual user is very different, and from a deeper understanding of the individual it follows that, if a product is suitable for one person, it may not necessarily be suitable for a group of people. In addition to individuality, age, gender, education, individual differences also arise from work, culture, life experience, social status, family status etc. For special systems, such as voice interaction systems, individual language skills and accents can also have an impact on interaction. We want to design products that are accessible to as many people as possible and that meet the needs of as many people as possible. In this case, it is important to understand individual differences. In other cases, the product may be used by more than one person, so it is important to understand this common group.

Age and gender differences are two of the most fundamental differences. Some abilities differ between ages, and products designed and developed for children cannot be based on the physiological, psychological and cognitive abilities of adults. Similarly, products designed for younger people may not be suitable for older people, and the other way around. This is because reaction time, eyesight, hearing and muscle strength are all deteriorating with age. Furthermore, many studies have shown that teenagers, between the ages of 18 and 25, are not mentally mature enough to cope with the road environment when driving and are therefore more likely to be involved in car accidents than other age groups. They also have different interests than adults and may be more interested in novelty and excitement. Similarly, with the advent of an ageing society, researchers in many Western countries are studying the characteristics of older drivers, hoping to extend their driving experience and improve their safety through interaction design.

In many ways, gender differences are also evident. In driving tasks, this difference may not be immediately obvious. However, in the use of infotainment, or some peripheral products, differences may arise. When someone studied why a European car sold so well in Europe in the 1970s but not so well in the US, it was thought to be because there was no small mirror on the back of the sun visor above the driver's seat. Because women in America, when preparing for work in the morning, may not have time to put on their make-up and need to do this in the car. The absence of a mirror can be a deciding factor in whether they buy the car or not!

The educational background and nature of the user's work can have a significant impact on interaction design. For instance, while users with an engineering background may expect the system to provide more logical explanations of the interaction, users without such a background may not want to know every technical detail. Good design boils down to a deep understanding of people. An understanding of the user starts with the following.

1. Understanding the user's strengths and weaknesses, the main emphasis here being on their cognitive abilities and limitations. What are they used to, what are they good at. What are they not good at, what don't they understand how to do, etc. This is described in more detail in Chap. 3.
2. Designing to help people do things the way they are used doing them: this is important if a design is trying to change the way people are currently used to do things, emphasising them to learn a completely new way of doing things, which can evoke resistance from users. Many years ago much research was done into the arrangement of keys on the keyboards of typewriters and it was concluded that the current arrangement was not the best design, but this research did not result in redesign of the keyboard, because people were already used to the current alphabetical arrangement of keyboards and did not accept the new arrangement.
3. Any design mediates a user experience. The design is not just a list of various functions that the user needs, stacked there. Through an in-depth understanding of the user, mastering the user's operational characteristics, the design can better guide the user to operate the relevant functions, so that the user feels that the product is designed for him.
4. Meet the needs of the user and, if possible, even involve them in the design. There are theories that the person who knows best is the person themselves. So, there are theories that involving users in the design will make the product more

responsive to their needs. This idea is difficult to implement in practice, especially for companies. However, it can be a very effective way to give designers a deeper understanding of user needs through user involvement in early design.
5. Use a tried and tested user-centred approach.

2.8 A Short History

When it comes to interaction design for cars, there are a few terms that often come up: ergonomics, human factors and interaction design. What is the difference?

Wojciech Jastrzebowski coined the term 'Ergonomics' in a philosophical narrative in 1857, from the Greek words ergon (work) and nomoi (natural law), to refer to the technique of optimising the design of a product to make it more user-friendly. In the early 1900s, industrial production still relied heavily on human/motor power and ergonomic ideas were being developed to improve worker productivity. Scientific management (as developed by Frederic Taylor) to increase worker efficiency through improved workflow was popular. However, in a strict sense, design in this period was still a design that 'made people fit the job', which is not the modern concept of 'ergonomics'. The modern concept is that we design work to meet the characteristics and needs of people. The prototype of this concept was born in the Second World War. There was a huge amount of modern weaponry being used in the war. It soon became apparent that many soldiers were not killed by the enemy but injured by their own weapons. This was because the weapons were not designed in accordance with human mechanics, anthropometry, human physiology and so on. This led to a great deal of research into the human being. To date, ergonomics still focuses on the effects of human mechanics and anthropometry, human physiology, environmental physiology, etc. on human performance. This is an important part of the design of automobiles, which includes the study of the comfort of car seats, seat belts, steering wheels, air conditioning, interior sound, lighting, the layout of physical buttons and so on.

Human Factors Engineering, on the other hand, is a discipline that has developed more on the basis of human cognitive psychology. The rise of this discipline was due to the development of large-scale industries and the emergence of many complex systems that brought complexities that, at that time, were beyond the scope of human cognitive psychology, such as the emergence of large industrial control rooms. The study of human factors has led to a broad understanding of the characteristics and limitations of human behaviour and cognition. Interaction design, on the other hand, emerged from the rapid development of electronic systems as computers became widespread. Interaction design is, in a way, a part of ergonomics and human factors engineering.

References

Hassenzahl M, Tractinsky N (2006) User experience—a research agenda. Behav Inf Technol 25(2):91–97

Kahneman D, Fredrickson BL, Schreiber CA, Redelmeier DA (1993) When more pain is preferred to less: adding a better end. Psychol Sci 4(6):401–405

Maslow AH (2013) A theory of human motivation. Simon and Schuster

Nielsen J (1994) Heuristic evaluation. In: Nielsen J, Mack RL (eds) Usability inspection. Wiley, New York

Norman DA (1988) The design of everyday things. Basic Books, New York

Norman DA (2013) Design of everyday things, Revised. Basic Books, New York

Sharp H, Rogers Y, Preece J (2019) Interaction design—beyond human-computer interaction, 5th edn. Wiley

Part II
Driving Psychology

Part II
Driving Psychology

Chapter 3
Basic Psychology

Abstract In this chapter main concepts from cognitive psychology are introduced in relation to driving, to provide a theoretical foundation for automotive interaction design. Insights concerning topics such as human information processing, human perception, attention, memory, learning and decision making are summarised. Going beyond cognitive psychology, we also summarise insights concerning emotions in relation to driving.

As we mentioned earlier, one of the most important foundations of interaction design is cognitive psychology, and this is overlooked by many people working in interaction design, even equating interaction design with aesthetics. Because of this, it is possible to make some very superficial mistakes, and it is also difficult to understand why some designs are good and some are not.

When it comes to cognitive psychology, we must mention Christopher D. Wickens and his book "Engineering Psychology and Human Performance" (Wickens et al. 2013). C. D. Wickens is a leading American cognitive engineering psychologist. His book "Engineering Psychology and Human Performance", first published in 1992 and now in its fourth edition in 2013, is a fundamental textbook for anyone working in human factors engineering. The book describes the foundations of cognitive psychology as it relates to design. It is highly recommended for anyone interested in interaction design research. Here, we will only give a brief overview of some of the important and commonly used knowledge that is closely related to interaction design in cars. The following content is based on the 2013 edition of the book, in which a number of examples related to car driving are used. This book is more suitable for those of us who work in automotive interaction design. The following, if not specifically stated, leans heavily on this book.

© China Machine Press 2023

F. Chen and J. Terken, *Automotive Interaction Design*, Springer Tracts in Mechanical Engineering, https://doi.org/10.1007/978-981-19-3448-3_3

3.1 Human Information Processing

Before we look further into cognitive psychology, we need to understand information processing. A schematic representation of how human information processing proceeds is shown in Fig. 3.1.

The model describes a series of different stages of information processing, and it characterises the flow of information as humans perform a task in four steps: sensory processing, perception, working memory (including thoughts and decisions) and response (selection and execution). Information from the external environment is first acquired and processed by our senses (visual, auditory, haptic, etc.). For example, when a car is approaching an intersection, the driver sees traffic lights, passing cars, and other road users, and may hear sounds from the car's audio system and the conversations of the passengers. However, the acquisition of information by the senses is not the same as perception. The process of perception involves establishing a connection between the information from the senses and knowledge in our long-term memory, resulting in the classification of the information from the senses as entities that have a certain meaning. For instance, in the case of visual perception, the collection of visual shapes acquired by the sensory system is connected with the knowledge in long-term memory, making us see objects and events, such as cars and bicycles that are heading in a certain direction with a certain speed, traffic lights that are in a certain state with a certain meaning etcetera. Our long-term memory contains the knowledge that we have acquired throughout our lifetime, consisting of factual knowledge, knowledge about categories, knowledge about rules and how to do things, personal memories, images, and so forth.

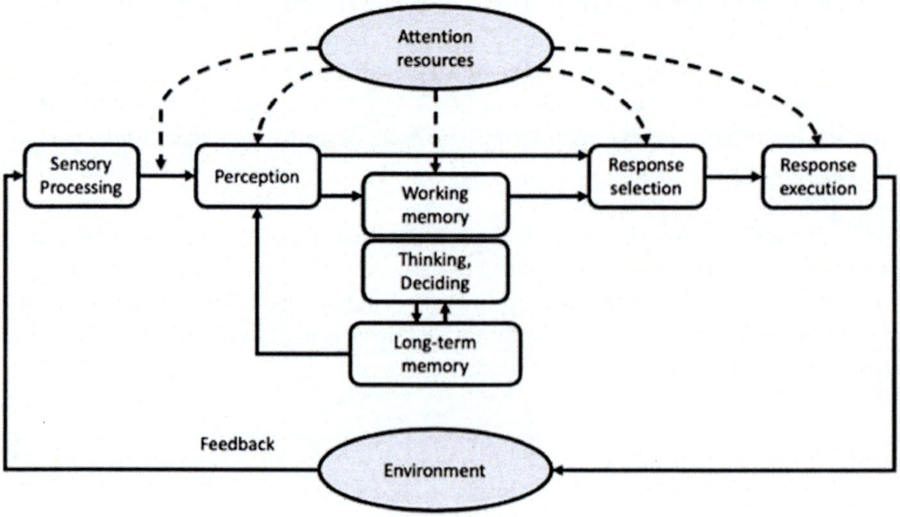

Fig. 3.1 Model of the human information processing process (after Wickens et al. 2013)

After perception, our information processing usually operates in one of two ways. At the lowest level, after perceiving the environment, one situation is that it immediately triggers a choice of a response. For example, if a driver sees a yellow traffic light, he can choose to step on the accelerator or step on the brakes, and this decision is based on a number of factors. For example, the driver may observe that the yellow light has been on already for a long time and will turn red immediately, so that, if he pushes the throttle, he is likely to encounter a red light. So, he chooses to step on the brakes, a decision that must be made quickly. Then, after the response choice, the next step is the execution of the response phase. It is thus clear that our sequential responses involve not only the muscles but also the brain's control of them.

However, perception of acquired information and understanding does not always lead to an immediate response. The driver can use working memory to temporarily retain the acquired light (yellow) state while scanning the road ahead and other information in the intersection (for example, approaching vehicles or other road users) to make a comprehensive judgement about the next action required. In fact, in many cases, the final action may not follow people's previous perceptions at all. For example, when you attend a scientific seminar, you may hear a scholar explain an important theory, but you choose not to take notes, but to reflect on the content and compare and integrate it with existing relevant knowledge in long-term memory. This means using working memory to store that information in long-term memory. It can then be used in the future in relevant situations. The role of working memory is therefore two-fold. It serves as a register to temporarily hold information. And it refers to the cognitive processes that work on this information, enriching the information, evaluating it in terms of our goals, and using the information for decision-making.

At this point, we note that the processes of perception and working memory are not two distinct and clearly separate boxes. The boundary between them is blurred, so that the stage from sensory memory to working memory is often described as the 'cognitive stage', the process of understanding the external information acquired through the senses. This process can be as quick as understanding and responding to a traffic light, and sometimes it can be slow, as in the case of a research conference.

Within this four-step (sensory memory, perception, working memory, response selection/execution) model, there are two other crucial elements: feedback and attention. Firstly, the execution of a response changes the original internal and external environment and therefore creates a new state, which is registered by the senses, and this is what the feedback loop shows. For example, when a driver starts to accelerate in order to change lanes on a motorway, his senses also start to pick up new information (e.g., that another vehicle is accelerating behind him in the left lane) and the driver may need to make further judgements and responses to this, either by accelerating further and completing the lane change, or by slowing down and waiting for another vehicle to pass before re-accelerating. Secondly, attention is an important tool that helps with much of the information processing. Attention plays two roles (Wickens and McCarley 2008). Firstly, attention acts as an information filter. In the process of moving from acquisition of sensory input to perception, attention focuses on certain elements for further processing, but at the same time blocks other information that it does not consider relevant from entering the perceptual process, so that the perceptual output is smaller than the sensory input. Continuing with the example above, the driver at an intersection focuses his attention on the traffic lights and ignores the communication between the occupants of the car and him. Secondly, attention acts as the fuel that provides power for our cognitive processes, and the mental resources or energy required for the various stages in information processing are provided by attention. Some stages require more resources, and certain tasks require more of a person's mental resources than others. For example, perceiving a traffic signal in fog will require more effort compared to when the weather is bright. But there is a limit to the supply of our attentional resources, and if the overall resources required for a task exceed the amount that can be provided by human capacity, this will overload the system and possibly result in failure in task performance.

While Fig. 3.1 provides a useful conceptual framework for understanding information processing, it should not be taken literally. Neuropsychological research has found that although some operations are associated with certain specific brain structures, this association is not clear at this stage of research. Also, these stages are not strictly sequential. After all, we may initiate an activity out of some inspiration, thought or intention that originates in long-term memory, flows into working memory, and then leads to a response (or we might rather say action), possibly without perceptual input. Furthermore, acquisition of information from the senses, perception, processing information in working memory and acting all go on in parallel. However, the information processing model is useful for task analysis, rationale descriptions, solution recommendations, and theoretical research in engineering psychology.

Now that we have this model of information processing, the subsequent treatment of cognitive psychology will follow the sequence of this model. The first is an understanding of the human senses. There are several major organs in the body that receive information from the outside world: the eyes (vision) and the ears (hearing) are two of the most important, followed by the nose (smell), the mouth (taste) and the skin of the body, especially the hands (touch). In the context of driving, smell and taste are less relevant, so we will focus on sight, hearing and touch.

3.2 Vision

For the driver, vision is the most important organ for obtaining information inside and outside the vehicle. Almost 80% of all driving-related information is currently obtained visually. Therefore, there are many studies on driver vision.

Figure 3.2 shows the anatomy of the human eye.

The entire eye is wrapped in a sclera, which is like the black box of a camera and is divided into two sections: the anterior and posterior. The anterior part of the eye is the light concentrator and is made up of the cornea, pupil, ciliary muscles and lens. Their function is to regulate and concentrate the light coming in from outside. Incoming light passes through the cornea, through the pupil and through the crystalloids, and is concentrated on the retina in the back of the eye, made up of photoreceptor cells. The

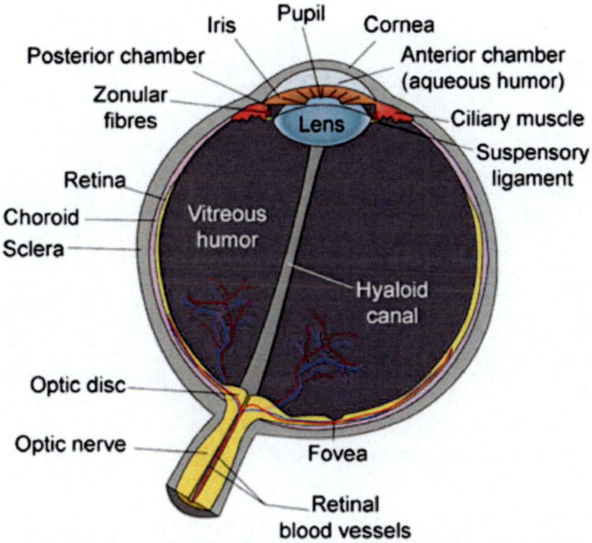

Fig. 3.2 Anatomy of the human eye. *Source* https://commons.wikimedia.org/w/index.php?curid=29385884. Licensed under Creative Commons Attribution-Share Alike 3.0 Unported. Author: soefm

pupil is a light-transmitting opening that adjusts its circumference according to the intensity of the incoming light. In darkness, the diameter of the pupil dilates, allowing more light to enter. In well-lit situations, the diameter of the pupil contracts so that the amount of light entering the eye is not too strong. With the pupil and crystalloid working together, the eye can receive light from a variety of sources: strong, weak, distant and near. The stretching of the ciliary muscle in the eye deforms the crystalloid and thus adjusts the refraction so that light can be focused on the retina to form an image. The pupil of the human eye may not change very much, but if we look at a cat's eye, we will see that in bright light, its pupil will shrink into a line, while in the dark, its pupil will become large and round.

The retina in the posterior section is made up of two types of photoreceptor cells, named rod cells and cone cells because of their shape. Their function is to convert the light produced by focusing the crystalloids into electrical signals that are sent to the brain. The cone cells are sensitive to colour (they can distinguish between the colours red, green and blue, which are also called primary colours, and other colours result from the combination of these three colours and black and white), and the rod cells are sensitive to variations in brightness (light/dark). The cones and rods are not evenly distributed on the retinal surface (see Fig. 3.3). The cone cells are mostly concentrated in a small central area of the retina called the macula ("yellow dot"), which has a diameter of about 5.5 mm. The central area of the macula is called the fovea, which has a diameter of about 1.5 mm. In the fovea, the concentration of cones is highest, so that, if our task demands that we see visual detail, we need to turn our eyes such that the centre of the area of interest is projected onto the fovea. The rods can be found mostly outside the macula. Furthermore, as can be seen in Fig. 3.3, the farther away we go from the fovea, the distribution of the rods decreases. Therefore, the farther away from the spot where we focus our visual attention, the progressively

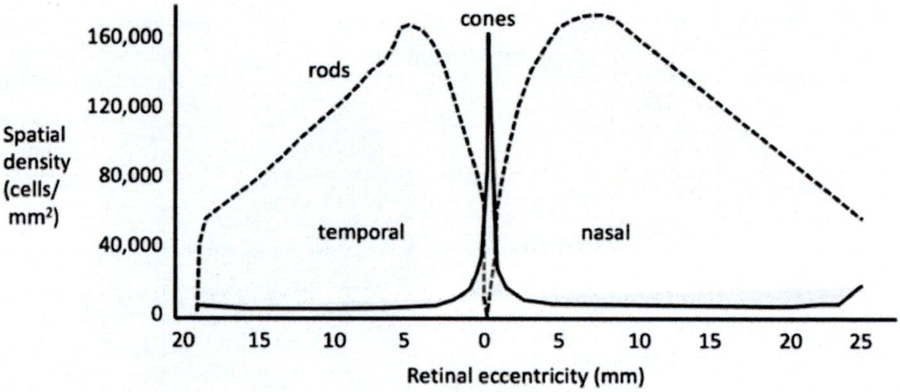

Fig. 3.3 Distribution of rods and cones across the retina. After: R. Milo and R. Phillips, Cell biology by the numbers (2015) http://book.bionumbers.org/how-big-is-a-photoreceptor/Fig. 2

less detail we perceive, and in the periphery of our visual field we only perceive gross shapes. Thirdly, since the rods are sensitive to variations in brightness, they are instrumental in perceiving movement in the periphery of our visual field.

The retinal nerve fibres converge in the optic nerve, which transports the electrical signals from the retina to the brain for further processing.

The ability of the eye to distinguish the size of objects is also called visual acuity and is divided into static and dynamic vision. Static vision is the driver's visual acuity at rest, while dynamic vision is the driver's visual acuity while the car is in motion. For example, at a speed of 60 km/h, a driver can see a traffic sign at 240 m; at a speed of 80 km/h, a driver can only see a traffic sign at 160 m (of course, this is also related to the size of the sign). Visual acuity is also related to brightness, which decreases as a function of distance and time of day; in particular, the light at dusk is most detrimental to the driver's ability to see. In addition, there is an adaptation process for vision from dark to light or light to dark, during which visual impairment can occur. This knowledge is important in designing traffic indication messages. If important information is expected to be visible to drivers from a greater distance, it should be designed according to the dynamic visual acuity in relation to the speed limit of the road.

Viewing information inside the vehicle is somewhat different to viewing information outside the vehicle. During driving, the relative speed between the person and the display system is static and the distance is relatively constant, regardless of the speed of the vehicle. A possible influence is the blurring of vision due to some vibrations caused by the high-speed movement of the car, but this should not be significant. Another factor is that when driving manually, the human eye cannot be taken off the road for too long (2 s is the accepted safety limit) and therefore the information that can be viewed is limited (NHTSA 2010-0053).

Since our eyes are positioned at the front of our head, we can only see part of the world. This is called the field of view (Fig. 3.4). Within the field of view, there is the useful field of view (UFOV, Wood and Owsley 2014). This is the part of the

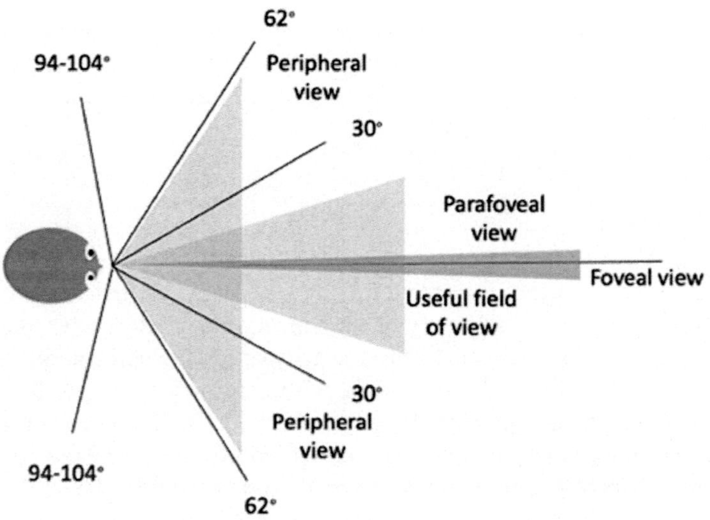

Fig. 3.4 Field of view

field of view from which information can be extracted in a brief glance without head or eye movements; that is, within this area, a person can extract information by simply looking. The width of the UFOV generally decreases with poor vision, and when giving effort to ignore distraction; these effects may be due to the fact that the UFOV is inversely correlated to foveal task demands (see below): the harder we work to process the information that we focus upon, the narrower the UFOV. UFOV also decreases with age, due to slower processing ability. Furthermore, the size of the UFOV is inversely correlated to the speed of the vehicle; as the speed of the vehicle increases, the driver's UFOV becomes significantly narrower. For example, at a speed of 40 km/h, the UFOV is 90°–100°; at a speed of 80 km/h, the UFOV is 60°. Finally, the UFOV has been shown to inversely correlate with vehicle crash risk, obstacle collisions and the propensity to fall.

Within the UFOV, there is an area where we see sharp detail. This is the foveal view. It is a narrow area of the field of view, and measures about 2 arc degrees. Next to the foveal view, there is the peripheral view. Within the peripheral view, we may distinguish the near peripheral view, the mid peripheral view and the far peripheral view. As mentioned above, the further away in the periphery, the more blurred the visual information. However, the periphery plays a very important role when driving. Much information from the environment is obtained through this peripheral view. Even in the far periphery, the visual system is sensitive to fast movement and bright light.

The way people perceive and experience different colours is called colour perception. People react differently to different colours, for example, red is highly visible and irritating, making people alert; yellow light is the brightest and has the highest intensity of reflected light, easily attracting people's attention; green light is softer, giving people a sense of calm and security. Therefore, traffic engineering will use red light as a no-go signal, yellow light as a warning signal, green light as a pass signal. Because this definition of colour in traffic is so well understood, the same colours are used to convey the same meaning in the design of in-car systems.

3.3 Hearing

In a broad sense, 'hearing' has two levels of meaning. The first level refers to the perception of sound, i.e., the ability of the sensory nerves to receive sound, which is innate and is mainly related to the integrity and development of the auditory system, i.e., whether the auditory system and related structures are histologically, anatomically and physiologically properly developed. The second level refers to the recognition or interpretation of sound, i.e., the ability to understand sound, which is based on the first level and results from the processing of the auditory system at all levels, which includes complex mental processes such as comprehension and memory, and therefore requires acquired knowledge.

Physiologically speaking, sound is transmitted mechanically through the earwax, external auditory canal and tympanic membrane in the external ear, the auditory chain, the eustachian tube in the middle ear and the internal and external lymphatic fluid in the inner ear to the special auditory cells located in the cochlea (Fig. 3.5). The auditory cells receive the mechanical vibrations of the tympanic membrane caused by the sound waves and transduce them into neural electrical activity containing sound

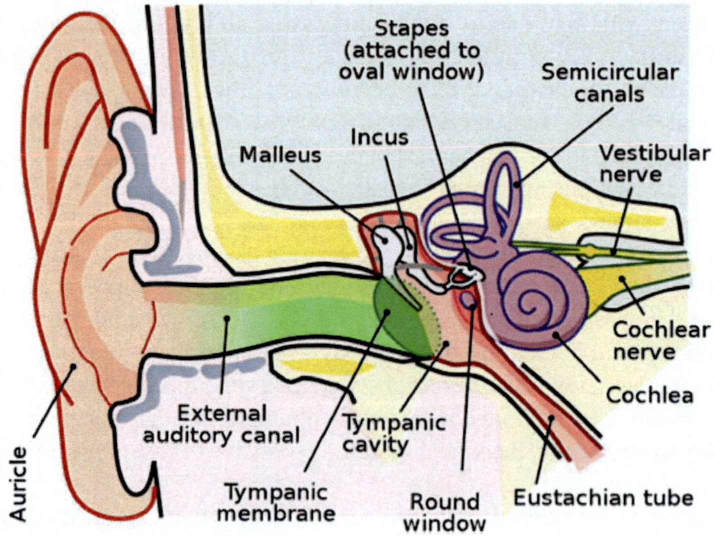

Fig. 3.5 Diagram of the ear. *Source* Chittka L, Brockmann A (2005) Perception space—the final frontier. PLoS Biol 3(4): e137. Licensed under the Creative Commons Attribution 2.5 Generic license. https://commons.wikimedia.org/wiki/File:Anatomy_of_the_Human_Ear.svg

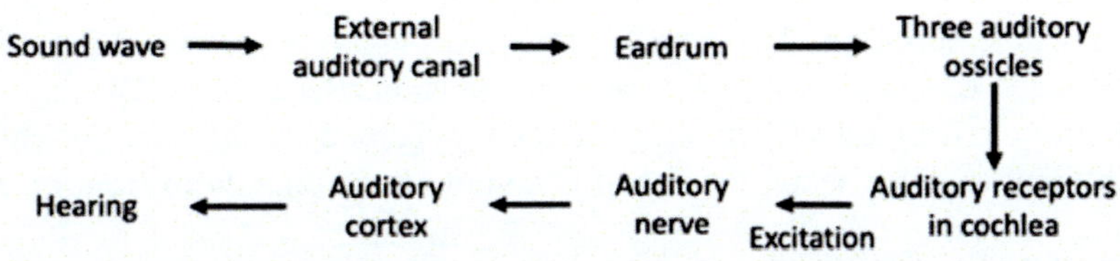

Fig. 3.6 Schematic representation of the process of auditory transduction

information, which is encoded in the form of different frequencies and combinations of nerve impulses, which are then re-encoded by relay neurons at various levels and transmitted to the auditory centre. The sensitivity of the auditory system changes with age, as the sensitivity of the nerves in the cochlea to different frequencies of sound waves decreases with age. Figure 3.6 shows the process of auditory transduction.

For interaction design, we are most interested in the perception of sound. The auditory perception of sound has a certain pattern. Firstly, the auditory nerve in the ear picks up the mechanical vibrations of sound waves, which determine the presence or absence of sound. This is followed by the elicitation of auditory attention. Generally speaking, content that is meaningful to the listener is likely to attract auditory attention. For example, if your name is mentioned at a very noisy cocktail party, you will perk up your ears, even if the sound is not too loud, and this is the "cocktail party effect". Our sense of hearing can also identify the direction of the source of sound. When any sound reaches the two ears, the distance between the ears and the shape of the head can cause a slight difference in the timing and intensity of the sound reaching the two ears: the Interaural Time Difference (ITD) and the Interaural Intensity Difference (IID) (Fig. 3.7). The intensity difference arises because, for sounds that do not originate from sound sources precisely in the front or back middle, the head dampens the sound arriving at the far ear. The human brain analyses these differences to determine the direction from which the sound is coming.

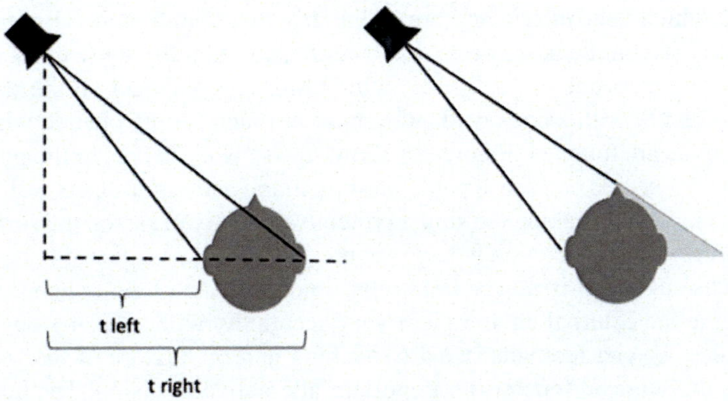

Fig. 3.7 Schematic representation of human localisation of sound. Left: Inter-aural time difference (ITD). Right: Inter-aural intensity difference (IID). After: Feinkohl (2014)

Of course, the human ear has some shortcomings in its judgement of direction and distance, and this is known as front-back confusion. Because the IID and ITD of sound arriving from directly in front and directly behind are the same for both ears, it is difficult to tell whether the sound is coming from directly in front or directly behind.

The ability of humans to judge the orientation of sounds has been increasingly developed and applied as an important part of interaction design. In automobiles, the ability to generate an alarm sound in the same direction as an emergency event is a clear example of this (Chen et al. 2007, 2008; Wang et al. 2012, 2017). Audible warning messages combined with orientation information will allow drivers to better respond to alarms appropriately.

Auditory recognition is the ability to distinguish similarities and differences in sound. It is the result of a combination of sensory and brain analysis. We listen to music, which is a perfect example of auditory recognition. We can distinguish sounds by their frequencies, peaks, timbres, etc., and can identify different sounds, their content, their emotions, etc. Timbre refers to the distinctive characteristics of sounds in terms of their waveforms. Different objects have different characteristics of vibration. Different sound-producing bodies have different timbres due to their different materials and structures. For example, a piano, a violin and a human being do not produce the same sound, nor does each person produce the same sound. Therefore, timbre can be understood as the characteristics of a sound. Auditory memory, on the other hand, is the storage of sound signals in the brain based on the recognition of sounds.

3.4 The Sense of Touch

The sensation produced by the skin's tactile receptors in contact with a mechanical stimulus is called haptic. The surface of the skin is scattered with touch-sensitive receptors, which vary in size and are irregularly distributed, generally most on the fingers and least on the head, back and lower legs, so that the sense of touch is most acute on the fingers and more diffuse on the lower legs and back. If the skin surface is touched lightly with a fine needle, the sense of touch is only elicited when certain specific points are touched. Figure 3.8 shows the sensory nerves in the human skin. These sensory nerves can sense pain, warmth, touch and vibration as well as certain complex sensations, making the skin a protective sensory organ on the surface of the body, but the sensory function is less specific.

The sense of touch is a very important function of the human hand. Often, we can perceive the external environment very accurately without using our eyes, and in particular, we can feel what we are touching and the texture of the surface. For the blind, the sense of touch is an important alternative to vision. The human hand is distinguished from the animal by its sensitivity and dexterity. In cars, the majority of operations and information input, before voice interaction became commonplace, was done by hand. As the driver's vision needs to be focused to the greatest extent

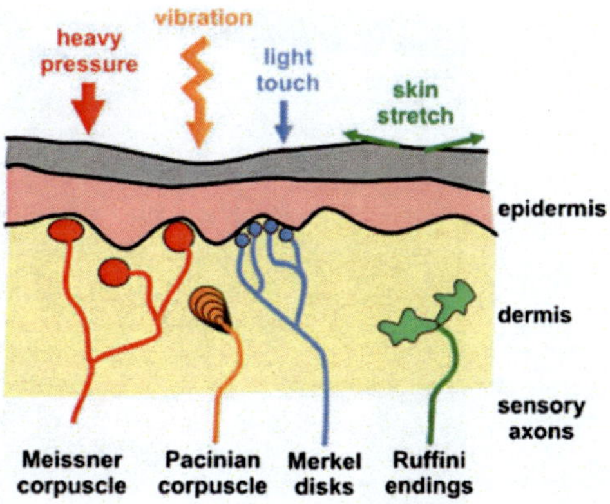

Fig. 3.8 Tactile receptors in the skin. *Source* http://www.pc.rhul.ac.uk/staff/J.Zanker/PS1061/L6/PS1061_6.htm. Reprinted with permission from J. Zanker

possible on the information outside the car, the design of the information input system in the car should rely as little as possible on the assistance of the eyes when operating by hand, so to speak. Many of the buttons in the car are designed so that the driver can operate them "blindly", without having to look at them, just by touching them.

In fact, the manipulation of the human hand is the result of instructions from the brain acting on the muscles of the hand and arm, while the tactile nerves in the skin perceive the manipulation and gather feedback to the brain in time, thus creating a circuit. Repetition of the same action makes this circuit more fluid and faster, eventually forming a reflex arc that is less dependent on the higher levels of brain activity. This is why a skilled driver can steer a car quickly, accurately and naturally, without much thought. There are many other examples of the development and application of haptics, for example, touch screens, as in Fig. 3.9.

There are two parts of the human cerebral cortex that are most elaborate, the mouth and the hand, which correspond to our ability to speak and to manipulate our hands. Figure 3.10 is an analogy of the little man in the human brain, where the size of the body parts reflects the sensitivity for touch.

In connection with the haptic sense, also proprioceptive perception needs to be mentioned. Proprioceptive perception refers to our perception of the position of the parts of our body, mediated through the action of nerve cells in our muscles, tendons and joints. When reaching for something, the movements of our hands and arms may be guided by our eyes (hand–eye coordination), or, when we reach for familiar locations, the reaching may be controlled by the proprioceptive information in coordination with haptic information once our hand reaches the object. For instance, when reaching for a button on the mid console, the driver may keep his eyes on the road, while the movement of his arms and hand are directed by proprioceptive information. Once his hand reaches the object, the haptic feedback may lead to fine adjustments in the position of the hand and fingers. Also, when turning buttons, we can feel resistance.

Fig. 3.9 Haptic interaction

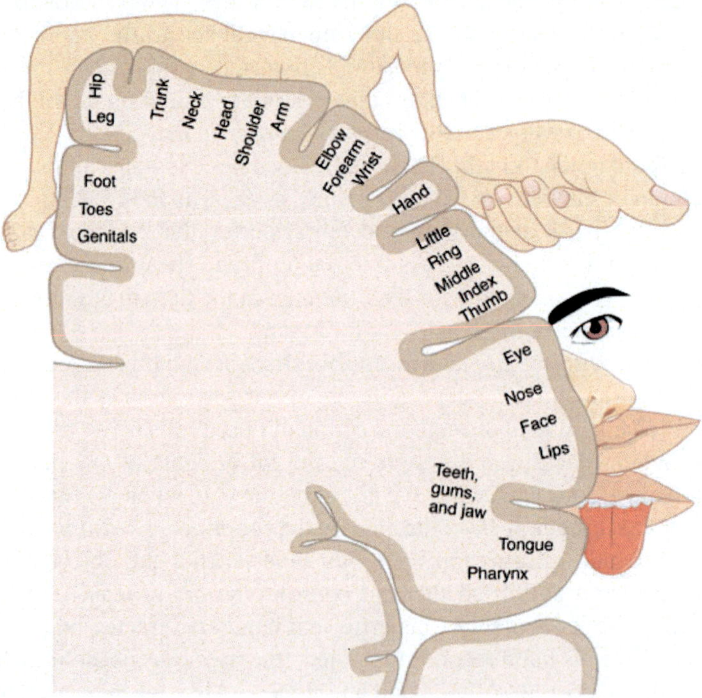

Fig. 3.10 Penfield's homunculus. From: anatomy & physiology, connexions web site. https://openstax.org/books/anatomy-and-physiology/pages/14-2-central-processing. Author: OpenStax College. Licensed under the Creative Commons Attribution 3.0 Unported

3.5 Attention

Driver attention is a critical factor in safe driving. A survey in the USA showed that 40,000 people are killed in traffic accidents each year, more than half of which are caused by distractions (Lee et al. 2009). While driving, drivers must focus their attention on the road and in the vehicle, selecting and focusing on what is most relevant to safety. Referring back to Fig. 3.1, attention can be directed by two processes. In the first place, attention can be guided by the goals that derive from the driving task we are performing, and which are held in working memory. In this case, we focus our attention on the road, and, moving our eyes across the visual scene, actively scan the environment for information that is potentially relevant to the driving task, such as the presence of other vehicles, bicyclists, pedestrians, traffic lights and traffic signs, line markings and so forth. In the second place, as we have seen in Sect. 3.2, sudden movements and very bright objects draw our attention, even if they are in the periphery of the visual field. This is what we call perceptual salience (the same applies to sounds: loud and high-pitched sounds are perceptually salient). Once potentially relevant information is identified, the visual system quickly decides whether attention should be given to this object or event. Unfortunately, perceptually salient objects and events in our environment may attract our attention also if they are irrelevant to the driving task, thus diverting our attention from the driving task (e.g., billboards on the roadside, a fancy car parked along the street, an accident in the opposite lane on the highway, etc.). Billboards on the roadside are usually intentionally designed to be perceptually salient and attract our attention.

Attention can take different forms. Selective attention refers to the situation where we direct our attention for a longer time to a single channel of information, such as in the case of reading. Divided attention, on the other hand, refers to the situation where we divide our attention across different channels of information, either in quick succession (such as when looking at the road and checking the speedometer and the navigation display) or at the same time (such as in the case of driving in combination with listening to music or engaging in a conversation with a passenger). The latter is also called multi-tasking, which will be discussed in more detail in Sect. 4.3. From the information processing model in Fig. 3.1, we know that attention affects many aspects of information processing, so we will discuss the issue of attention in relation to different aspects.

Selective visual attention

Within the visual field, selective visual attention can be engaged in six different task types (Wickens et al. 2013).

1. Scanning back and forth in the visual area looking for task-relevant information, e.g., the driver is constantly looking at the front and side of the car and the rear and side mirrors.
2. The line of sight is scanned back and forth over a specific path, supervising the control to ensure that certain dynamic variables such as line markings are controlled to be within range. If they are out of range, some form of manual

control is performed to restore them to their original state, for example lane keeping in driving. This task is highly goal-oriented.
3. Attention, including monitoring, especially in response to certain unexpected events. (Such events are not included in the supervisory control tasks to be performed.)
4. Search for a specific, often predefined, target. For example, driving while looking for the person you want to pick up at a specified location.
5. Reading.
6. Confirming that certain controls are in place (e.g., handling feedback).

Many tasks are clearly a mixture of some of the above. For example, reading diagrams or interpreting maps when following instructions to operate equipment usually involves some combination of searching and reading.

There are four factors that determine how visual attention will be directed (Wickens et al. 2013):

1. (Perceptual) Salience: The extent to which an object or event stands out from the environment, by virtue of its size, colour, intensity, contrast or movement, etc. Objects that stand out from the environment contain information that is potentially relevant to the task, and therefore salient objects have a tendency to draw an individual's attention.
2. Expectation: We tend to focus more on areas where much information is expected. While Salience is a bottom-up process, guided by environmental input, Expectation-guided direction of attention is a top-down process.
3. Effort: Moving the focus of attention from one location to the other takes time and effort. The larger the movement, the more effort it takes. This is also known as Information Access Effort (IAE).
4. Value: This indicates the usefulness (importance) of the information (i.e., the relevance of the information to the task, weighed by the relative importance of the task). Attention will be preferably directed to channels with high information value.

Since there is the possibility of salient objects or events catching visual attention, there is also the possibility that visual attention should be drawn but is missed for various reasons, this is known as change blindness (Wickens et al. 2013): it is used to describe situations where changes in the environment go unnoticed. Traffic accidents often occur because drivers suffer from change blindness and fail to notice relevant information on the road, leading to incorrect actions.

When referring to visual attention, the concept of 'area of interest' (AOI) needs to be introduced. The AOI is an external physical area in which one can find information relevant to the task. With the development of eye-tracking technology, there has been an increase in the study of visual attention. The scanning of the environment may seem easy, but it is not without energy consumption. As mentioned under 3 above, the movement of one's visual attention from one AOI to another requires effort and digests energy. With effort, comes fatigue. Similarly, the distance between two AOIs determines the amount of effort expended (IAE).

The visual angle that the fovea (Fig. 3.2) spans is between approximately 2 arc degrees. If the AOI is less than 4 degrees, the IAE paid is minimal. If the visual angle of the two AOIs is greater than 4 degrees, inspecting both AOI requires an eye movement (saccade). If the distance increases further, in particular when the visual angle is larger than 90 degrees, a rotation of the head is required, ultimately in combination with a rotation of the body. It follows that the further apart the two AOIs are, the more energy is required for visual scanning. Furthermore, the driver needs to be aware of the information on the dashboard and mid console as well as capturing the road information. As the AOI on the road and the AOI on the dashboard are about 30 degrees from each other, the driver only needs to turn his or her eyes, whereas the AOI on the mid console requires an assisted head turn. The larger the mid console, the further the AOI in the mid console is from the AOI on the road, and the greater the head turn, which may even require an assisted body turn. As a result, the more energy is expended and the longer it takes to move the eyes between the two. The advantage of a head-up display (HUD) is to reduce the distance between the AOI on the road and the AOI in the display.

Visual search

The search for a target through our visual selective attention is called visual search. Generally speaking, the object to be searched is predetermined in advance. This search function of human vision is indispensable in our everyday life and has been extensively studied at the end of the last century and the beginning of this century, especially in the field of driving (Ho et al. 2001). Visual search is usually carried out more or less systematically by eye movements within the Useful Field of View (UFOV). The size of the UFOV affects the effectiveness of the visual search. In general, experienced drivers have a larger UFOV than novice drivers (Owsley et al. 1998), demonstrating that UFOV is susceptible to training.

When it comes to visual search, there is a question of efficiency and speed. We won't go into the details of the factors that influence this. One well-known factor that has a direct impact on search efficiency is the layout of the display and presentation of the target being searched. In general, the efficiency of visual search is improved by grouping related and similar information. Figure 3.11 shows a typical example. Here it can be seen that it is easier to search visually in image (a) than in image (b).

In traffic, the visual scene is often complex and contains many elements, and when we need to identify information of interest from this multifaceted world, our vision employs two parallel processes. (1) Preattentive processing: This is a bottom-up process, which automatically groups features in the environment into objects, establishes relations between objects such as nearby-far away, and groups similar objects into groups of objects. Features that play a role in this process are captured by Gestalt laws such as Figure-Ground, Proximity, Closure and so forth. (2) Attention-guided visual search. This is a top-down process, using working memory resources. It works on the output of the pre-attentive processes and inspects the output emerging from the pre-attentive processes for relevant entities. Since the pre-attentive processes are susceptible to the perceptual salience of elements in the visual field, the attention-guided visual search can be facilitated if the target is perceptually salient. In that case,

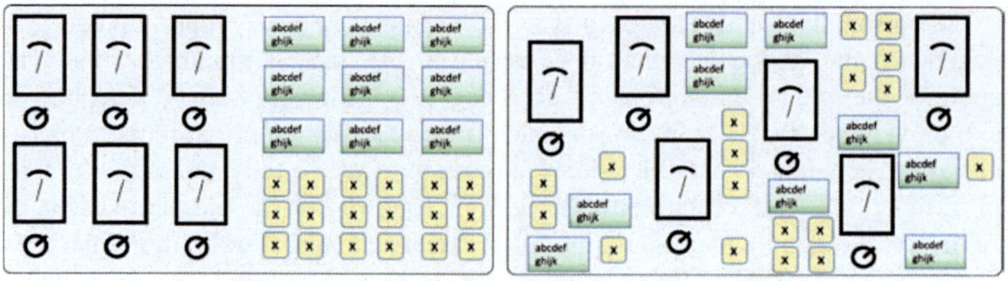

(a) (b)

Fig. 3.11 Layout of information display in relation to visual search. Diagram **a** groups similar information together in clusters. In diagram **b** the information is dispersed in a less orderly and structured manner

(a) (b)

Fig. 3.12 Arrays for visual search. **a** No perceptually salient elements. **b** The targets are given a different colour, making them perceptually salient

the attention-guided visual search will not inspect the set of objects emerging from the pre-attentive processes one by one, but will immediately jump towards the salient objects. Compare for instance the task of searching for the '4'-s in Fig. 3.12a, b. If the '4'-s stand out from the environment by their colour, the visual search can be completed much faster (of course, at the expense of missing '4'-s that are not yellow). Putting information in the environment to facilitate information processing and remembering is known as "external cognition".

The same process works in traffic. In Fig. 3.13, the search for bicyclists on a rural road at night is much facilitated if the bicyclists use reflective materials and lighting. Note that the second bicyclist, not using reflective materials and lighting, will be missed easily.

Auditory attention

Hearing differs from vision in three important ways: (1) hearing is omni-directional: sound can come to the ear from any angle; (2) hearing is full time: unlike the eyes, which may close so that we cannot see, our ears do not close; (3) auditory input is volatile: a spoken word, a sound, comes and goes with time. It cannot be retained, unlike vision, where objects may stay present in the environment.

Again, as with visual attention, pre-attentive processes and attentive processes work at the same time. The sounds of the environment are picked up and processed by the ears, but we do not 'hear' all the sounds. Instead, our auditory attention selects

Fig. 3.13 Searching for bicyclists on a rural road at night. *Source* https://www.fietsenloix.be/slim-op-de-fiets/119-zien-en-gezien-worden-op-de-fiets

one channel that contains information that is relevant to our task. For instance, we may focus on a conversation, while "suppressing" the sounds from the radio. The pre-attentive processes code information such as timbre and the spatial location from which the sounds come, and this information is used by our auditory attention to tag the relevant sound stream. However, again, perceptual salience comes into play. If there is a sound from the environment that is perceptually salient, e.g., through its amplitude or frequency, it may draw attention, and the person may switch attention from the sounds that s/he was focusing on to the sound that just arrived. The fact that people generally distinguish/identify sounds by several different sound characteristics such as timbre and spatial location, provides a good basis for the design of sound interaction.

Whether visual or auditory, if the designer does not want the user to process two or more pieces of information at the same time, then all but the primary information can be seen as a distraction. Of course, in everyday human life, information is not received in a single modality, and hearing and vision often work together. For example, when driving, the driver's eyes look ahead on the road, while the electronic map gives a spoken indication of the road. This multimodal interaction will be discussed in more detail in Sect. 9.2.

3.6 Memory

As shown in Fig. 3.1, human memory can be divided into working memory and long-term memory. Working memory is a more active type of memory, temporary in nature, used to store and process new information. It acts as a workbench, examining, evaluating, comparing and transforming the information acquired by the senses. This part of memory is like our own self-awareness, representing the active part of the brain, and it is also the process of "encoding" the acquired information and storing

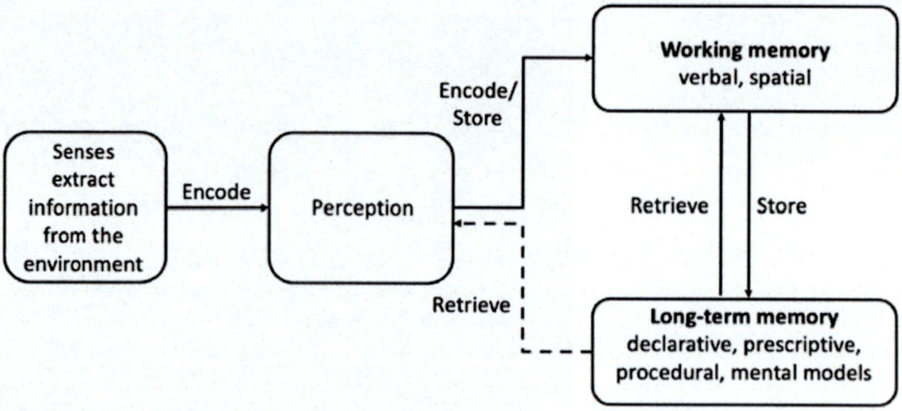

Fig. 3.14 About the process of remembering

it in long-term memory. Long-term memory is where we store knowledge about the world and ourselves, and about how we do things (Wickens et al. 2013, p. 197).

Our memory process can be schematically represented by 3 stages: encoding, storage and retrieval. This is shown in Fig. 3.14.

Working memory There are different models for working memory, sometimes also called Short-term Memory, as opposed to Long-term Memory. Here, we use the model proposed by Baddeley (Baddeley 1986, 1995; Baddeley et al. 2009). In this model, a distinction is made between two registers for storing information, and the Central Executive. The registers are the Verbal register for storing verbal-linguistic information such as text and speech, and the Analogue-Spatial register for storing visual and spatial information. The Central Executive performs the mental operations on the information stored in the two registers, such as extracting the gist from a verbal message stored in the Verbal register (or just rehearsing the information such as a number that we want to memorise) and extracting parameters like speed and direction of moving objects stored in the Analogue-Spatial register.

Baddeley sees four roles for the central executive: (1) Coordinating performance on multiple tasks; (2) Retrieving information from long-term memory; (3) Temporary preservation and manipulation of information retrieved from long-term memory; (4) Selectively responding to stimuli. The different activities performed by the Central Executive are all guided by the goals of the tasks at hand, and these goals are also contained in Working Memory.

A further characteristic of Working Memory is that it has limited capacity, as opposed to Long-term Memory, which is generally assumed to have (near) unlimited capacity. There have been many experiments exploring how much information can be held in working memory, how long it lasts and how forgetting occurs. We won't go into the details of how these studies have been done, but summarise the main insights. In case of rehearsing information (repeating them in our mind), the number of independent items, like digits or words, that can be held in working memory is about seven. If the number increases further, forgetting will occur. Also, this capacity varies from person to person, and is influenced by matters such as fatigue (hence, the famous 7 ± 2 rule).

As we talk about "independent items" here, we need to understand what these are. For example, five unrelated numbers, letters, or words can be thought of as five independent items. A mobile phone number is 11 numbers (including the country code but excluding the leading zeros). If we apply grouping, we can reduce the number of independent items making a telephone number easier to remember. For instance, we often divide the phone number into groups of 3 or 4, forming a sequence of $3 + 4 + 4$, making it easier to remember. Furthermore, a string of letters may form a word, so that it becomes one independent item. A typical application of this is the Swedish license plate number, which usually consists of 3 letters and 3 numbers, for example "MUM 226". This makes 2 groups of 3 "independent items" each, where the first group can be pronounced like a single syllable and then constitutes one independent item. Likewise, digits in telephone numbers can be replaced by letters, so that 1-800-265-5328 becomes 1-800-COLLECT, which is much easier to remember. Thus, when presenting information, we also need to consider how we can help the user to group the information in a way that it is easy to remember.

However, most tasks in real life do not concern rehearsing, but processing the information for further use. This involves processes like extracting the meaning of a verbal message, integrating it with information from long-term memory, maintaining a model of the world around us, in short, deciding upon the meaning of the state of the world in view of our goals, and deciding upon actions. Again, Working Memory has limited capacity. For this type of tasks, the limitations of working memory have been less well explored, but in general, it can be said that the limited capacity of Working Memory is evident from the existence of forgetting: if new information comes in, it overwrites older information. If we listen to a conversation, the literal text of a sentence that we heard just a few seconds ago is already lost, and only the gist of it remains. And as the conversation continues, fewer and fewer details from the earlier parts of the conversation ore retained. For visual memory, the situation is somewhat different: if we are shown a large number of pictures, and are show a sample of pictures after a short while, some of which were shown before, we still have very good memory of whether a particular picture was in the original set or not. But this is not because we kept all pictures in Working Memory. Instead, apparently our visual Long-term Memory works differently than our auditory memory.

The above provides the theoretical basis for the subsequent discussion on how to design interfaces so that the presentation of information matches the way information in working memory is coded. This is the well-known SCR compatibility principle (Wickens et al. 2013, p. 202), which specifies the optimal association of display formats with working memory codes. In SCR, S (stimulus) denotes the display format (auditory and visual), C (central processing) denotes the two possible central processing codes (verbal and spatial), and R denotes the two possible response formats (manual and acoustic). Figure 3.15 shows the best matching codes between information content, access modality and central processing or cognition.

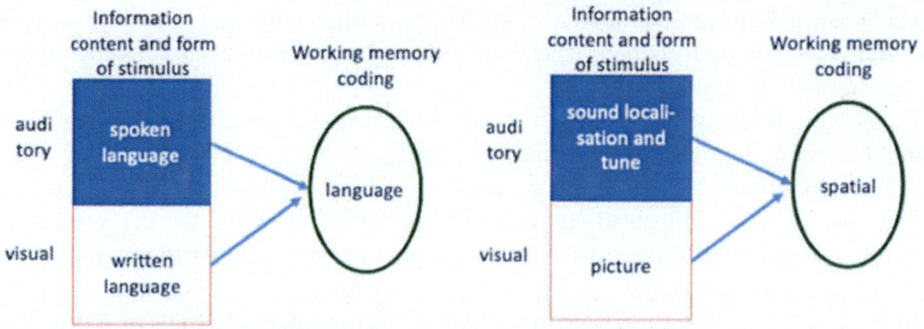

Fig. 3.15 Best-fit model between information content, mode of access and central processing or cognition

Any information (stimulus modality) in the environment can be divided into two encoded forms: verbal, or spatial. And the modalities of presentation are also two: visual or auditory. This gives rise to four possible modalities: textual information expressed in a visual mode is written, in an auditory mode it is speech. Spatial information is expressed visually as a picture, while spatial information is expressed aurally as a sound with orientation and pitch information. Figure 3.16 shows the best matching patterns. For example, a map is a better description of the location of a geographical space than a textual description. At the same time, a short message that needs to be remembered in words is best represented by speech, which is easier to remember than written words (Nilsson et al. 1977). This principle has a very important practical value for designers. When a short text message needs to be conveyed to the driver, the best method is through speech so that the information is not lost in the auditory senses at the point where the message is received. However, if it is a longer message, then it is better to write it in text that can stay longer, or to repeat the speech message.

Long-term memory acts as a repository of our knowledge, both about the world and about our own past history. The latter is also called episodic memory, as opposed to the

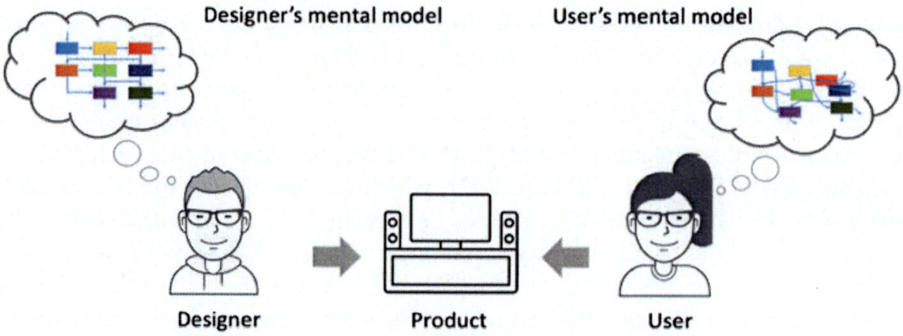

Fig. 3.16 The designer's mental model versus the user's mental model

former type of knowledge, which is sometimes called semantic memory. Knowledge about the world is stored as declarative, prescriptive and procedural knowledge. Declarative memory is factual information that can be represented as statements ("declarations"). Prescriptive knowledge concerns our knowledge about rules and prescriptions, such as applying to traffic signs. Procedural memory is knowledge about how to do things. In addition, researchers of human long-term memory have distinguished a type of knowledge called schemas, which consist neither of precise declarations or prescriptions nor of procedural knowledge. A schema is a generalized form of knowledge about a particular type of situations. A well-known schema is the restaurant schema, which captures the knowledge about what happens when we visit a restaurant. For instance, we may know that, when we enter a fancy restaurant, a waiter may welcome us and take our coats, then guides us to a table, asks whether we want something to drink and will bring the menu. Then, after a while, s/he asks about our choice of dishes. Then the dishes will be brought. After dinner, the waiter will bring the bill, we pay, get up and wait for the waiter to bring our coats. This knowledge facilitates our visit to the restaurant, as it is the basis for expectations about what will happen. In other cultures, other schemas may exist for restaurant visits.

Knowledge can be highly automated, so that retrieving and applying the knowledge requires little effort, or, in other words, hardly uses working memory capacity, while other knowledge is not automated, so that retrieving and applying the knowledge requires much effort, or, in other words, uses much working memory capacity. For instance, for experienced drivers, retrieving and using knowledge of how to drive (braking, steering and so forth) is highly automated and requires little working memory capacity, while for novice drivers, performing the same tasks takes much working memory (at the expense of monitoring the traffic and deciding upon actions related to manoeuvring through traffic).

Forgetting Forgetting information may be the result of two different mechanisms. One mechanism is decay of memory traces. This mechanism seems to apply more to information stored in Long-term memory. Information that is not used (activated) for a long time, appears to degrade and get lost in the end. This applies for instance to a foreign language that we learned in secondary school and did not practice throughout our adult life. In later life, most of our knowledge about this language has been lost. The same may apply to our knowledge about the meaning of traffic signs. Traffic signs that we had to learn for our driver's license may no longer be remembered after a while, if they are not encountered in real life. It should be noted, however, that there also other forms of "forgetting" information from Long-term Memory, which relate more to a failure to access the information. This applies for instance when we are looking for a name or a word, where we know we know it, but it "won't pop up", although it may be "at the tip of our tongue". Usually, after a while it suddenly pops up, providing evidence that we did not forget it. The second mechanism is "overwriting" information, which appears to apply more to Working Memory. As stated above, the capacity of Working Memory is limited. If it "full", e.g., with a string of digits, and further digits are added for memorisation, part of the information that is already stored in Working Memory is overwritten by the new information and is lost.

3.7 Mental Models

A mental model represents a person's knowledge about a part of the world. In the context of interaction design, a mental model represents the user's knowledge about a system. A mental model is an internal cognitive construct formed in the human mind on the basis of interactions with the external world, which can help people make predictions about the outcome of actions. Therefore, mental models also determine how people react to a particular state of the world. Mental models are created spontaneously by users through their interaction with the system and involve both unconscious and conscious processes. They can be built up through repeated use or training (Moray 1999). Mental models are often used to explain the use of technologies, especially when new technologies or products are first introduced, and are built up through continuous practice.

The formation of mental models is a process of accumulating experience through the interaction with external physical systems. A distinction can be made between structural and functional models. Structural models contain knowledge about how the system exactly works ("under the hood"), functional models contain knowledge about how to achieve particular goals through interacting with the system. Since models (in particular functional models) derive through interaction with the system and different people will have different interaction histories with the same system, different people may have different models of the same system, and therefore, mental models may be incomplete or incorrect, containing wrong assumptions about how to achieve our goals. For instance, when we get home in winter after a day in the office and want to bring the temperature to a comfortable level as quickly as possible, we may set the heating to a high setting, thinking that it will get warm faster. However, this applies only if the heating works with valves. A modern system with a thermostat works in an all-or-none fashion, so that temporarily setting the target temperature to a higher value does not matter. Similarly, when we are waiting for the lift, we may press the lift button twice, or even several times in a row, hoping that it will make the lift come faster. These observations show that it is very common for people to use the wrong mental model to guide their actions. It should be noted, however, that a wrong mental model may not prevent people from reaching their goal. Usually, the worst that may happen is that the interaction becomes less efficient, e.g., because the user performs more actions than needed to reach the goal.

The concept of a mental model is important to interaction designers to explain why some designs are easy for users to operate and why others are difficult (Fig. 3.16). The designer has a mental model of how the system being designed should be operated, and he or she designs the product according to this model. The user also has her own mental model of how the system operates. The more similar these two mental models are, the better the product will work, whereas if they are farther apart, the worse the product will be. Or, otherwise said, the more the designer, through good design, helps the user to build the correct mental model, the easier it will be for the user to operate the system. This is why we need to do user research to understand the user's mental model when we design a product.

Studying users' mental models and the way mental models are constructed means studying their problem-solving styles and operating habits. Sometimes, for the sake of convenience, the designers study their immediate colleagues, on the grounds that "they also drive, they are also drivers", and we can see from Fig 3.16 why they are not representative of users. In the next section, we will look in more detail at how knowledge is built about how to interact with systems.

3.8 Learning

In learning psychology, usually a distinction is made between intentional learning and incidental learning. In the case of intentional learning, people study learning materials and try to memorize them explicitly, e.g., for reproduction at a later time during a test. In the case of incidental learning, people explore the environment and, while doing so, process the information presented by the environment in Working Memory. At a later time, it turns out that some of this information was stored in Long-term memory and can be reproduced if desired, e.g., to tell what a novel that one read was about.

In the context of learning to interact with a system through a user interface, both forms of learning apply. People may explicitly study a manual or tutorial to understand and memorize how to operate the interface. However, in case of consumer applications and systems it has been found that people are not very fond of reading manuals or going through structured tutorials and prefer to learn how to operate the system while using and exploring it. The way people learn to work with systems was studied by Rasmussen, who made a distinction between three forms of behaviour: Knowledge-based behaviour, Rule-based behaviour and Skill-based behaviour (Rasmussen 1986)

.

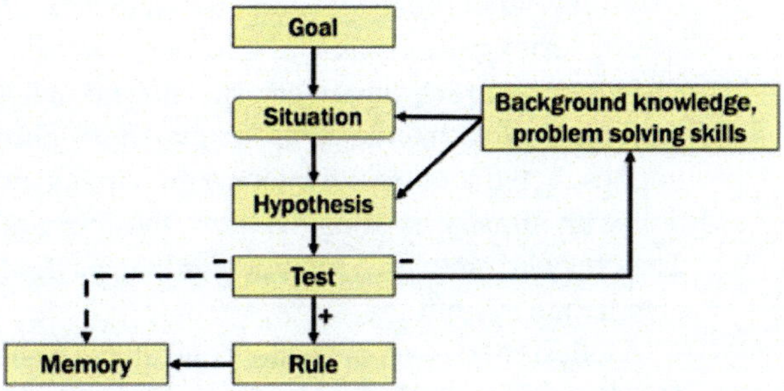

Fig. 3.17 The process of exploring a user interface based on background knowledge

(1) Knowledge-based behaviour (KBB) applies when the user is confronted with a new system and tries to figure out how to work with the system. The situation resembles a problem-solving situation, and the user uses the information that the user interface provides in combination with his background knowledge to generate hypotheses about how to achieve certain goals with the system. This process may be represented as in Fig. 3.17. The user has a goal, e.g., to find out how to operate the system. Based on inspection of the User Interface, the user generates a hypothesis about what to do, or what the effect will be of a certain action. In generating this hypothesis, the user applies his background knowledge. The user then performs the action, testing the hypothesis. Either the outcome is as expected, meaning that the hypothesis was correct. Or the outcome is not as expected, meaning that the hypothesis was not correct, and a new hypothesis needs to be generated. If the hypothesis was correct, the user may derive a rule stating that "If I perform action X, the result will be Y", or alternatively "In order to achieve Y, I need to perform action X", and store this rule in memory. Note that, as long as the action or a sequence of actions results in the desired outcome, the hypothesis will be confirmed, even if there would be more efficient ways to reach the same goal.

The notion of 'background knowledge' is rather vague, but can be understood as follows: Users have knowledge about the world from past interactions with the world and bring this knowledge to try to make sense of the situation. This knowledge is rather varied. E.g., we know that, in order to achieve something, we have to act. If we don't do anything, usually nothing will happen. In case of electrical devices, we know that they need power, so, if we are confronted with a new electrical device, we look for ways to turn on the system. In the case of electronical devices with screens, we know that we have to enter commands through a keyboard, we may know the function of a mouse from previous occasions, and we may know that icons on the screen can be touched and that that will activate an application. Also, we may know the meaning of certain terms such as File, Edit, View, Tools etcetera from previous occasions, and know that we can move a cursor to these interface elements and click on them to activate a menu from which we can choose a menu item.

The background knowledge that people bring to the situation differs between people, depending on their previous experiences. Some people may bring only very general background knowledge, while other people may bring very specific background knowledge. For instance, people who are experienced smart phone users, have very specific knowledge that helps them to quickly find out how to interact with a new smart phone that they bought. This also implies that, the more the interaction with a new system resembles the interaction with already existing systems, the more relevant background knowledge users may bring to the situation, and the easier it becomes to figure out how to operate the system.

As designers, we have no control over the background knowledge that people bring with them. However, it is not unusual to make certain assumptions about people's background knowledge. For instance, in the context of smart phones, it is usually assumed that people already have knowledge about how to work with smart phones. For people who don't, there are manuals to explain the working of the system. On the other hand, designers may design the information that the user interface provides such that it helps the users to figure out how to operate the system. As was already mentioned above, assumptions may be made about the background knowledge that people bring, and if the design of the user interface complies with these conventions, the user's task of figuring out how to operate the system is much facilitated. Secondly, the principles of Mapping, Affordance and Constraints may be applied (see Sect. 2.6). Technically speaking, we may say that the process of hypothesis generation in Fig. 3.17 is informed by two sources: one the one hand the user's background knowledge, and on the other hand the mapping, affordance and constraints represented in the user interface. If the designer appropriately applies these concepts, the user receives helpful clues about how to achieve his goals.

2. Rule-based behaviour (RBB). As mentioned above, the process of exploration leads the user to discover a number of rules about what to do in order to achieve certain goals (incidental learning). Accordingly, if the user at a next occasion needs to interact with the system, he may activate these rules in order to determine what to do. The collection of rules makes up the user's mental model.

   According to Rasmussen, Rules may also be obtained in other ways, bypassing Knowledge-based interaction. For instance, the user may consult a manual, or ask someone who is knowledgeable about the system for help. In the latter case, the most effective help consists obviously of helping the user reflect on what to do or by telling him user what to do, instead of taking over the interaction and explaining what to do. In the latter case, the user may not learn anything.

3. Skill-based behaviour (SBB). If a user frequently interacts with the system, the rule-based interaction will become automated, and the user has become skilled at the interaction with the system. Skill-based behaviour involves highly automated action patterns, which do not require complex conscious thought. The user does not need to think about what to do in order to achieve a goal, but immediately knows what to do.

Since skill-based behaviour arises from frequent interaction, it also means that for functions that the user only needs infrequently, the process from knowledge-based through rule-based to skill-based interaction may not be the most useful route. Instead, forms of interaction that take the user by the hand, such as wizard-based interaction, may be more meaningful.

3.9 Decision Making

We have to make all kinds of decisions every day, such as waking up early and what to wear. We have to consider the occasion we are going to, the specifications and requirements of the meeting we are attending, the weather conditions and so on. Many factors have to be taken into account. Do you drive yourself to work in the morning, or do you take the bus or a taxi? When making this decision, you need to consider the time of day, the accessibility of transport, the convenience of parking at the office if you drive, the cost of a taxi, and whether you can get a taxi in the morning rush hour. These daily decisions may have little impact on a person's day, but some of our decisions may have life-threatening consequences, endangering ourselves and others.

In the course of driving, people violate traffic rules, intentionally or unintentionally. Many times, this behaviour does not put you or others in danger, but sometimes a momentary indiscretion may lead to irreversible results. Some years ago, a tragic story happened in northern Sweden. In the winter in northern Sweden, the rivers were so thickly frozen that people could skate on them and many cars took the opportunity of a shortcut and drove over the ice instead of bypassing the rivers, so that the frozen rivers were often rutted. One father, driving home with his two daughters, did what many people did and drove over the icy river, only to have the ice break up and sink into it, drowning all three people in the car. The two remaining members of the family, the son, unable to cope with the tragedy, chose to die by suicide, leaving the mother, who eventually became schizophrenic.

There are more than one million major road accidents worldwide each year, and the number of people who die in road accidents each year is in the top 10 of causes of death. For every traffic accident, there is a series of decisions and processes behind the decisions made by those involved. Sometimes, if we look at individual decisions, they may feel right or not so badly wrong, but the combination of factors may have irreversible consequences. One of the key aims of interaction design is to help drivers make the right decisions. Therefore, we need to understand the decision-making process. In this section we will discuss the issues of decision making in a systematic way.

3.9.1 Decision-Making Process

Decision-making is a process of responding to the information obtained, and in the process of decision-making the following characteristics emerge (Wickens et al. 2013, p. 247): (1) Uncertainty: Almost any decision contains varying degrees of uncertainty. This uncertainty results in decisions that produce outcomes that may not be expected. Sometimes, the outcome is even unpleasant and can be costly for the bearer of the consequences. This is a condition we call risk. (2) Temporality: Time plays two important roles in the decision-making process. One is that at a certain point in time we have to make a decision, for example which way to go when driving. And then there is the time pressure. That is, the decision has to be made and implemented within a limited time frame. For example, when the light turns yellow, do I accelerate through the light, or do I slow down and stop? This decision must be made and executed before the light turns red. (3) Familiarity and expertise: Our decisions will change as we gain experience. An experienced driver is able to respond to different traffic situations with ease, almost instinctively, making a satisfactory decision. In contrast, a novice driver may be unable to make a satisfactory decision in time because of his inexperience, which is why we call a novice driver a road killer.

As we have said before, decision-making is done in working memory and is an important part of the basic functions of working memory. When a decision needs to be made, the first thing a person needs to do is to look for relevant cues in the environment. Often, however, these cues can be confusing, ambiguous, contain a lot of uncertainty and can even be misinterpreted. Selective attention plays a key role in this process by filtering the cues acquired by the senses on the basis of previous experience and knowledge (stored in long-term memory and acquired through learning). The selected information is further processed to create situation awareness (perception, understanding of the present situation, prediction of the future situation—see Sect. 5.1), which allows the decision maker to make further assumptions: what will be the consequences if I make such a decision? A typical example is electronic navigation. On the digital map, a traffic jam may be shown on the road ahead, and the system predicts how long it will take to get through the traffic jam. At the same time, the system may suggest a change of route and tell the driver what the consequences of changing the road might be. In this way, the driver may make the decision that a change of road may save time but may require a long additional drive. The user then needs to weigh which alternative is more cost effective, the time saved or the extra miles driven.

Of course, the decision-making process is not always a one-off process, it is more often than not a **multi-iteration** process. The assumptions of a decision may lead the driver to search for more, or other, clues to help make the decision. Take the previous example. Perhaps the driver has several different destinations in mind for the trip, so instead of changing the route as suggested by the navigation, would it not make more sense to change the order of destinations and adjust the time plan for the whole day?

3.9.2 *Choice of Action*

Once a decision has been made, the next step is to act. Often there are several different actions that can be taken to implement the same decision, so how do we choose our actions? For example, let's suppose we have decided to replace our mobile phone because it is currently two years old, has been dropped a few times and is not working well. Once this decision is made, we are faced with different choices of action: which brand of phone to buy? What price point? Where to buy it, etc. Different choices may produce different results, and the greater the uncertainty, the greater the risk.

In choice behaviour, it is often assumed that our goal is to make the best decision and maximise the benefits on the basis of a cost–benefit analysis. However, there are two types of decision makers. One type is indeed called the maximiser. They do a comprehensive consideration of the various alternatives, scoring their importance, their costs and benefits, and then perform a weighting calculation. Of course, in daily life, much of this is done more or less intuitively, but the net result is that maximisers, once they have arrived at the best alternative, may be hesitant to implement the decision, because they feel that there may still be a better alternative they are not yet aware of. The other type is called satisficer. These people have a (loosely defined) criterion for what they find acceptable or satisfactory, and once they have identified an alternative that meets this criterion, the decision-making process stops. In everyday driving, the difference between maximisers and satisficers may apply primarily to higher-levels goals (strategic goals), and less to the concrete operational goals of throttling and braking. However, for novice drivers, for whom the decision-making process at the tactical and operational levels has not yet been automated, the difference between satisficing and maximising may also apply to lower-level goals. For instance, when accelerating, maximisers may be in doubt when to shift to a higher gear. And when crossing an intersection, they may take a long time to decide whether an available gap is wide enough to cross the intersection safely.

Talking about the choice of action inevitably requires talking about the difficulty of the choice and the reaction time. Some of the choices we make are almost automatic, requiring little mental effort and time, while other choices may involve more complex reasoning, requiring more effort and time. For example, the amber light in traffic lights is a warning that the light will soon turn red. If the driver cannot brake and stand still before the stop line except by risking a head–tail collision with the following car, he may drive on; otherwise, he should stop. Thus, the process of choosing an action involves estimating the distance to the stop line and the distance needed to come to a standstill given the speed of the car and the braking power of the vehicle. Experienced drivers do this almost routinely (although there may be differences between drivers concerning their inclination to brake for the amber light; some drivers may bypass the choice process and routinely speed up when seeing the traffic light turn amber). On the other hand, if there is a traffic jam on the highway and there is an exit giving access to a possible alternative route, the choice process may involve more conscious reasoning involving a comparison of the time that will probably be lost in the traffic jam (depending on the length of the traffic jam) and the time lost by taking the alternative route (which may include rural roads), and an estimate of the likelihood that many other drivers chose the alternative route as well, causing slow traffic, and possibly other considerations as well.

We may classify reaction time (RT) as simple reaction time and choice reaction time. Simple reaction time refers to the reaction time of a person when given a simple stimulus. This is a rare occurrence in our life and is mostly found in the laboratory. Simple reaction time is one of the fundamental steps in the study of reaction time. Laboratory studies have demonstrated that simple human response times to

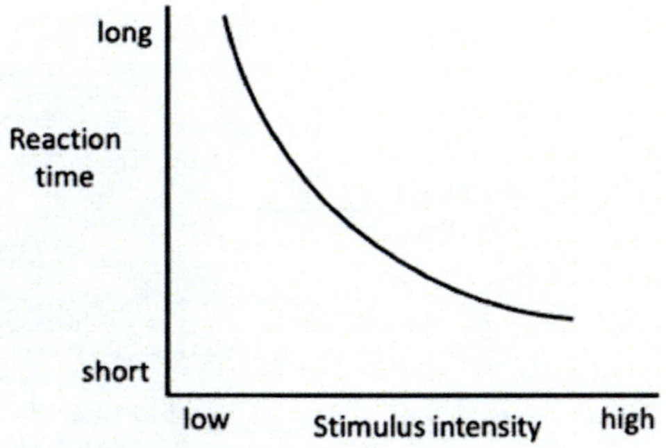

Fig. 3.18 Reaction time versus stimulus intensity. After Wickens et al. (2013), Fig. 9.1

sound stimuli are 30–50 ms (milliseconds) and to visual stimuli 130–170 ms (Woodworth and Schlossberg 1965). Therefore, the RT is different for different modalities. Considering that the propagation of sound is not limited by orientation and that the response time to an acoustic stimulus from different orientations is the same, whereas visual stimuli are closely related to the orientation of the stimulus and must be in the human field of view in order to be received, alarms in the real world often use sound as a warning.

RT is not only related to the modality of the stimulus, but also to the intensity of the stimulus. The stronger the stimulus, the faster the reaction (Fitts and Posner 1967). This is illustrated in Fig. 3.18.

It follows that the first factor influencing RT is the modality that produces the stimulus and the second factor is the intensity of the stimulus. The third factor influencing the speed of RT is temporal uncertainty. The time interval between two stimulus signals is called the inter-stimulus interval (ISI). If this ISI is a constant value, e.g., 0.5 s, then the observer can predict when the next stimulus will occur, so that the RT can even be reduced to 0 s. However, if the ISI is variable and the observer is uncertain when the next stimulus will occur, then the RT is prolonged. It has been shown that in the case of variable ISI, RTs are longer if the mean ISI is relatively short and shorter if the mean ISI is relatively long (Fitts and Posner 1967). If the ISI is too short, it results in subjects not having enough time to prepare. This feature has a good application in traffic design. When the traffic light turns red to green, in some countries there is a yellow light transition, which allows the driver to be prepared and ready for action when the light turns green. But if the duration of the amber interval is too short, the driver is not yet prepared when the light turns green. The fourth influencing factor is human expectation. As in the previous case, when the ISI is relatively long, people are expecting the next stimulus to be generated. This is why the RT is short.

There are many additional factors that may affect reaction speed. For example, repetition of a stimulus can speed up a reaction. If a reaction is performed by two separate hands, it will also be faster than having the same hand deal with two stimuli. Of course, training can also speed up the reaction. There are two other important factors, one being stimulus–response modality compatibility, which is discussed in detail under Multimodal Interaction. The other factor is the mutual matching of

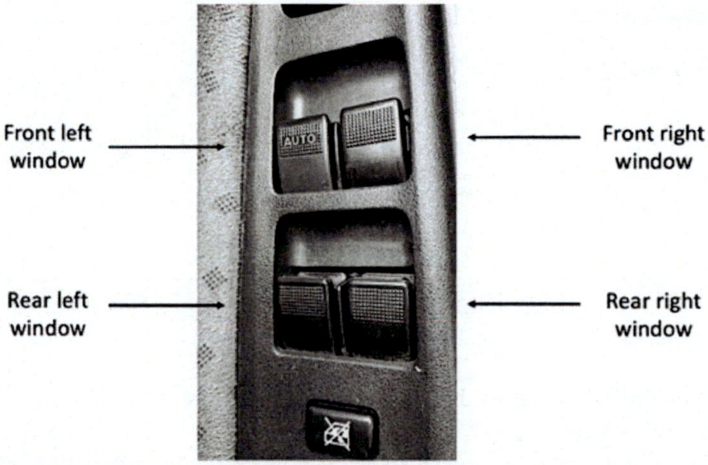

Fig. 3.19 Matching spatial layout. The four keys in the driver's left-hand door of the car control the windows on the car. The spatial layout of their arrangement corresponds to the spatial layout of the windows

spatial locations (mapping). The best example of this is the arrangement of the controls for the windows on the car doors, as shown in Fig. 3.19. Proper application of the Mapping principle results in fewer errors and faster action.

A further factor affecting reaction time to a stimulus is described by bottleneck theory. As described in Sect. 3.5, perceptual processes operate mostly in parallel. However, the capacity for working memory to process information in parallel is much more limited; instead, activity in working memory, such as deciding and selecting appropriate actions, is characterised by serial operation. Thus, when there is a series of stimuli that appear in quick succession, requiring a response to each one, the observer's response to the stimuli arriving later is delayed. Figure 3.20 illustrates this theory. The perceptual analysis of stimuli S1 and S2 is conducted in parallel,

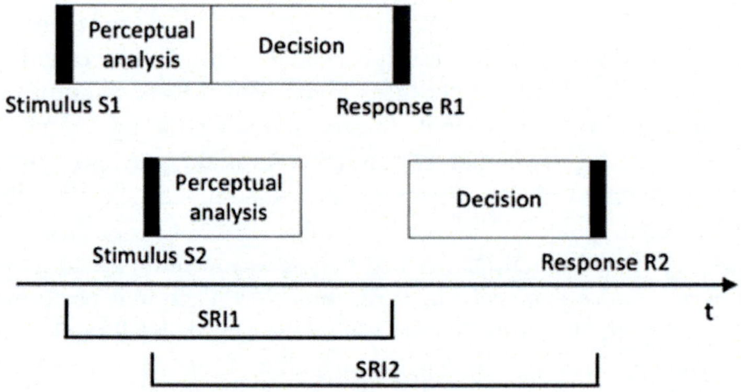

Fig. 3.20 Bottleneck theory of information processing (Wickens et al. 2013, Fig. 9.10)

Fig. 3.21 Relationship between reaction time and accuracy

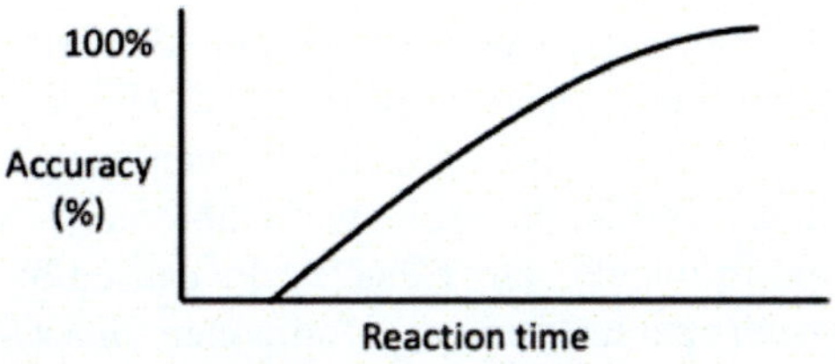

but deciding upon an action for S2 needs to wait until the decision making for S1 is completed. Accordingly, the stimulus–response interval (SRI) for S2 is longer than for S1.

Finally, there is an inverse relationship between the speed of the response and the correctness of the response. In general, the faster the response, the less accurate the response action is likely to be; and the more accurate the response, the longer it takes to respond. The speed-accuracy operating characteristic (SAOC) can be expressed as in Fig. 3.21 (Pachella 1974).

Figure 3.21 shows an interesting insight about the consequences of the level of accuracy that is required. To demand 100% accuracy at all times would mean an excessive increase in human response time, and the efficiency of the work would be reduced. In order to increase efficiency, a certain level of accuracy would need to be sacrificed. This is often referred to as the Speed-Accuracy trade-off. Where this value is set for a specific situation will depend on the nature of the job, i.e., the tolerance of the job to error, and also on individual differences. In traffic, some drivers may prioritize accuracy (correctness) of decisions and actions, driving more carefully and taking more time at intersections. Other drivers may put more weight (relatively speaking) on efficiency and take more risk. Interestingly, "slow-down", is a form of strike by which employees continue doing their work but do everything by the book and follow all guidelines, causing a drop in production and losses for the employer. Again, this shows that high accuracy is at the expense of efficiency.

3.9.3 Error Classification

It is commonly believed that over 90% of all traffic accidents are caused by human error. In a user study, an elderly driver told that if everyone on the road obeyed the rules of the road, there would not be so many traffic accidents. Here we will not go into the specifics of traffic accidents, but return to cognitive psychology. In general, human error means inappropriate behaviour. Errors are generally divided into two categories, namely, mistakes or violations, and slips or lapses. A violation is defined as a failure to form the correct intention (e.g., running a red light). A slip is defined as an error in action when the intention is correct (e.g., failing to notice a traffic signal). Slips are usually caused by deviations in the operation of the series, or by a lack of concentration, or by the fact that the correctly operated button is somewhat similar to another button, causing a mistake, or that two buttons are too close together, etc.

Accordingly, much of automotive HMI design aims at reducing the likelihood of slips and mitigating the effects of slips that occur.

We often say that to err is human, so that any system that requires people to operate it has a reliability problem. In this context, we are referring to errors caused by human performance, rather than errors caused by the quality of the system. If we consider each operating unit as a component of a system with a reliability of 0.9, it means that a human being will make 10 errors in 100 operations. Whether these components are connected in series or in parallel has a different impact on the confidence level. A series relationship means that after the operation of one component is completed, the operation of the other component is started, and the output of the first component is the input to the other component. When two components are connected in series, the errors cascade; if the confidence level of each is 0.9, the confidence level of the system is $0.9 \times 0.9 = 0.81$. When two components are connected in parallel, the system will only go wrong if both components go wrong, which means that the error rate of the system becomes $0.1 \times 0.1 = 0.01$, so the confidence level of the system becomes 0.99. The so-called fault-tolerant design, or redundant design, is to avoid the cascading effect.

In summary, good design may help to prevent errors and to mitigate the consequences of errors that cannot be prevented.

3.10 Emotions and Driving

When you get bad news, how do you react? Anxiety, sadness, anger or ignoring. How long do these negative emotions affect you? A few hours, a day? Emotion also plays an important role in the interaction between people, and most people have the ability to quickly detect the emotional state of the person with whom they are communicating, whether it is anger, joy, sadness or indifference, by the expression on their face, body language and tone of voice. This enables them to respond with an appropriate, corresponding emotional and verbal phrasing. The importance of emotions in communication is also witnessed by the fact that many multimedia interactive systems now offer emoticons that allow people to express their emotions.

Would you expect a system to be designed to take your emotions into account? At this point in the development of artificial intelligence, the systems we design are still far from being able to respond to the user's emotions in the same way as a human. There is much research being done to explore whether certain emotions produce certain patterns of behaviour. For example, does anger make people more focused? Does excitement cause people to do dangerous things? At the same time, what attitude should the system take in response to the user's emotions? Should we design interfaces that try to put the user in a happy state all the time?

Emotion and driving has been a topic of interest to researchers for many years (Hu et al. 2013), and it is well known that road rage is threatening traffic safety. Emotional drivers may not be able to focus on important information at the right time. There are many studies that show that drivers' emotions have an impact on their driving.

Angry driving often results in dangerous driving behaviour that affects not only the driver himself but also other road users and can lead to serious accidents. Emotion detection with AI technology is already quite accurate (Park et al. 2017). Therefore, the development of interactive systems within vehicles that can respond to driver emotions, especially road rage, has become a hot topic.

Some studies have shown that anger can cause people make irrational decisions. Angry drivers may not be aware of their driving risks (Jeon 2015). Also, angry drivers have an 'illusion of control', are more likely to drive in a risky and aggressive manner (Shamoa-Nir and Koslowsky 2010), and to break traffic rules. When drivers are angry, certain specific driving behaviours can be identified (Deffenbacher 2016; Jeon 2015; Pêcher et al. 2011), such as increased acceleration and speeding. They may drive at maximum speed; increase throttle pressure; and increase steering wheel usage. They increase overall driving errors, increase lane deviations, and reduce safety levels and take higher risks. They sound the horn, yell and display hostile postures, tail other vehicles or make harmful movements and end up leaving the car for verbal assault or physical violence because they are unable to communicate.

Currently, devices that can sense and determine the driver's emotions in a vehicle are multi-parameter, multi-sensor devices. The driver's emotion is determined by facial expression analysis, corresponding to relevant physiological data, changes in driving behaviour and other parameters. Face analysis to determine the emotional state of a person is well established in the field of artificial intelligence and robotics. However, designing systems to respond to different driver emotions is not a simple matter. Some in-vehicle systems use different methods of emotion regulation, such as changing the colour of the display, the style of the music, etc., in order to be able to control the driver's emotions. Research by Lisetti and Nasoz (2005) suggests that emotionally intelligent car interface systems can enhance driving safety by improving understanding of the driver. AI algorithms could be used to analyse driver psychological data and design intervention interfaces with different strategies, such as turning on the radio, opening the windows or playing music; this is the solution that most have adopted, but because driver data on vehicle driving is limited and most of this research has been done in laboratory-heavy settings, it is questionable how effective it will be. Perhaps the use of more active interactions, such as car systems that actively talk to the driver and certain safety assistance systems that automatically assist the driver in manoeuvring the car, would be a better solution. This is an area where a great deal of research is still needed. In Chap. 7 we will come back to this topic.

References

Baddeley AD (1986) Working memory. Clarendon Press, Oxford

Baddeley AD (1995) Working memory. In: Gazzaniga MS (ed) The cognitive neuroscience. MIT Press, Cambridge, MA, pp 755–784

Baddeley AD, Hitch GJ, Allen RJ (2009) Working memory and binding in sentence recall. J Memory Lang 61:438–456

Chen F, Qvint G, Jarlengrip J (2007) Listen! There are other road users close to you—improve the traffic awareness of truck drivers. In: Lecture notes in computer science (including subseries Lecture notes in artificial intelligence and Lecture notes in bioinformatics), vol 4555, LNCS

Chen F, Hagernäss J, Lindh M, Jarlengrip J (2008). Listening to the traffic—introduce 3D-sounds into truck cockpit for traffic awareness. In: 2nd International conference on applied human factors and ergonomics 2008, Las Vegas

Deffenbacher JL (2016) A review of interventions for the reduction of driving anger. Transport Res f: Traffic Psychol Behav 42:411–421

Feinkohl A (2014) Psychophysical experiments on sound localization in starlings and humans. Doctoral dissertation, Carl von Ossietzky Universität Oldenburg

Fitts PM, Posner MA (1967) Human performance. Brooks Cole, Pacific Palisades, CA

Ho G, Scialfa CT, Caird JK, Graw T (2001) visual search for traffic sign: the effects of clutter, luminance and aging. Hum Factors 43:194–207

Hu TY, Xie X, Li J (2013) Negative or positive? The effect of emotion and mood on risky driving. Transport Res f: Traffic Psychol Behav 16:29–40

Jeon M (2015) Lessons from emotional driving research. In: 19th Triennnial congress of the international ergonomics association, Melbourne, Australia

Lee JD, Young K, Regan MA (2009) Defining driver distraction. In: Regan Ma, Lee JD, Young K (eds) Driver distraction: theory, effects and mitigation. CRC Press, Boca Raton, FL

Lisetti CL, Nasoz F (2005) Affective intelligent car interfaces with emotion recognition. In: Proceedings of 11th international conference on human computer interaction, Las Vegas, NV, USA

Moray N (1999) Mental models in theory and practice. In: Gopher D, Koriat A (eds) Attention and performance XVII: cognitive regulation of performance. MIT Press, Cambridge, MA, pp 223–258

NHTSA-2010-0053 (2010) Guidelines for reducing visual-manual driver distraction during interactions with integrated, in-vehicle , electronic devices, Version 1.01

Nilsson LG, Ohlsson K, Ronnberg J (1977) Capacity differences in processing and storage of auditory and visual input. In: Dornick S (ed) Attention and performance VI. Erlbaum, Hillsdale, NJ

Owsley C, Ball K, McGwin G, Sloane ME, Roenker DL, White MF, Overley ET (1998). Visual processing impairment and risk of motor vehicle crash among older adults. J Am Med Assoc 279:1083–1088

Pachella RG (1974) The interpretation of reaction time in information processing research. In: Kantowitz BH (ed) Human information processing. Erlbaum, Potomac, MD, pp 41–82

Park BJ, Yoon C, Jang EH, Kim DH (2017) Physiological signals and recognition of negative emotions. In: 2017 International conference on information and communication technology convergence (ICTC). IEEE, pp 1074–1076

Pêcher C, Lemercier C, Cellier JM (2011) The influence of emotions on driving behavior

Rasmussen J (1986) Information processing and human-machine interaction: an approach to cognitive engineering. North Holland, New York

Shamoa-Nir L, Koslowsky M (2010) Aggression on the road as a function of stress, coping strategies and driver style. Psychology 1(1):35

Wang MJMJ, Li YCYC, Chen F (2012) How can we design 3D auditory interfaces which enhance traffic safety for Chinese drivers? In: Proceedings of the 4th international conference on automotive user interfaces and interactive vehicular applications—automotiveUI '12, (c), 77. https://doi.org/10.1145/2390256.2390268

Wang M, Lyckvi SL, Hong C, Dahlstedt P, Chen F (2017) Using advisory 3D sound cues to improve drivers' performance and situation awareness. In: Proceedings of the SIGCHI conference on human factors in computing systems. https://doi.org/10.1145/3025453.3025634

Wickens CD, McCarley JM (2008) Applied attention theoryitle. CRC Press, Boca Raton, FL

Wickens CD, Hollands JG, Banburry S, Parasuraman R (2013) Engineering psychology and human performance, 4th edn. Pearson, Boston

Wood JM, Owsley C (2014) Useful field of view test. Gerontology 60(4):315–318. https://doi.org/10.1159/000356753

Woodworth RS, Schlossberg H (1965) Experimental psychology. Holt, Rinehart & Winston, New York

Chapter 4
Mental Load and Fatigue

Abstract This chapter discusses two important issues relating to driving. The first concept is mental load. Mental load is a central concept to understand driving performance: if drivers experience high mental load, driving performance may deteriorate and safety may be jeopardised. On the other hand, if drivers experience low mental load, they may become inattentive or drowsy, and again safety may be jeopardised. In this chapter, we consider the concept of mental load, look at the relation between mental load and psychological stress, and look at how mental load can be measured. The second important concept is fatigue. Main insights concerning fatigue are summarised, and its effects on driving. Also, we look at methods for detecting fatigue, how drivers may cope with fatigue and how systems may be developed that support drivers in coping with fatigue.

Workload can be differentiated into physical load and mental load. For people working in front of computers, as well as for drivers, physical load is usually a minor problem, or no problem at all, so the main problem comes from psychological load. Here we use the terms psychological (work-)load and mental (work-) load to mean the same. The issue of mental load is probably the most frequently encountered problem in human factors research work, and one of the most studied. When studying the operational behaviour of operators and designing for them, many questions are often asked: How busy is the operator? How complex are his tasks? Is there capacity for additional tasks? Is the operator able to cope with unexpected events? How does he feel during the task? Mental load refers to the extent to which a person's operations place a demand on the information processing resources in the human brain. Our brain has limited resources to process information and when its capacity is exceeded, overload occurs. In this chapter, we will talk more about mental load.

A related topic is driving fatigue. Both low and high mental load and fatigue may have adverse effects on driving performance, but the relation between mental load and fatigue is indirect and far from clear.

© China Machine Press 2023

F. Chen and J. Terken, *Automotive Interaction Design*, Springer Tracts in Mechanical Engineering, https://doi.org/10.1007/978-981-19-3448-3_4

4.1 About Mental Load

Mental load directly affects a person's ability to operate at work. Figure 4.1 illustrates the relationship between task demands, performance and mental load and good or bad handling.

In Fig. 4.1, the X-axis represents the mental resources required for a task and the Y-axis represents the mental resources that are supplied by the human brain. There is a maximum to the mental resources that can be supplied. In the left side of the graph, where the task demands are less than the maximum mental resources that can be supplied (in the case of relatively simple tasks), high performance can be achieved (as indicated by the dashed line) with low mental effort, and the driver still has reserve capacity left to perform other tasks. This part of the curve is called the underload region. However, if the driver decides to engage in an additional task, the combined mental load imposed by both tasks will increase and may increase to the extent that the performance for one or both tasks may suffer. We will come back to this in Sect. 5.2 about multitasking.

If the task becomes more complex, the driver spends more resources on the task and mental load increases; still, he can achieve high performance. This is the case up to a certain point: if the resources demanded by the task fully absorb the resources that the driver has available, the user is fully engaged in the task; mental load is at its maximum and no reserve capacity is left for other tasks. If the task gets even more demanding and the demand for resources exceeds the person's capacity, i.e., at the right side of the graph, the driver is in a state of overload, and task performance will decrease, becoming evident from errors. As can be seen, however, the performance already starts to degrade before the resources supplied reach the maximum. In particular, if drivers need to perform close to the maximum of their ability for longer periods of time, they will experience stress and start to make mistakes.

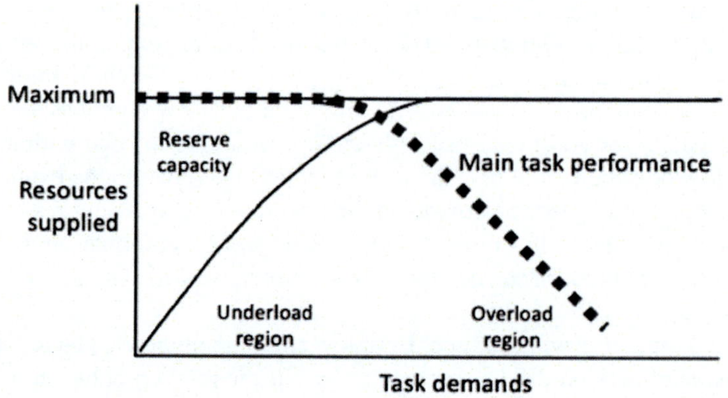

Fig. 4.1 Relationship between task demands, performance and mental load. After: Wickens and Hollands (1999)

For a skilled motorist, motorway driving with modest traffic density and good weather conditions is when he is comfortable and the demand on his mental resources is on the left side of Fig. 4.1. In this situation, he has reserve mental resources for other tasks, such as making phone calls. The amount of these reserve resources depends on the complexity of the road, the traffic density and his driving proficiency. If there is an accident or road construction ahead and he needs to change lanes or exit the motorway, these scenarios will require his full attention and his driving task may reach a critical point in terms of mental load, and he will probably not answer the call even if it comes in. If there is an emergency on the road at this time, the resource demands of the driving task may enter the right-hand side of Fig. 4.1. If the driver has no resources left to deal with the more complex driving task because he is answering a phone call, and the task of answering a phone call takes up some of his mental resources, his control of the vehicle decreases and his ability to react to unexpected situations such as a car in front of him suddenly braking may break down, and the car may rear-end the car in front of him. Fortunately, most drivers prioritise safety and will disengage from the phone call if the driving task requires most or all available resources.

4.2 Measurement of Mental Load and Reserve Capacity

So far, we have talked loosely about the amount of resources demanded and supplied, but to be more precise, resources and psychological load should be measured. There are three commonly used methods: behavioural measures of primary task performance, measures of secondary task performance, and measures of subjective workload estimates.

Behavioural measures: These are direct measures of how well the operator performs the task, which include the correctness, accuracy, effectiveness, time to task, speed of response, frequency of operation, error rate, etc. of the operation itself. In the context of driving, typical measures are SDLP (the Standard Deviation of the Lane Position), Speed Maintenance and Distance to a leading vehicle. However, as can be seen from Fig. 4.1, if the task requires fewer resources than available to the user (or, in other words, if the task is in the "underload" range), the performance will tend to be perfect, regardless of the difficulty of the task: If the task becomes more complex, the person may simply supply more resources to the task ("work harder"). This means that, if the task is in the underload range, behavioural measures are in general not sensitive to variations in task demand. Conversely, if the task requires more resources than available to the user, the performance will decline, and the degree to which the performance declines is a good indication of the task difficulty, or the extent to which the operator is able to handle the task.

Secondary task performance: If a person's task requires fewer mental resources than he has available, he may operate another task at the same time, which is called a secondary task, and this method is used to estimate the amount of resources demanded by tasks in the underload region. For drivers, we call the tasks directly related to driving a car (controlling the steering wheel, stepping on the accelerator, braking, obtaining safety aids, etc.) primary tasks, and other tasks such as adjusting the car's temperature, listening to music, making a phone call or texting secondary tasks. The performance on the combination of a primary and secondary task may be a good indication of the task difficulty of the primary task. The assumption is that, if the difficulty of the primary task increases, there will be less reserve capacity left for the secondary task, so that either the performance of the secondary task will suffer, or the performance of both tasks will suffer. The difficulty of the primary task, and hence the resource demand of the primary task, is then inferred indirectly by measuring the performance of the primary and secondary task. If the secondary task is done well, the load on the primary task is low (and the ability of the operator to handle the primary task is high). For example, if we design a section of road of varying complexity in a driving simulator and ask the driver to read out the information displayed on other screens, the speed at which the information is picked up and read, the correctness rate, etc., may be used to measure the performance of the primary task, and infer from that the resources demanded and mental load induced by the primary task. It is important, however, that the experiment is not set up in such a way as to put the cart before the horse and allow the driver to ignore driving and focus on reading the information. For test situations such as in simulators, special tasks have been devised such as Probe Reaction Time, where drivers, while driving, react to a visual stimulus that is presented at irregular moments. The reaction time and the number of misses provide information about the mental load induced by the primary task. The disadvantage with secondary tasks is that they induce extra workload, so that they are not practical in realistic driving situations. Instead, in realistic driving situations, so-called "embedded" secondary tasks may be used, such as the frequency of inspecting the rear mirror and the side mirrors, or the ability to engage in a fluent conversation. This is also what examiners do during driving tests. Typically, when task demand is high, such as for novice drivers, the first thing that suffers is these embedded secondary tasks and the ability to engage in a conversation.

Subjective scales: Generally, when the above two methods are used to measure psychological load, they are combined with subjective scales. A subjective scale is a questionnaire used to measure psychological load. There are many off-the-shelf scales that have been developed and gained widespread acceptance and application. Researchers are advised not to design such subjective scales themselves as a last resort, but to use off-the-shelf methods whenever possible. The most commonly used is the NASA-TLX (task load index) (Hart and Staveland 1988) and from the NASA-TLX has evolved a method specifically designed to measure driver psychological load called the DALI (Pauzie 2014).

4.3 Mental Load and Stress

We are all under different kinds of stress at different times in our lives. Stress is often seen as a state of emotional arousal that can affect a person's operations and behaviour and, if severe, can disrupt behaviour and have a negative impact on health. However, stress is not always negative and can also act as a motivator to motivate a person to operate. Distinguishing whether different stress conditions weaken or strengthen a person's cognitive abilities or their ability to act is one of the many challenges of stress research.

Stress is studied in different fields, such as biology, medicine, psychology and sociology. In different disciplines, stress is defined differently and studied in different ways. In engineering psychology, it is generally analysed in terms of comparing the behavioural manifestations of stress and the absence of stress. Stressors may include environmental influences such as noise, vibration, heat, lighting, speed, and psychological factors such as anxiety, fatigue, frustration and anger, as well as organisational factors such as severe punishment. Figure 4.2 is a diagram of the relationship between stress and information processing. It can be seen from this diagram that stress affects the information processing process in a number of ways.

Stress generally has three effects: (1) an emotional experience, where stress can make a person feel excited or depressed; (2) sympathetic arousal, where reactions such as increased heart rate and reddening of the skin can be observed; and (3) an effect on the person's ability to process information, which can become faster or slower. Causes of stress and their effects may vary. For example, exposure to whole-body vibration (environmental stressor) affects truck driver discomfort, but also results in fatigue, decreasing the truck driver's vigilance. Occupational demands

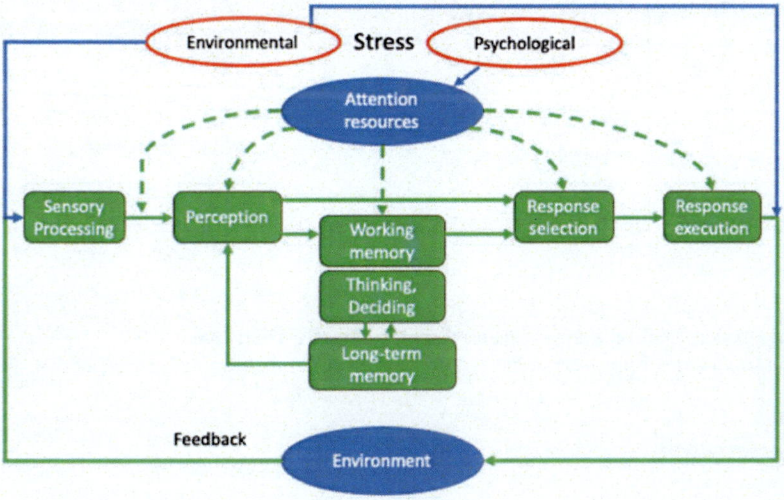

Fig. 4.2 Schematic representation of the relationship between stress and information processing processes

(especially in the case of conflicting demands) may induce stress and elevate at-fault crash risk.

Stress and mental workload are strongly related. As we have seen in the previous section, increased mental load, especially for prolonged periods of time, tends to cause stress. If we are under pressure due to high workload, then, if we need to print a document and the printer fails, we will feel stressed. In turn, prolonged feelings of being stressed will result in feelings of high mental load.

Yerkes and Dodson published Yerkes Dodson's Law in 1908, as shown in Fig. 4.3. We can see that the curve is an inverted U-shape. When a person's arousal level is in the climbing phase on the left side, an increase in stressor intensity can increase the arousal level and thus the person's ability to operate. For example, motorways with curves and up and down hills may make drivers perform better, compared to a straight motorway. This is because curves and up and down hills increase the tension and driving stress of the driver, thus increasing their arousal level and their ability to drive the car. If the road is straight like an aeroplane runway, the driver may get drowsy because driving is less difficult and stressful and arousal level is lower. However, when the stress level is already at a high level, further increasing the stressor intensity will reduce human performance. For instance, if the road is already challenging, then exceptional situations such as road construction sites may increase the stress level beyond the optimal point and result in overlooking relevant information. Furthermore, complex tasks generate more stress than single tasks, so the curve reaches its peak earlier with complex tasks than with simple tasks.

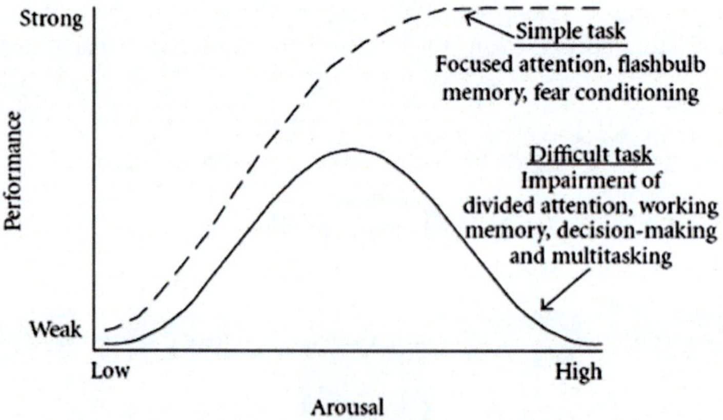

Fig. 4.3 Yerkes Dodson's Law. Licensed under *Creative Commons CC0 1.0 Universal Public Domain Dedication.* https://commons.wikimedia.org/wiki/File:OriginalYerkesDodson.JPG

4.4 Physiological Correlates of Mental Load

Any mental activity has a physiological basis. The British physiologist and neuroscientist Charles Sherrington, proposed as early as 1890 that the basis of mental load was the activity of the brain (Roy and Sherrington 1890). Sherrington was awarded the Nobel Prize in Physiology or Medicine in 1932. He argued that blood flows through the aorta to the brain to provide the oxygen needed for brain activity. When a person's cognitive activity takes place in a certain part of the brain, the oxygen consumption in that part of the brain increases, and therefore the blood supply increases. This theory was the basis of neurophysiology until 100 years later, when Positron emission tomography (PET) and functional magnetic resonance imaging (fMRI) were developed. A large number of findings from fMRI techniques have revealed the response of cognitive activities in the brain. However, the equipment associated with PET and fMRI is too expensive and not readily available to study the response of cognitive activity in the brain during various driving manoeuvres, and therefore cannot be used to measure the mental load during different manoeuvres. Here, we present some other operational techniques, and some of the results produced by applying these techniques to human brain activity.

Electroencephalography (EEG) records the electrical activity of the brain through electrodes placed on the subject's scalp. Spectral power bands at different EEG frequencies have been found to be sensitive to increased working memory (WM) load and to changes in attentional resources. EEG can be used to assess mental load in an operational setting, but it can be somewhat difficult due to noise. One of the problems is that EEG can be disturbed by eye movements and muscle movements. There are already a number of algorithms that can separate valid data from interfering waves, so the laboratory use of EEG is becoming increasingly sophisticated.

Event-related potentials (ERPs) represent the neural responses of the brain to specific sensory, motor and cognitive events. ERPs are analysed by averaging the EEG over time periods locked to specific stimuli or response events. At present, ERPs occupy a somewhat unique position among the tools of cognitive neuroscientists. This is because they offer the only neuroimaging technique with high temporal resolution and millisecond performance in terms of temporal accuracy, unlike techniques such as PET and fMRI, which are slower because they track cerebral haemodynamics. Researchers often use the millisecond precision of ERP to examine the neural mechanisms associated with cognitive processes at corresponding time periods. For example, recording temporal information from ERPs provides critically valid evidence that attentional regulation is neurally activated approximately 100 ms after stimulation (Hillyard et al. 1998).

Heart rate is another physiological measure for mental load. Backs et al. (2003) measured three different heart rate measures during simulated driving on easy to difficult curves and found that they were affected differently by curve radius. They therefore concluded that the different effects suggest that the resource demands of driving can be distinguished between central and peripheral neural activity.

Finally, the use of pupil diameter as a measure of mental load has been explored. Researchers have observed that pupil diameter correlates closely and accurately with the mental demands of cognitive activity. In addition to pupil size, blink frequency and the duration of sustained gaze at a target can be used to assess the cognitive load of the task (Marquart et al. 2015). As the workload induced by the driving task increases, the pupil enlarges, blink frequency increases, and gaze forward time increases significantly, indicating that the driver is highly focused on the driving task. Pupil measurement can be highly sensitive, although it is not yet possible to diagnose the type of workload demand. It provides a comprehensive picture of the mental demands in an information processing system. However, changes in ambient lighting must be monitored during the measurement, as they can affect the pupil diameter. Also associated with the autonomic nervous system, the measurement is also susceptible to changes in mood.

4.5 Psychological Load Testing Methods

There is a long history of research into methods of assessing psychological load, which are summarised in Table 4.1.

Mental load is a complex concept, and measuring it is complex as well. Preferably, different methods are combined. For estimation of mental workload in real driving conditions, insights from research may be applied to predict which situations will cause high workload, taking into consideration driver characteristics such as fatigue and level of expertise. If such additional factors are not taken into consideration, the estimates may be too crude to be useful.

4.6 About Driving Fatigue

Driving fatigue is an age-old subject and a difficult problem to solve, due to the following reasons.

(1) The onset of fatigue is not easily predictable
(2) The degree of fatigue is not easily detected in advance
(3) The effects of fatigue on driving vary from person to person
(4) There is no good method of waking up from fatigue

In interaction design, it is highly desirable to use various technologies, mainly eye-tracking and vehicle behaviour monitoring, to determine whether the driver has a degree of fatigue that may result in a reduced ability to cope with road emergencies, making driving behaviour potentially dangerous. After detecting the presence of fatigue, the car may take a number of steps to push through a range of possible ways to alleviate fatigue. Here, we will systematically review the existing research findings on driving fatigue to bring together and integrate fatigue-related factors and

Table 4.1 Summary of measures of psychological load

Method name	Description	Advantages	Disadvantages
Measures of primary task performance	Measures such as SDLP, Lane Keeping performance, Distance maintenance	The measurement method is straightforward and effective and can be combined with physio-psychological indicators, and, if carried out in a driving simulator, with secondary task techniques	It is sometimes difficult to establish a quantitative relationship between the performance and the level of psychological load. If the task demand is in the underload region, primary task measures are not accurate measures of mental workload
Secondary task techniques	Measures such as probe reaction time Need to be carried out on a driving simulator	Sensitive to changes in mental load in the underload region	The subtasks need to be designed with great care as they interfere with the main task If the primary and secondary task employ different sensory modalities, the sensitivity may be affected (Multiple-resources theory—see Sect. 5.3.3)
Physiological measurements	Heart rate, ECG, EEG, eye movement and other measurements	Sensitive to mental load, but if combined with a secondary task, it measures the mental load generated by the primary and secondary tasks together. It is continuously measured data and can be used in real-world driving	The data can be easily interfered with by other factors, and the method may be intrusive, as the measurement electrodes come into contact with the human body. Also, the data analysis requires specific expertise
Subjective scales	NASA-TLX, DALI (Driving Activity Load Index) SWORD (Subjective Workload Dominance (Vidulich 1989)	Easy to administer, quick Can be used in conjunction with other measures	It's a post-task method. Hence, it is not sensitive to variations in task workload over time due to low temporal resolution NASA-TLX uses 6 subscales and weights, so is more time-consuming

coping methods: the definition of fatigue, the causes of fatigue, the effects of fatigue on driving, the detection of fatigue and coping methods.

Definition of fatigue: Definitions of fatigue may be very confusing, lumping many different issues together, and making it difficult to measure and compare relevant studies (Phillips 2015). There are many factors that contribute to 'fatigue' that, if not clearly understood, can have an impact on the discussion that follows, such as exhaustion, sleepiness, stress, anxiety, burnout or boredom, and the term 'fatigue' often becomes an elephant in the room that encompasses many related phenomena (Hancock et al. 2012).

Fatigue is not just a subjective feeling; it has a psychophysiological basis. The determination of fatigue is generally made in 3 ways: physiologically, psychologically (subjectively), and behaviourally. The measurement of only one aspect reduces accuracy. For example, in the case of driving behaviour, lorry drivers indicate subjectively that they experience fatigue after 11 h of continuous driving, while measures of driving behaviour do not yet show signs of fatigue (Hamblin 1987). On the other hand, when measured with cognitive tasks, changes in cognitive ability can be measured after perhaps five minutes (Dinges and Kribbs 1991). Furthermore, psychological research has shown that human behavioural motivations can influence the results.

We quote the definition of fatigue proposed by Phillips in 2015: "*Fatigue is a suboptimal psychophysiological condition caused by exertion. The degree and dimensional character of the condition depends on the form, dynamics and context of exertion. The context of exertion is described by the value and meaning of performance to the individual; rest and sleep history; circadian effects; psychosocial factors spanning work and home life; individual traits; diet; health, fitness and other individual states; and environmental conditions. The fatigue condition results in changes in strategies or resource use such that original levels of mental processing or physical activity are maintained or reduced.*" (Phillips 2015).

The definition illustrates that fatigue is the degree of change relative to an optimal subjective or objective state, where the optimal state is the average of a healthy individual in a fully recovered, rested state, or a group of fully recovered, rested and healthy individuals. Secondly, there are corresponding fatigue dimensional characteristics in terms of psychological and physical aspects. For example, the psychological aspects are indicators of cognition, habits, emotions, the ability to self-regulate and sleepiness. The physiological aspects are various biochemical indicators, such as electrocardiography, electrodermal, electroencephalography, etc. Of course, fatigue can also be observed externally, for example in facial expressions and various eye changes. The definition also suggests that fatigue can also be judged through motor behaviour and performance outcomes.

Fatigue is a dynamic and multidimensional concept. The relationship between fatigue and sleepiness is worth bringing up separately. Sleep can relieve fatigue, but sleepiness does not always result from fatigue. For example, driver drowsiness may be due to the time of day when they drive, e.g., late at night or early in the morning, or after a meal; it may also be due to the lack of stimulation in monotonous driving situations and the cardiovascular resonance caused by the vehicle during travel (Connor 2011).

It is important that sleepiness is integrated into the concept of fatigue. Firstly, the psychophysiological state of sleepiness partially overlaps with the psychophysiological state of fatigue. The perception of the sleepiness state (perceived need for sleep) is a sub-dimension of the experiential dimension of the fatigue condition. The cognitive, adaptive, affective and self-regulatory dimensions of fatigue will also have dimensions related to narcolepsy.

Secondly, (Borbély 1982) used homeostatic and circadian factors to describe the effects of sleep on driving and also how this contributes to driver fatigue. In a sense, fatigue may be caused by the driver trying to stay awake. It is attributed to the effort of staying awake and trying to maintain normal driving in a drowsy state, in addition to any other form of exertion (May and Baldwin 2009). Importantly, fatigue varies according to the intensity of the exercise, the 'time of day', but also the 'time of day' due to the 'effort to stay energised' (Tepas and Price 2000).

4.7 Causes of Fatigue

Almost everyone who drives or rides in a car has experienced that sitting in a car, or in a vehicle that moves at a constant speed (including planes, trains, boats, etc.) makes you sleepy easily. Although a number of factors can affect fatigue, three are crucial: sleep, continuous wakefulness, and circadian rhythms. And fatigue itself affects one's judgement of one's level of fatigue.

Fatigue is a psychophysiological state that occurs long before one has even entered a state of sleepiness. Fatigue has a negative impact on many human abilities, such as one's reaction time, one's ability to concentrate, and one's ability to judge road conditions correctly. Apart from unpredictable psychological factors such as emotions, the two main categories of factors that contribute to fatigue are: internal physiological factors and task-related factors (Thiffault 2011). Physiological causes include.

1. Individual differences in fatigue sensitivity that may be related to sleep disorders, other medical conditions or physiological variability.
2. Circadian rhythm, with early morning (e.g., 4:00 am to 7:00 am) being the highest risk time.
3. Sleep time in the most recent period, including primary sleep time and naps.
4. Sleep inertia, and staying awake, arousal habits.
5. The time that has elapsed since the last major sleep; especially if it is more than 16 h, independent of work or specific work activities.
6. General health status and recent behaviour, i.e., diet and exercise.
7. Caffeine intake.
8. The use of prescription and over-the-counter medication.
9. Bright and dim light.
10. The rhythm of the car's movement, resonating with the human blood circulation

The main task-related factors are the duration of the operational task, the complexity of the task and the monotonicity of the task. Hours-of-service (HOS), a factor that has a direct impact on fatigue, has received widespread attention in many countries, and different countries have developed different rules governing it. The rules on driving hours for commercial vehicle drivers contain specific regulations relating to the driver's schedule. These include the minimum daily off-duty time, the maximum daily driving time, the maximum number of working hours (for truck drivers this is the limit on total working time), the regularity of schedules, the maximum weekly working time, the rest period between assignments, the rest time required for driving and the use of the sleeper berth in the truck (including the "sleep" needs).

Studies have reported that the average long-haul commercial vehicle driver drives more than 14 h a day and gets only 5–6.5 h of sleep per night, compared to the 7–8 h of sleep that the average person needs. Therefore, drowsiness while driving due to lack of sleep is the cause of 70% of traffic accidents involving long-haul drivers. Staying awake for 17 h has the same impact on driving ability as a state with 0.05 mg/ml of alcohol in the blood. A driver who drives without sleep for 24 h has the equivalent of 0.1 mg/ml of alcohol in his blood (Williamson and Feyer 2000).

Also, insomnia is common among professional drivers. One type of insomnia arises from a condition called Obstructive Sleep Apnoea (OSA). People with this insomnia have a driving accident rate that is 3 to 4 times the rate of average drivers. In addition, studies have shown that driving ability is reduced due to drowsiness, and this varies by age. For younger people (20–25 years old), sleepiness may result in a 1.9-fold reduction in driving ability, while for older people (52–60 years old) there is little effect (Philip et al. 2005).

In addition to fatigue due to high task loads, task-related fatigue can also be due to monotonous and underloaded tasks, in particular as the driver's role and tasks change with increased levels of automated driving (May and Baldwin 2009). Studies have shown that drivers are more likely to fall into fatigue under automated driving conditions (Neubauer et al. 2012). The monotony of the task and the lack of task engagement are the main causes of passive fatigue. Passive fatigue occurs as a result of lack of stimulation or lack of physical exertion due to the driver's overly monotonous and underloaded tasks. Passive fatigue can lead to consequences such as a reduced level of driver alertness to the point of missing warnings from the system or taking over in the wrong way after a warning.

4.8 Effects of Fatigue on Driving

A number of studies (Dinges and Kribbs 1991; Philip et al. 2005) have shown that fatigue affects driving behaviour in specific ways. Typically, after 2–3 h of continuous driving, drivers become tired and exhibit reduced steering control. Widely documented effects of fatigue include.

- Longer reaction times: fatigue increases reaction times in emergency situations.

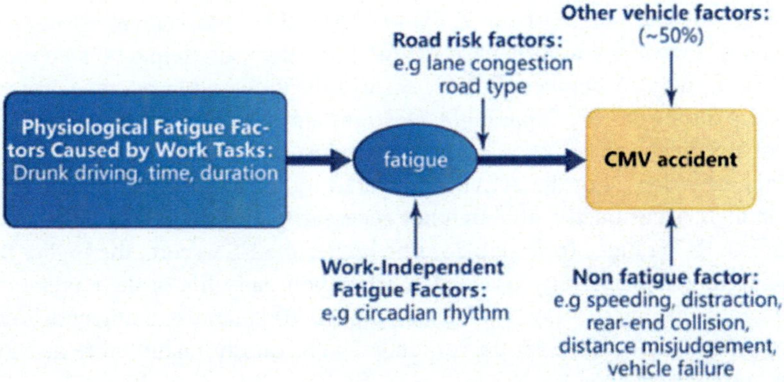

Fig. 4.4 Commercial vehicle accident causation model. Based on Sharwood et al. (2011)

- Reduced alertness: subjects perform less well on attention-based tasks when sleep deprived (e.g., tired drivers are slower to notice impending hazards, such as road construction or working conditions such as roadblocks ahead).
- Reduced information processing capacity: fatigue reduces both information processing and short-term memory accuracy (e.g., a fatigued driver may not remember the first few minutes of driving).

It should be noted in particular that one of the categories of fatigue and drowsiness is called Microsleeps, a type of sleep that lasts only a few seconds, which is particularly dangerous because the driver is not aware that he or she is drowsy or has fallen asleep. These few seconds are enough to cause a driving accident. In truck accidents, the effects of fatigue are even more pronounced. Figure 4.4 summarises the factors that contribute to commercial vehicle (CMV) accidents (Sharwood et al. 2011).

4.9 Fatigue Detection Methods

When introducing the definition of fatigue, we mentioned that the detection of fatigue is generally divided into three aspects: physiological, psychological and behavioural. The detection of fatigue is usually closely related to the results of the detection of sleepiness. The main methods of detecting sleepiness include self-assessment scales and objective measures.

Self-assessment scales: Common self-assessment scales are the Karolinska Sleepiness Scale (KSS, Akerstedt and Gillberg 1990), the Stanford Sleepiness Scale (SSS, Hoddes et al. 1972) and the Epworth Sleepiness Scale (ESS, Johns 1991). The Karolinska Sleepiness Scale and the Stanford Sleepiness Scale measure how sleepy or drowsy a person feels at a given point in time. For instance, the KSS requires the subject to rate how they feel within five minutes prior to the test. Administered on a 9-point scale ranging from 1, "very alert" to 9, "very sleepy, struggling with drowsiness", the KSS is often used in studies to track participants' sleepiness over time. The Epworth Sleepiness Scale measures the individual's average propensity to fall asleep during the day. It has eight questions and asks subjects to assess their usual level of sleepiness or falling asleep while performing eight different activities on a 4-point scale (0–3). For the activities covered by the scale, most people engage in them at least occasionally, although not necessarily every day. The ESS score (sum of 8 items, 0–3) ranges from 0 to 24. The higher the ESS score, the higher the individual's average propensity to sleep (ASP) in their daily life or their average degree of sleepiness during the day, with a score above 10 generally being considered as a higher propensity to sleep for the respondent. The questionnaire takes no more than 2–3 min to answer.

Objective measures: The most commonly used objective measures are electroencephalography (EEG) to measure brain activity, psychomotor vigilance testing (PVT) and electrooculography (EOG) to measure eye movements. EEG and EOG are common objective measures of sleepiness for laboratory and naturalistic driving studies. For the detection of fatigue, the most accurate method of measuring fatigue is the direct monitoring of brain activity. EEG is commonly used as a pathway for EEG monitoring, in which four waveforms are generally monitored to determine driver activity: beta waves (generated when the brain is performing a task), alpha waves (generated when the brain is at rest), theta waves (generated when the brain is fatigued), and delta waves (generated when the brain is asleep). Studies have shown that the β-wave component increases when cognitive load is increased; when fatigue levels increase, the α/β/θ-wave decreases and the δ-wave component increases (Barua et al. 2017; Jap et al. 2009). However, EEG monitoring is not suitable for monitoring in realistic driving situations due to its complexity and invasive nature, and is therefore mainly used in simulator experimental studies.

Apart from changes in brain activity, the most obvious symptom of fatigue appears in the eyes. According to research, the time difference between a visual stimulus and its response is one of the main methods of determining perceptual activity. The task used to measure this time difference is often referred to as the Psychomotor Vigilance Task (PVT). This time difference shows how quickly a person can respond to a visual stimulus, and the time difference is strongly affected by fatigue. Furthermore, studies have shown that there is a close relationship between this time difference and the percentage of eyelid closure over a given period of time. The change in the PERcentage of eyelid CLOSure over time is known as PERCLOS (Wierwille and Ellsworth 1994; Wierwille et al. 1994). The PERCLOS is therefore another main indicator of fatigue detection, and subjects are generally considered to be in a state of fatigue when the PERCLOS is >80%.

Yet another measure is the length of the eye closure cycle. An eye closure cycle is generally divided into three phases, mid-eye closure, eye closure and eye opening, where the speed of eye closure (~350 mm/s) is slower than the speed of eye opening (~150 mm/s) (Jin et al. 2013). Under normal conditions, a complete eye closure cycle is typically within 200 ms. When a person enters a tired state, the eye closure cycle

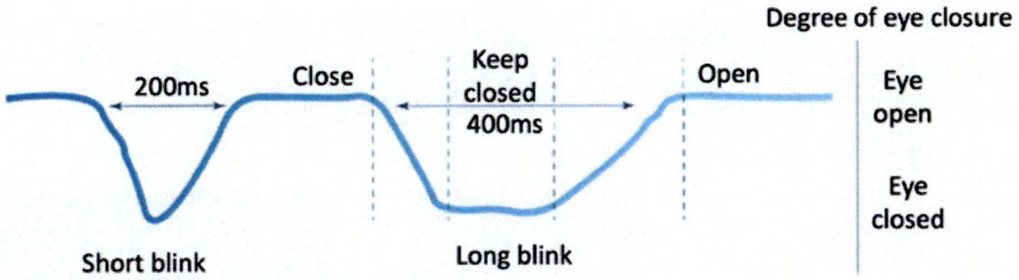

Fig. 4.5 Schematic diagram of the eye closure cycle

is extended to 300–600 ms. And the eye closure cycle in a sleepy state is extended to more than 600 ms (e.g., Fig. 4.5). It is worth noting that these reference values are averages and the individual differences in the eye closure cycle are large, and the corresponding detection thresholds need to be corrected for each individual, so that it is not useful for detecting driver fatigue in practice.

Finally, behavioural measures can be used. The behaviour of alert drivers is characterised by frequent small corrections of the position of the vehicle in the lane, measurable from the steering wheel. Once drivers get tired, they are less alert on deviations of the position in the driving lane, and corrections become less frequent, but therefore also larger corrections are needed. This difference can be used for classification of the driver state.

4.10 Coping with Fatigue

Notwithstanding in-car monitoring devices that are currently in place to detect early on whether a driver is fatigued or drowsy, and outside detection devices to see if a driver is driving in a non-safe manner, or even rumble strips on both sides of the road to wake up a drowsy driver, these measures are only likely to give the driver some indication and help in case of imminent danger, and do not prevent fatigued driving. Effective ways to prevent fatigue are:

1. Sleep well, eat sparingly, avoid alcohol and drugs that may cause sleep
2. Lower the temperature in the driver's seat and ensure good air circulation
3. Upright driving position, allowing the legs to bend at 45°
4. Take breaks every 2 h and wash your face
5. Sing to yourself, talk to passengers, listen to the radio to stay awake

As for how to deal with driver fatigue that is already present in drivers, so far, the focus has been on the following approaches.

1. The usual practice is to listen to the radio, wash your face, open the windows, talk to passengers
2. Changing driving speed, in particular increasing the speed. This creates more challenge and arousal. However, its effect may be short-lived.

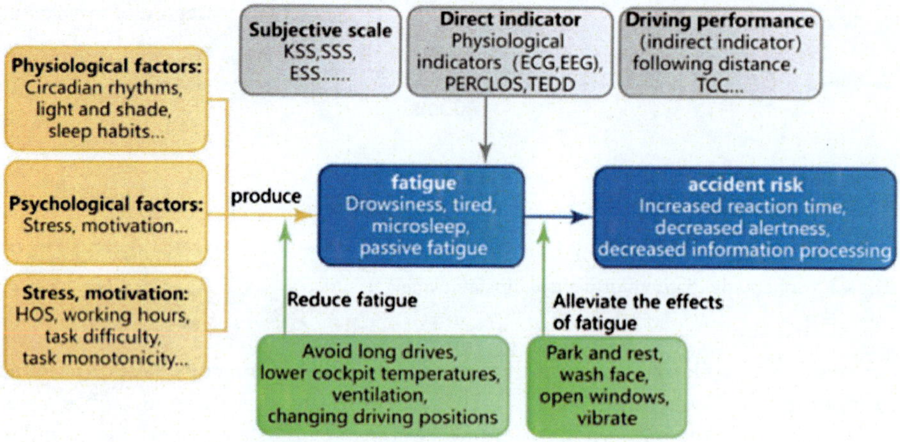

Fig. 4.6 Summary of fatigue studies

Once sensors are implemented detecting the driver's alertness using measures such as PERCLOS or behavioural indicators such as the frequency of steering wheel corrections, the output can be used to steer systems such as Driver Alert System or Driver Attention Alert, that may give warning signals and remind drivers that they need a break. Other solutions include steering wheel vibration. As micro-sleep is the most common occurrence, where the driver is prone to changing lanes due to problems with vehicle control, steering wheel vibrations generated by lane change alarms can be effective in preventing accidents and waking the driver up.

Figure 4.6 summarises the results of studies about fatigue.

References

Akerstedt T, Gillberg M (1990) Subjective and objective sleepiness in the active individual. Int J Neurosci 52:29–37

Backs RW, Lennerman J, Wetzel JM, Green P (2003) Cardiac measures of driver workload during simulated driving with and without visual occlusion. Hum Factors 45:525–538

Barua S, Ahmed MU, Begum S (2017) Classifying drivers' cognitive load using EEG signals. Stud. Health Technol Inform. 237:99–106

Borbély AA (1982) A two-process model of sleep regulation. Hum Neurobiol 1:195–204

Connor J (2011) The role of driver sleepiness in car crashes: a systematic review of epidemiological studies. In: Verster JC, George CFP (eds) Sleep, sleepiness and traffic safety. Nova Science Publishers Inc., New York

Dinges DF, Kribbs NB (1991) Performing while sleepy: effects of experimentally induced sleepiness. In: Monk TH (ed) Sleep, sleepiness and performance. Wiley, Chichester

Hamblin P (1987) Lorry drivers' time habits in work and their involvement in traffic accidents. Ergonomics 30(9):13–23

Hancock PA, Desmond PA, Matthews G (2012) Conceptualizing and defining fatigue. In: Matthews G, Desmond P, Neubauer C, Hancock PA (eds) The handbook of operator fatigue. Ashgate Publishing Ltd., Farnham, Surrey

Hart SG, Staveland LE (1988) Development of NASA-TLX (Task Load Index): results of empirical and theoretical research. In: Hancock PA, Meshkati N (eds) Human mental workload. North Holland, Amsterdam, pp 139–183

Hillyard SA, Vogel EK, Luck SJ (1998) Sensory gain control (amplification) as a mechanism of selective attention: electrophysiological and neuroimaging evidence. Phil Trans Royal Soc London-Ser B: Biol Sci 353:1257–1270

Hoddes E, Dement WC, Zarcone V (1972) The history and use of the stanford sleepiness scale (abstract). Psychophysiology 9:150

Jap BT, Lal S, Fischer P, Bekiaris E (2009) Using EEG spectral components to assess algorithms for detecting fatigue. Expert Syst Appl 36:2352–2359

Jin L, Niu Q, Jiang Y, Xian H, Qin Y, Xu M (2013) Driver sleepiness detection system based on eye movements variables. Adv Mech Eng 5

Johns MW (1991) A new method for measuring daytime sleepiness: the Epworth sleepiness scale. Sleep 14(6):540–545

Marquart G, Cabrall C, Winter JD (2015) Review of eye-related measures of drivers' mental workload. In: 6th International conference on applied human factors and ergonomics (AHFE)

May JF, Baldwin CL (2009) Driver fatigue: the importance of identifying causal factors of fatigue when considering detection and countermeasure technologies. Transp Res F: Traffic Psychol Behav 12(3):218–224

Neubauer C, Matthews G, Langheim L, Saxby D (2012) Fatigue and voluntary utilization of automation in simulated driving. Hum Factors 54(5):734–746

Pauzie A (2014). A method to assess the driver mental workload : the driving activity load index (DALI), May. https://doi.org/10.1049/iet-its

Philip P, Sagaspe P, Taillard J, Valtat C, Moore N, Åkerstedt T, Bioulac B (2005) Fatigue, sleepiness, and performance in simulated versus real driving conditions. Sleep 28(12):1511–1516

Phillips RO (2015) A review of definitions of fatigue—and a step towards a whole definition. Transp Res F: Traffic Psychol Behav 29:48–56. https://doi.org/10.1016/j.trf.2015.01.003

Sharwood LN, Elkington J, Stevenson M, Wong KK (2011) Investigating the role of fatigue, sleep and sleep disorders in commercial vehicle crashes: a systematic review. J Australas CollE Road Saf 22(3):24–34.

Wierwille WW, Ellsworth LA (1994) Evaluation of driver drowsiness by trained raters. Accid Anal & Prev 26:571–581

Wierwille WW, Ellsworth LA, Wreggit SS, Fairbanks RJ, Kirn CL (1994) Research on vehicle-based driver status/performance monitoring: Development, validation, and refinement of algorithms for detection of driver drowsiness. Final Report. US Department of Transportation. NHTSA. DOT HS 808 247

Chapter 5
Situation Awareness, Multi-tasking and Distraction

Abstract In this chapter, we analyse the different levels of control involved in driving. Furthermore, we discuss the concept of situation awareness. Maintaining good situation awareness is central to safe driving. Finally, while situation awareness is central to the driving task, drivers may engage in other activities ("multi-task"), in particular when mental load is moderate. However, such additional activities may cause distraction and therewith impede situation awareness.

Before considering situation awareness, multi-tasking and distraction, we briefly analyse the driver's task and introduce concepts that were introduced by Michon (1985) and will be useful later on. When driving, usually the driver has a certain destination in mind. S/he may do the daily commute of going to the office or, when going home from the office, drive to the kindergarten to pick up the children from the day care or drive to the supermarket to buy groceries. On Saturday or Sunday, s/he may take the family for the trip in the countryside. This is called the level of **Strategic control**. It involves deciding about the destination, deciding about the type of route—either the fastest route or a scenic route, or a route that avoids traffic jams. It also involves choices concerning the desired time of arrival and the associated cruising speed. At the **Tactical** (also called Manoeuvring) level, the driver is concerned with manoeuvres such as obstacle avoidance, gap acceptance, making turns, overtaking and so forth. The actual manoeuvres are influenced both by the strategic goals and the traffic situation. While lane changing manoeuvres to avoid obstacles derive from the exigencies of the actual traffic situation, lane changing manoeuvres may also be initiated to satisfy Strategic goals, e.g., in order to overtake slow traffic. In turn, the Tactical level feeds into the **Operational** level, which concerns the actual operations such as steering, throttling, braking (Fig. 5.1).

While driving, the driver needs to perceive the events and objects constituting the traffic situation, use this information to build a model of the traffic situation that needs to be updated continuously and used to predict the course of events so that the driver can anticipate future events (Situation Awareness, see Sect. 5.1). Furthermore, the driver may temporarily shift his attention from the driving task to other tasks/activities, either tasks supporting the driving task, such as consulting the screen of the navigation system, or other tasks, such a accepting a mobile phone call

© China Machine Press 2023

F. Chen and J. Terken, *Automotive Interaction Design*, Springer Tracts in Mechanical Engineering, https://doi.org/10.1007/978-981-19-3448-3_5

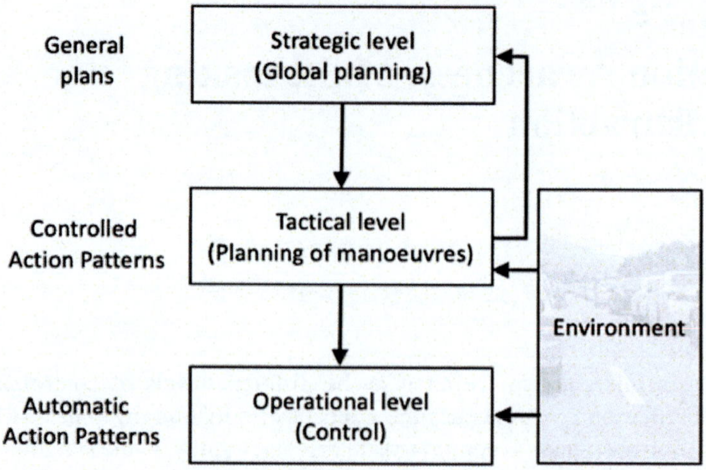

Fig. 5.1 Three levels of control involved in the driving task (after Michon 1985)

(see Sect. 5.2). Thus, other tasks may distract the driver from the driving task (see Sect. 5.3).

5.1 The Concept of Situational Awareness

Situational awareness (SA) is generally defined in terms of a specific purpose, a particular object, a particular function, or even a particular job, and is about people's awareness of what is happening around them, understanding the meaning of the information and what it means for the future (Endsley 1995b). The concept of situational awareness first arose from military activities and was originally defined as (Endsley 1988): "*The perception of the elements in the environment within a volume of time and space, the comprehension of their meaning, and the projection of their status in the near future.*" SA is a necessary condition for almost every decision to be made. The relationship between SA and decision making is shown in Fig. 5.2. The understanding of the current situation and the expected evolution of the situation in the near future form the basis for decision making and acting.

Building and maintaining Situation Awareness comprises three phases.

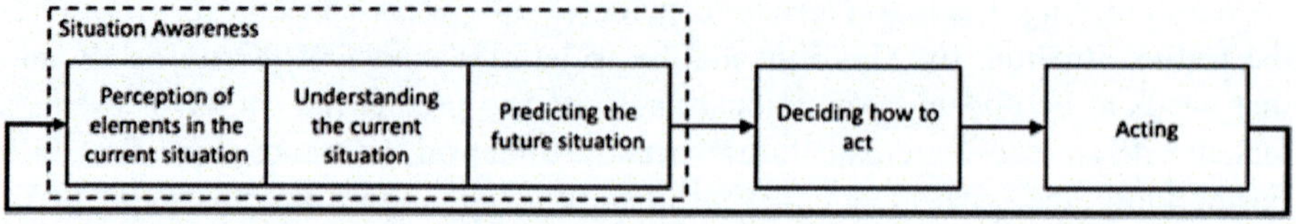

Fig. 5.2 The relationship between situational awareness and decision making (after Endsley 1995b)

Step 1: Perception of the elements of the current situation
Step 2: Understanding the current situation
Step 3: Prediction of the future state

Stage 1: Perception of the elements of the current situation.

These elements include the current state of the internal and external environment, attributes, dynamic changes, etc. The content of the elements is different for different jobs, functions and purposes. For example, a driver driving on a motorway needs to perceive the relative position of the cars in front, to the side and behind him, their speeds, how far away he is from the next exit, etc. At an intersection, the driver needs to know the nature of the intersection (whether he is on a main road or a secondary road, and the location, direction and speed of other vehicles, bicyclists and pedestrians, and additional relevant information, such as traffic signs, traffic lights and pedestrian crossings). This information can be obtained primarily through the sense of sight, although occasionally other senses may come into play as well. In many areas, accidents are often caused by elements of the situation not being fully perceived. Many traffic accidents are caused by distracted drivers, or road traffic that is so complex that drivers do not catch the critical information in time to make a correct decision. Endsley's research also demonstrated that 76% of SA errors by fighter pilots were due to problems with Stage 1 (Endsley 1995b).

Stage 2: Understanding of the current situation.

In stage 2, the information obtained from stage 1 is integrated into a model of the current state of the world. For example, when a car is travelling on a motorway, the driver notices that the brake taillights of the car in front of him are on and that a car to the left behind him is reducing the distance to him. From these signs, he develops an understanding that the car in front is slowing down and the car on his left is accelerating. It has been shown that 19% of SA errors by fighter pilots are due to Phase 2 problems (Endsley 1995b).

Stage 3: Prediction the future state of the world.

From an understanding of the current state of the world, the evolution of the current state into the near future can be predicted. On the basis of information about the location, speed and direction of other vehicles, bicyclists and pedestrians as contained in the model of the world, the driver predicts how the situation will evolve and for instance anticipates whether a gap between two vehicles on the main road will be sufficient to cross the main road. When predicting the future state of the world from stage 2, it helps if people have good knowledge and experience of what needs to be done. An experienced person is able to predict the future state with relative ease, so that they can anticipate the future state and make the right judgement about their next move. For example, an experienced driver's judgement of traffic conditions can be very helpful in driving and responding to various emergencies that occur on the road. A novice driver who has just obtained his driving licence, on the other hand, is often prone to errors in judgement, and a large part of the reason for this is that this prediction is not well made. As in the previous case, an experienced driver can accurately determine that, if he does not slow down, he may hit the car in front of him, or if he tries to change lanes, he may collide with the car on his left.

There are many factors that influence SA, starting with temporality. Time plays a large role in SA, both in terms of time itself and in terms of the various elements associated with time. The most typical question is "How much time do we have left?" The world is constantly changing with the passage of time, and in this dynamic situation SA is also a dynamic process. For car driving safety, getting the critical information, making the right judgement and taking the right action at the right time is of paramount importance. The second element is perception and attention. For the various situational elements in the first stage of SA, the features may be scattered across the environment, but due to our attention allocation problems, much of the information may not be perceived and the wrongly missed information makes our SA less than complete and accurate. The third factor is working memory and long-term memory. The second and third stages of SA occur in working memory, while long-term memory provides a store of previous knowledge and experience for comprehension and prediction. Since mental resources for managing information in Working Memory are limited, high workload has an adverse effect on Situation Awareness. The fourth factor is the mental model. Mental models help a person to determine what information is important, and also to understand the meaning of information and make predictions about the future. If a person does not have a mental model of what is at hand, he or she can develop situation awareness through various analyses and knowledge, but this process is slow and resource-intensive.

As explained before, human information acquisition is generally driven by two different mechanisms, a purposeful search and a direct stimulus-driven capture of attention. Salient information from the external environment, such as alarm sounds, or lights that are flashing, is quickly captured by the human attention. At the same time, people will consciously select information according to their goals. Setting the right goals is crucial, and goal setting leads to the selection of a correct mental model, which dominates people's attention to capture and process the relevant and correct information. The goals play a key role in interpreting this information and producing predictions about the future. Similarly, a person's expectations lead a person to look for the information he expects and thus also dominate the process of understanding and prediction. So, expectation building also plays a large role in SA.

During driving, the driver needs to pay attention to many things, such as the driver's route location, road curvature, the location of nearby traffic and pedestrians, speed and so on. This information needs to be integrated into a model of the current situation, that is used to predict the evolution of the situation in the near future, to serve as a basis for decision making. This process is achieved both through automatic, pre-attentive information processing and information processing that involves selective attention, and hence demands the allocation of attention resources. Stage 1, perception, involves mainly automatic pre-attentive processes that occur subconsciously and have little demand on cognitive resources. Stages 2 and 3, maintaining and updating a model of the world and predicting the future situation, involve controlled processes that demand cognitive resources. As may be expected, in particular for stages 2 and 3, but also for Stage 1, driving experience plays a major role. Deriving the speed of other vehicles (stage 1) and using this to construct a model of the world (stage 2) and predicting the evolution of the location of vehicles in the next few seconds (Stage 3), places a high demand on the cognitive resources of novice drivers, and becomes faster and more automated once drivers become more experienced. Also, experienced drivers know better what elements of the environment to attend to.

5.1.1 Measuring Situation Awareness

The concept of Situation Awareness has been applied to many different fields, including teamwork, health care, aviation and driving. Research into factors influencing SA and into the effect of driver assistance systems on building and maintaining SA requires methods to measure SA. Since SA is a mental construct, it cannot be measured directly in an objective manner. This means that the research needs to rely on self-report methods, i.e., methods that involve (verbal) reports or judgements by participants in experiments, or on derived objective measures, from which inferences have to be made about the driver's SA. In the next paragraphs, we go into methods for measuring SA in the context of driving.

Self-report methods that are most commonly used for measuring situation awareness in the context of driving include SART and SAGAT. SART (Situation Awareness Rating Technique, Taylor 1990) requires participants to fill a post-test questionnaire containing ten items addressing Instability of the situation, Complexity of the situation, Variability of the situation, Arousal, Spare mental capacity, Division of attention, Concentration of attention, Information quantity, Information quality, Familiarity of the situation. Participants score each item on a seven-point rating scale ranging from Low to High. The scores on the ten items are then grouped into three dimensions, demands on attentional resources (D), supply of attentional resources (S), and understanding of the situation (U). Situation Awareness can then be analysed through inspecting the scores for the three dimensions, and through considering the composite score, $SA = U - (D - S)$.

SAGAT (Situation Awareness Global Assessment Technique, Endsley 1995a) is a freeze probe technique that is applied in driving simulator contexts, where the scenario is frozen at random moments and participants are asked about their perception of the situation and expectations about what will happen next. The questions can be verbal questions, or schematic drawings with specific assignments, such as "what is the position of [object x]?" that can be administered through a screen. The questions are derived from an analysis of what information should have been extracted from the situation, in terms of the three SA stages proposed by Endsley, perception, comprehension, prediction. The answers to the questions are compared to the objective scenario, and are scored for correctness, giving an indication of the participant's SA. Experiments have shown that the outcomes of the SAGAT method have high validity, and can also be used to evaluate how a design concept contributes to improve or degrade SA. Disadvantages are that the freeze method interferes with the sensation of presence in the scenario and with the driving task, so that it can be used only in experiments with driving simulators. A method called DAZE, attempting to deal with these disadvantages, was developed by Sirkin et al (2017) for measurement of SA in natural driving situations in automated driving. Participants are probed about particular events by questions presented on a touchscreen and respond by clicking Yes or No. The method may still interfere with the main task and cause distraction, but is set up such that the interference is reduced as much as possible.

Yet other self-report methods have been applied, including critical incident interviews. In this method, participants are exposed to critical events, and are asked what happened and where they were focusing their attention just before the event. These reports can then be compared to the actual scenario and to the outcomes of an analysis of what information a driver should have extracted.

Objective methods for measuring SA include eye tracking, but the results are mixed. In the first place, the correlation with outcomes of the self-report methods is usually low. In the second place, eye tracking primarily reflects the first stage of SA, perception, and not the other two stages. In the third place, the method requires a dynamic world model, against which the outcomes of the eye tracking can be evaluated. Currently, this makes eye tracking as a method for measuring SA suitable only for use in driving simulator experiments. With improvements on sensor technology and artificial intelligence, automatic understanding of real-life situations may evolve sufficiently to make the use of eye tracking for measuring SA also viable for real-life contexts. The question remains whether, by the time the technology has evolved so far as to achieve these requirements, drivers still need to build SA. It may well be that automation of driving systems will have evolved so far that people no longer need to drive themselves, making the need to remain situationally aware obsolete. In Sect. 16.4.3 we will come back to this question.

5.2 Multi-tasking

Multi-tasking is the interleaving of two (or more) different tasks. In the context of driving, multi-tasking may cause safety hazards. It may distract the driver from the driving task and cause accidents. For instance, a driver, while driving, is picking out the music he wants to listen to. Although he keeps his eyes on the road ahead from time to time, most of his attention is temporarily on the music he is picking out. At the same time, a pedestrian is walking and playing with his mobile phone. While at the verge of crossing the road, he takes a quick glance around and sees no danger. Unfortunately, these two people happen to meet at the same junction, and you can imagine what happens next.

When studying multi-tasking, we typically compare the performance on a task when a person is performing that task in combination with another task (multi-tasking) with the performance when the person is performing the same task in isolation. If the performance on the task degrades, this is called dual task decrement.

For example, in the case above, the driver's ability to drive in a dual task condition (driving and looking for music) is much lower than when focusing only on driving. When studying multi-tasking, the most common psychological factor we consider is divided attention, which affects several parts of the information processing process. There are three factors that play a role in the case of divided attention: (1) The mental resources demanded by the task, i.e., the amount of effort the user needs to put into completing the task. This typically depends on the difficulty of the task; (2) The allocation of resources to the different tasks. This typically depends on the priority of the goals for the different tasks, but also on the effort of switching between different tasks; (3) The structure of the resources needed for the different tasks. This typically depends on the similarity of the underlying cognitive processes and mental operations executed when performing the tasks. These three points will be expanded below.

(1) Even if parts of the driving task such as the concrete operations of steering and throttling are highly automated processes requiring little effort (Michon's Operational level), task execution typically involves Working Memory activity. In particular, we need to evaluate whether our operational activities are still in line with our goals and, if not, take corrective action. Also, we need to keep our plan in mind and prepare for upcoming turns and similar manoeuvres (Michon's Tactical level). This process of evaluation, taking corrective action and planning involves Working Memory activity. As mentioned above, the capacity of Working Memory (both Baddeley's Central Executive and the Verbal and Analog-spatial registers) is limited, and our mental resources for performing Working Memory activity (executing the Central Executive function and rehearsing and maintaining information in the two registers) are limited as well. This also affects our ability for multi-tasking. Typically, simple tasks use few resources and complex tasks use more resources. E.g., driving on the highway on a quiet morning in the weekend with bright weather is a relatively simple task. In such a situation there are only few objects to stay aware of, and the situation around us does not change a lot, so that maintaining the situation model does not require many resources. Also, the driving task is relatively simple: we may maintain more or less constant speed, and the only thing we have to do is to make small adjustments to our position in the lane and occasionally check our mirrors to see whether the traffic situation behind us changes. On the other hand, driving at evening rush hour in wintertime when it's wet in an urban environment is a complex task. This is a very dynamic situation, and we need to be aware constantly of changes in the traffic situation, like cars in front of us braking and bicyclists approaching from behind., so that maintaining and updating the situation model takes many resources. Also, we need to constantly adjust speed and position in the lane and prepare for turn manoeuvres. While the driving task in both cases has high priority, in the latter case it requires more resources (mental workload) and there will be fewer resources left for another task. If the driver still wants to perform another task, like occasionally

check his mobile phone for newly arrived text messages, task performance for the primary (driving) task will be impeded.

The ability of dealing with multiple tasks also depends on experience. This is because experience helps the user to deal with the complexities of the task. Consequently, a novice driver who has just learned to drive, may be afraid to talk on his mobile phone or do other things while driving even in situations that experienced drivers may consider simple. For the novice driver, in such a situation there is a high demand for mental resources. When he has been driving for a few years and has gone from being a novice to a skilled driver, especially when he is driving on a road that he is very familiar with and has become very good at coping with what may happen on the road, the demand for resources will be much lower compared to that of a novice, so that he has resources left for doing something else while driving. Of course, if he is driving in an area he is unfamiliar with or if the traffic situation becomes more complicated, e.g., when driving in rush hour in the city in winter when it's raining, the mental resources required for the driving task will increase again, and even an experienced driver may refrain from multi-tasking.

(2) Furthermore, research has shown that switching between tasks takes effort and that there is a certain resistance to interrupting the current task and switching to another task. Otherwise said, a task that the user is currently focusing upon is temporarily allocated a higher priority. This can be understood as follows (Fig. 5.3). At time t1, the user is focusing on task A. The contents of Working Memory can be represented as a set of concepts and relations between concepts, i.e., a structured representation of task state (we may assume that, in the case of driving, the situational model is held in the analogue-spatial register and that the Central Executive is concerned with processing information from the world, updating the situational model and anticipating the future situation). At t2, the user switches to task B that s/he was working on before. This involves storing the structured task representation for task A into Long-term Memory and retrieving the structured representation associated with task B from Long-term Memory into Working Memory. At t3, the user switches to task A again,

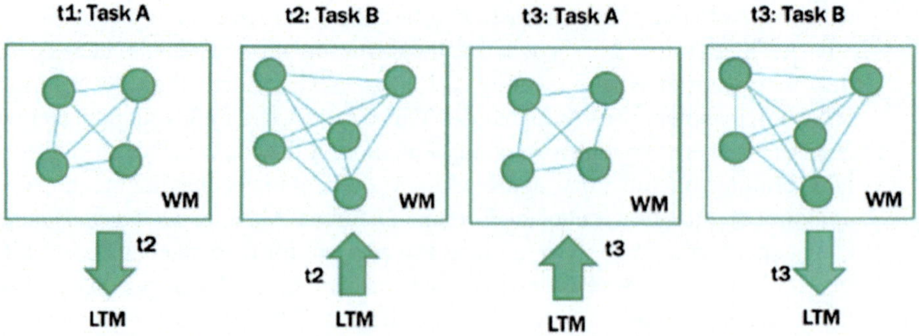

Fig. 5.3 The process of task switching

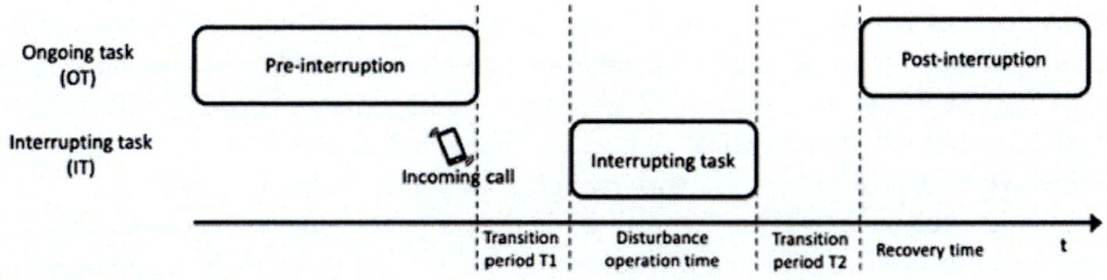

Fig. 5.4 Schematic diagram of multi-tasking transitions. After Wickens et al. (2013)

so that the structured task representation for task A needs to be retrieved from Long-term Memory into Working Memory and the structured representation associated with task B needs to be stored into Long-term Memory again. It can be easily imagined that this process of moving knowledge representations between different registers and rebuilding knowledge structures in Working Memory takes effort, and that users therefore dislike interrupting the current task. The issue of switching between tasks will be further elaborated below.

Even if this representation in correct in that it expresses that switching between tasks takes effort, there is also evidence, that this representation is simplifying. As said before, Baddeley distinguishes two registers in Working Memory, the verbal register and the analogue-spatial register, and there is evidence from psychological experiments that the resources for maintaining the knowledge representation in the two registers and operating on them are specialized, in the sense that resources for operating on the verbal register and resources for operating on the analogue-spatial register cannot be completely exchanged. The consequence is that, if one of the two tasks requires resources for operating in the analogue-spatial domain and the other task requires resources for engaging in a conversation, there will be less mutual intervention. On the other hand, the resources for the operation of the Central Executive are generic and are shared between the analogue-spatial and verbal register, so that evaluation of goal achievement and decision making will be impeded in a multi-tasking situation. Otherwise said, the situation shown in Fig. 5.4 only applies if task B also involves the analogue-spatial domain. If task B involves the verbal register, there is no need to clear the analogue-spatial register. However, the Central Executive would still have to switch from task A to task B. Furthermore, when two tasks employ the same register, e.g., the analogue-spatial register, such as driving and interacting with a touch screen, this will result in a drastic performance decrease for one or both tasks, depending on how the user sets the priorities.

5.2.1 *Switching Between Multiple Tasks*

While driving, the driver often operates additional tasks at the same time. This is especially true of infotainment systems. In the future, as cars become more autonomous, the design of infotainment systems is becoming more and more important and the concept of the car as a third space in addition to the home and office is becoming more and more accepted. However, for a long time to come, manual driving will still be a task for the driver. It is here that we need to understand what is happening cognitively when we switch from one task (driving a car) to another (choosing music).

We use the term OT (ongoing task) to denote a task that is in operation, e.g., the driver is driving. Then, a phone call comes in, and this call is an interrupting task IT (Interrupting task). This process can be represented as in Fig. 5.4. There is a transition period T1 between the user's OT and IT tasks, and another transition period T2 when the user returns to OT after completing IT. The lengths of T1 and T2 depend on the complexity of the tasks: the more complex the tasks, the longer T1 and T2 will be. The transition durations also depend on the design of the interaction. If the design specifies how to proceed, the transition periods will be shorter. Finally, if the frequency of switching between two tasks is higher, the transition duration will also be longer.

The relationship between OT and IT is influenced by several factors: (1) The driver's mental engagement and focus (engagement) on the task. For example, if the driver is driving on an unfamiliar road with a mix of road users, or a mountain road with a cliff on one side, it requires the driver's full attention and it is very difficult for any IT to take his attention away from the driving; (2) Modality, which has already been mentioned. If the mental resources required by the IT are the same modality as the OT, both requiring the driver's visual attention, then combining both tasks will be difficult, and the IT will strongly interfere with the OT. If the modality required by the IT is different from the OT, there will be less interference; (3) Dynamics: when the driver is driving manually, the driver's actions will vary depending on the state of the vehicle and the road. If the road is bumpy and the situation is unstable, the driver cannot be distracted from IT; (4) Priority: driving safely should always be the first priority in the driver's task. So, if IT is present, it may affect driving safety and the driver should be cautious with IT; (5) Targeted: if the OT's target has been accomplished, then the impact of IT intervention will be small.

The problem of multi-task interference is also influenced by other factors, one being the similarity of the interfering tasks: The more similar the two tasks are, the greater the interference will be. For example, if different voices are used to cue different meanings, but if the two voices are not very different, it is more difficult for a person to be able to tell the difference, and the interference between the two voices increases. Similarly, the same problem exists with the use of iconic cues. And, as already mentioned before, another factor is the difference between skilled and novice workers.

5.3 Driving Distraction

Driving distraction is a subject with an extensive history of research. In the 1990s, when mobile phones became commonplace and people began to drive and talk on the phone at the same time, and traffic accidents due to such distraction began to be reported, the number of publications on distraction exploded (with roughly 3000 hits in Google Scholar for the period 1980–1990 and more than 15,000 hits for the period 1990–2000).

5.3.1 Types of Driving Distractions

Driver Distraction is the diversion of the driver's attention from driving to other, competing activities, which can easily occur under assisted driving conditions. Related to distraction are the concepts of inattention and concentration loss, when a driver's distraction is due to thinking about something else. Inattention is defined as "not paying enough attention to keep driving safely" (Lee et al. 2009; Regan et al. 2011), and distraction is the result of inattention. In general, there are four basic types of distraction: visual distraction, auditory distraction, operational distraction, and cognitive distraction. Specific definitions of these types of distractions for drivers are as follows.

- *Visual distraction*: The driver's visual attention is on an object or information outside the traffic situation, for example, the eyes are looking at the mid console.
- *Auditory distraction*: The driver's attention is taken away from the driving task due to an auditory stimulus such as a bell or music.
- *Operational distraction*: The driver performs physical operational tasks outside of driving, such as manually adjusting the radio volume or selecting a song.
- *Cognitive distraction*: The driver's attention is taken away from the driving task due to a conversation or other activity, e.g., s/he is thinking about something else

The effect of distraction on driving depends not only on the type of distraction, but also on the frequency and duration of the task, meaning that even if a task is not a strong distraction, a driver who engages in it frequently or for a long time may increase the risk of causing an accident to a level comparable to that of performing more difficult subtasks less frequently.

The sources of driver distraction can be categorised in different ways, internal (distracted thoughts) or external (factors outside of oneself), technical (operating a piece of equipment in the vehicle) and non-technical (e.g., eating), self-inflicted (e.g., one is trying to call someone else) or caused by others (a friend calling), in-vehicle or out-of-vehicle. Table 5.1 gives a summary of common distractions in the car. Research by the European Road Safety Observatory (European Commision 2018) shows that, on average, between 20 and 30% of the time drivers are distracted by something unrelated to driving, and that a third of these distractions come from outside the vehicle and about a fifth from using devices such as mobile phones. For commercial vehicle drivers, distractions are responsible for 70% of accidents.

Age has a strong influence on distracted behaviour, with younger drivers being more likely to be distracted compared to middle-aged and older drivers. According to the Ministry of Transport of China in 2014, about 47% of simple traffic accidents

Table 5.1 Distractions commonly done in the car

Sources and types of distraction	Traffic related?	Self initiated?	Technology-related	Inside vehicle	Type of distraction
Phone	No	Yes	Yes	Yes	Auditory-cognitive
Passenger	No	Yes/no	No	Yes	Visual-auditory-cognitive
Music	No	Yes	Yes	Yes	Auditory-perhaps cognitive
Texting	No	Yes	Yes	Yes	Visual-cognitive-physical
Equipment handling	No	Yes	Yes	Yes	Visual-cognitive-physical
Enter destination in navigation system	Yes	Yes	Yes	Yes	Visual-cognitive-physical
Follow instruction navigation	Yes	Yes	Yes	Yes	Visual-auditory-cognitive
Reacting to warnings	Yes	No	Yes	Yes	Visual-auditory-cognitive
Looking at advertisements	No	No	No	No	Visual-cognitive
Eat, drink, reaching for object, facial care	No	Yes	No	Yes	Visual-physical
Daydreaming	No	Yes/no	No	Yes/no	Cognitive

Source European Commission, Driver Distraction, European Commission, Directorate General for Transport, February 2018

nationwide were caused by distracted drivers, amounting to 3.1 million; and about 38% of general traffic accidents were caused by distracted drivers.

5.3.2 *Effects of Distraction on Driving*

The extent to which distraction affects driving is influenced by factors such as age, fatigue, driving experience, personality and even a correlation with the co-driver. Regardless of the cause of distraction, the adverse effects include slower driving speeds, closer following distances, poorer route keeping, more driving errors and narrower visual focus. Figure 5.5 shows the factors that contribute to driving distraction and the associated effects.

Studies have shown that visual plus manual distraction tasks, such as entering a string of numbers by hand, present the highest risk of accidents. The results of

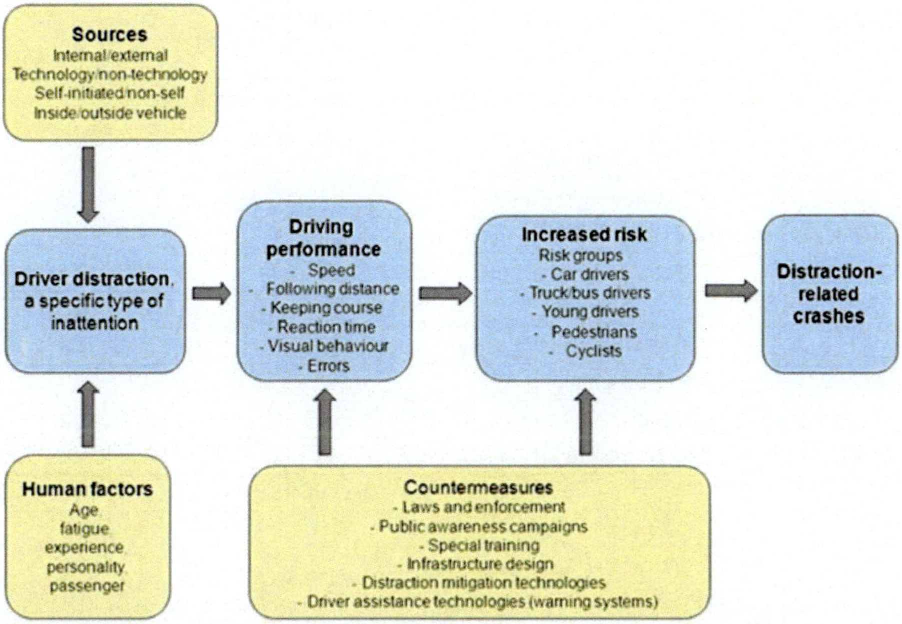

Fig. 5.5 Factors that produce driving distractions and associated effects. *Source* European Commission (2018), courtesy C. Goldenbeld

studies on the degree of risk of mobile phone use, however, are divergent, with the risk of answering a phone in naturalistic driving studies being much lower than in experimental studies, possibly due to the fact that in realistic situations, drivers will adjust their use to a less attention-demanding way of using the phone. Table 5.2 presents an assessment of risk factors for common distracting behaviours (Dingus et al. 2016), where the risk index is the probability of an accident when engaging in the corresponding distracting behaviour compared to the probability of an accident when not engaging in any subtask.

Studies have shown that visual distraction generally causes the greatest driving hazards (NHTSA-2010-0053 2010). This is because driving is primarily a visual task, and visual distraction can cause drivers to take their eyes off the road ahead, resulting in large and frequent lane departures, sudden swerving, and failure to react to the brakes of the vehicle in front of them.

Cognitive distraction leads to a lack of sensitivity to traffic information and the road environment, resulting in "seeing but not noticing". Studies have shown that cognitive distraction leads to a longer reaction time of 130 ms on average (Horrey and Wickens 2006), and that cognitive distraction also leads to a reduction in the ability to extract information from the peripheral vision. For auditory distraction, answering a phone or talking to a passenger result in at least a reduction in driving speed; increased following distance; longer reaction times; and reduced lane keeping performance.

Table 5.2 List of risk indices for each common distracting behaviour

	Distracted behaviour	Risk index
Equipment in the car	Other in-vehicle equipment (e.g., use of the central control screen)	4.6
	Regulating air conditioning	2.3
	Adjusting the radio	1.9
Mobile phone related	Dialling (handheld operation)	12.2
	Reaching for another object	9.1
	Typing	6.1
	Reaching for the phone	4.8
	Browse mobile	2.7
	Call	2.2
Other	Read or fill in	9.9
	Looking at a passenger outside the vehicle for a long time	7.1
	Eating	1.8
	Drinking water (non-alcoholic)	1.8
	Personal care (e.g., fixing hair)	1.4
	Swaying to music	1.0

Of the four types of distraction, visual distraction is directly related to driving operations. Drivers taking their eyes off the road ahead for longer than 2.0 s significantly increase the risk of a crash/near crash. When drivers take their eyes off the road ahead for a total of more than 2.0 s in a six-second period, this also increases the risk of an unsafe event occurring compared to normal driving. We have summarised the effects of various types of distraction on driving operations.

- The various visual-manual tasks associated with the use of different devices (texting, entering numbers, entering destinations, operating music devices) result in reduced driving ability, i.e., more frequent and longer periods of time that the eyes are away from the road, more objects missed on the road, poorer lateral position control, slower reaction times and more conflicts with other road users.
- The use of mobile phones and hands-free calling seems to have a similar negative impact on driving in terms of task performance.
- Talking to passengers appears to have less impact on the driving task, as the passenger can assist the driver with the driving task and adjust the speed of speech and the complexity of the communication to respond to the changing demands of the driving task.
- Eating and drinking while driving can lead to greater deviations from lateral position control with the vehicle and reduced speed.
- Roadside advertising may affect driving behaviour. The effects that have been found are reduced speed and greater deviations in lateral position. Advertisements attract the driver's visual attention, increase reaction times and lead to more driving

errors. Billboards located in the middle of the field of view or at street level are particularly distracting for drivers.

However, research into the effects of distraction on driving has come mainly from laboratory work, which presents several problems, such as.

- The effects of distraction on driving performance have been obtained mainly through experimental studies, particularly driving simulator experiments, and the conclusions drawn from these studies are not necessarily the same as in practice.
- There is not always a direct correlation between driving behaviour and the risk of driving accidents. Few data are available to document both driving behaviour and its impact on accident risk.

5.3.3 Resource Allocation and Workload

In order to further address the mechanisms by which distractions arise, a discussion of the allocation of attentional resources is required. Human attentional resources are limited. When a person is operating in a multi-tasking situation, attention is divided between different tasks. According to Wickens' Multiple Resources Theory (Wickens 1984, 2002), the distraction of a secondary task from the primary task is determined by the resource requirements of the secondary task. As was mentioned in Sect. 5.2, when the total attentional resource demands required for multiple tasks exceed the human attentional resource limit, competition between tasks arises, resulting in reduced performance on all or some tasks. However, this depends also on the nature of the resources. According to Multiple Resources Theory, there are specialised resources for verbal information and for analogue-spatial information. Competition is more intense between tasks that deploy the same type of materials. The driving task involves primarily analogue-spatial information, and accounts for almost 90% or more of the demand for visual attentional resources, which is why distractions involving visual perception and analogue-spatial information have the greatest impact on driving performance. Any in-vehicle subtask that involves visual-manual operation creates competition for attention resources with the driving task. If the secondary task is an auditory-speech operation, there is less competition for attentional resources with the driver's primary task and, therefore, less interference. However, such tasks may create cognitive distraction. In addition, because of the immaturity of speech technology, the high error rate, and the fact that the sound information is volatile, auditory-speech interaction is often accompanied by a visual presentation. This reduces the value of speech-based interaction.

In addition to the direct competition for attentional resources, the performance impact of multi-tasking is also reflected in the cost of switching between tasks and the time taken to do so. When a driver shifts from the primary driving task to a secondary task, the disruption in state understanding of the primary task (reduced situational awareness) results in more attentional resources being consumed when switching from the secondary task back to the primary task, resulting in longer reaction times and more operational errors.

Attentional demands are not uniform during a driving task, and in some situations (e.g., unfamiliar road turns) the workload of the driving task can be high, and if another task interferes with attention at such a moment, this can lead to driving task malfunctions. The same interfering task may cause less disruption if the attentional demands from the driving task are lower, such as when driving at a steady pace on a quiet highway.

Workload comes from a combination of factors such as the time demands of the task, the number of activities, and the complexity of the activity. In general, basic driving tasks (e.g., controlling a vehicle, scanning for hazards, planning a route, etc.) impose different workloads on the driver, which increase or decrease depending on the driving conditions (e.g., road complexity, weather, traffic flow, etc.) and the driver's state (fatigue, alertness, etc.). The driver meets the workload requirements across tasks by allocating resources in a way that is responsive to driving.

Due to the complexity of the factors affecting workload, it is difficult to monitor the workload associated with driving (or the driver's ability to handle additional tasks) in real time. Drivers themselves have a degree of resilience and are capable of handling higher workloads (e.g., by slowing down, increasing headway with the vehicle in front) and in order to cope with high load driving situations, drivers may resort to skipping or ignoring tasks that are not related to driving until the driving is finished or it is safe to stop.

In addition to the active management of the driver's workload, it is also possible to reduce the driver's workload for basic driving tasks by means of assistance systems. However, these systems may be mainly effective when applied in specific situations; for example, when driving in unfamiliar areas, the use of navigation systems can be effective in reducing the driver's workload.

5.3.4 Monitoring Distraction

Distraction is monitored in a similar way to fatigue, generally by means of cameras and eye movements, but the corresponding monitoring indicators are different. Firstly, there is a 2-s rule for visual distraction monitoring: the rule holds that taking your eyes off the road for more than 2 s poses a significant risk to driving safety and therefore should be avoided. However, this is not a definitive threshold; driving tasks are complex and are influenced by many surrounding circumstances, traffic flow and other factors such as light and visibility, so that 2 s may be too long or too strict, depending on the situation. Secondly, intervals of looking away for more than 2 s may be accompanied by turning the head. Head turns have therefore also proven to be a reliable indicator of visual distraction.

Eye movement monitoring

The eye-movement parameters used for visual distraction are the frequency of off-road gaze and the duration of gaze, which can be acquired using head-mounted glasses or a camera facing the driver. A driver is visually distracted when the eyes drift from the road ahead. If the eyes drift from the road ahead for more than 2 s, the driver is currently distracted in a way that is considered a driving risk.

Behaviour monitoring

Similar to the effect of fatigue, the effect of distraction can also be observed from the behaviour of the vehicle. Distraction results in fewer steering wheel manoeuvres to adjust the position of the vehicle in the lane and to larger deviations in lane position. Systems such as Attention Assist and Driver Alert Control monitor the driving input, either stand alone, or in combination with eye movement and eye lid monitoring, to identify driver distraction and driver fatigue, and warn the driver if a certain threshold has been passed.

Detection response tasks

The Detection Response Task (DRT) is a way of measuring distraction and fatigue and cognitive load using correct response rates and reaction times. The general design of the DRT requires the driver to respond to a randomly presented stimulus, including a visual stimulus such as a flashing light in the peripheral vision, and a haptic stimulus such as a vibration of the device. Specific standards for the operation of the DRT can be found in ISO standard 17488:2016.

The DRT can determine the current state of driver alertness by comparing it with indicators in normal conditions; it can also determine whether the secondary task load is too high by comparing the driver's keystroke hit rate and reaction time with and without subtasks. The DRT is simple to equip and easy to assemble, and can be used in naturalistic driving studies. However, it adds another task, which may cause distraction in itself and further increase the workload, so that it is not suitable for use in real-life driving contexts.

References

Dingus TA, Guo F, Lee S, Antin JF, Perez M, Buchanan-King M, Hankey J (2016) Driver crash risk factors and prevalence evaluation using naturalistic driving data. Proc Natl Acad Sci 2636–2641

Endsley MR (1988) Design and evaluation of system awareness enhancement. In: Human factors and ergonomics society annual meeting, August, 97–101. https://doi.org/10.1177/154193128803200221

Endsley MR (1995a) Measurement of situation awareness in dynamic systems. Hum Factors 37(1):65–84

Endsley MR (1995b) Toward a theory of situation awareness in dynamic systems. Hum Factors J Hum Factors Ergon Soc 37(1):32–64. https://doi.org/10.1518/001872095779049543

European Commission (2018) Driver distraction. European Commission, Directorate General for Transport, February 2018. https://ec.europa.eu/transport/road_safety/document/download/ab9ce97f-3801-49b0-9fd1-0957bc9d4464_en

Horrey WJ, Wickens CD (2006) Examining the impact of cell phone conversations on driving using meta-analytic techniques. Hum Factors 48(1):196–205

Lee JD, Young K, Regan MA (2009) Defining driver distraction. In: Regan Ma, Lee JD, Young K (eds) Driver distraction: theory, effects and mitigation. CRC Press, Boca Raton, FL

Michon JA (1985) A critical view of driver behavior models: what do we know, what should we do? In: Human behavior and traffic safety. Springer, Boston, MA, pp 485–524

NHTSA-2010-0053 (2010) Guidelines for reducing visual-manual driver distraction during interactions with integrated, in-vehicle, electronic devices. Version 1.01

Regan MA, Hallett C, Gordon CP (2011) Driver distraction and driver inattention: definition, relationship and taxonomy. Accident Anal Prevent 43(5):1771–1781

Sirkin D, Martelaro N, Johns M, Ju W (2017) Toward measurement of situation awareness in autonomous vehicles. In: Proceedings of the 2017 CHI conference on human factors in computing systems, May, pp 405–415

Taylor RM (1990) Situational awareness rating technique (SART): the development of a tool for aircrew systems design. In: Situational awareness in aerospace operations (AGARD-CP-478). NATO–AGARD, Neuilly Sur Seine, France, pp 3/1–3/17

Wickens CD (2002) Multiple resources and performance prediction. Theor Issues Ergon Sci 3(2):159–177

Wickens CD (1984) Processing resources in attention. In: Parasuraman R, Davies R (eds) Varieties of attention. Academic Press, New York, pp 63–101

Wickens CD, Hollands JG, Banburry S, Parasuraman R (2013) Engineering psychology and human performance, 4th edn. Pearson, Boston

Chapter 6
User Experience

Abstract Safety is a central goal for automotive design, which legitimises the focus on the cognitive processes involved in driving and the failures that may occur. However, drivers are not just information processing organisms; instead, they also experience emotions while driving. This insight is captured by the notion of user experience and creating positive user experiences has become a major goal for automotive design, adjacent to safety. In this chapter, we discuss the notion of user experience and summarise user experience theory. Finally, we look at recent trends that go beyond the user experience.

User experience and user interface are different concepts. A user interface is a tool designed by the designer to help the user talk to the system to complete a task, while the user experience is the feeling the user gets when using the interface to complete a task. Therefore, different experiences of the same interface can be very different and can vary from person to person. In this chapter, we look at the user experience in more detail.

6.1 In-Vehicle User Experience

In Chap. 2 we mentioned some of the concepts of user experience in general terms. Experience is defined as "the process of doing and seeing things and of having things happen to you" (Merriam-Webster Dictionary). User experience (UX), on the other hand, is related to people's experience of technology. UX can be defined as "the personal perceptions and responses that result from the use or intended use of a product, system or service".

In addition to safety considerations, the user experience for drivers and passengers is increasingly valued by the automotive industry, and this has become a reflection of the competitive advantage offered by car manufacturers. This means that, in addition to performance and appearance, car manufacturers now need to work on improving and enhancing the experience that using a car generates. One of the characteristics of the automotive industry is that change is slow and development cycles are several

© China Machine Press 2023

F. Chen and J. Terken, *Automotive Interaction Design*, Springer Tracts in Mechanical Engineering, https://doi.org/10.1007/978-981-19-3448-3_6

years long. This means that while UX may have made significant advances in other areas of technology, in the car it is still in its infancy. Furthermore, traditional methods used for HMI research, focusing on cognitive issues, may not be applicable to UX research, as UX research requires holistic, contextual and ethnographic considerations. A central issue at present is that much of the work labelled as automotive UX does not actually address important aspects of the automotive experience, but rather places more emphasis on the introduction of new features rather than considering issues of time, context and emotion. This may be due to the fact that much of the research labelled as UX has little or no connection to UX theory, but uses methods traditionally used in other areas of automotive research, and the results do not approach experience in a holistic way, but rather focus on certain pre-selected aspects of experience.

6.2 Pragmatism and User Experience

Since Norman et al. (1995) introduced user experience to the field of HCI (Human Computer Interaction) in the mid-1990s, the term "user experience" has become increasingly popular. The number of publications on the subject has also increased (at CHI 2015, 715 out of 936 articles at least mentioned the term "user experience").

The study of experience was central to Dewey's pragmatism, and the philosophical stance of pragmatism provides the epistemological basis for the modern study of user experience. In his seminal work Art as Experience (Dewey 1934), Dewey argued against the prevailing notion of the time that the value of art depended solely on the ultimate purpose of the artwork (e.g., painting). Dewey's original statement reads: "*An experience is a product, one might almost say bi-product, of continuous and cumulative interaction of an organic self with the world. There is no other foundation upon which aesthetic theory and criticism can build*". McCarthy and Wright examine Dewey's research in the context of HCI and note that "the experience that arises between the self and the object as felt by the attentive person is constituted by the relationship between what people do and the materials and tools they use. It consists of what people do and what the system does to people, their struggles and aspirations, and their feelings, including fears, beliefs, hopes, pleasures and fantasies. It records the experiences of life and activities of all kinds." In other words, experience is the emotion of each person, personal, subjective and unique (McCarthy and Wright 2004). Indeed, we humans experience our lives in this world. Experience is shaped through a dialectical relationship between the self, the artefact and the environment. Even if all other variables remain constant, different people will have very different experiences. Human experience cannot be described in terms of a priori truths and fundamental absolutes; instead, it is temporary, probabilistic, and dependent on changing, uncertain factors and characteristics. These insights have had a huge impact on the way user experience research and design is conducted (Wakkary 2009).

6.3 User Experience Theory

Several theorists have contributed to the concept of user experience. Here, we summarise the most important contributions.

1. ***Patrick Jordan's Four Pleasures (Pleasurable Design)***

   In 1996, Jordan (2002) proposed that designers should aim to design pleasurable products, where pleasure includes physiological, psychological, social and/or ideological aspects.

- Physiological pleasure is related to how satisfying the interaction with the product is in terms of its physical properties. Do the materials used feel good? Do they offer a luxurious feel or are they cheap? Do the controls feel good in the way they are handled? Are there no sharp edges, etc.?
- Psychological pleasure is related to whether it is easy for users to understand how to interact with the car and the in-car systems, and whether the interaction is exciting. Excitement may arise from novel interaction techniques. However, the effects of novelty are relatively short-lived. People quickly become accustomed to the novelty of the interface and no longer find it exciting. If that feeling passes and people still feel that this way of interacting with the system is 'essential', then the design takes on new meaning. Other determinants of psychological pleasure include the presence of surprise, which encourages exploration.
- Social pleasure is related to improving communication with others. Relatedness is one of the basic needs of people according to Self-Determination Theory, and this explains the popularity of using social media while driving, enabling communication with people elsewhere, even if it distracts drivers from the primary task. However, in essence, driving itself is a social activity, even though we often experience it as a highly soloist activity, leading to competition with other drivers and sometimes anti-social behaviour. Ways to enhance the ability to establish contact with other drivers and to enhance social behaviour have been explored by Wang et al. (2020).
- Ideological pleasure relates to supporting people's ideals. For example, the use of certain materials or technologies or applications that help to reduce energy consumption may help to achieve one's sustainability goals.

Specific applications may involve different types of pleasure. For example, an app that allows people to compare their fuel usage, or more precisely their eco-driving style with that of other users of the app, may enhance their efforts towards the ideal of sustainability on an ideological level, compare their scores with those of others on a social level, and make them psychologically aware that their scores are better than the scores of others, making them feel excited.

2. ***Theory of demand***

   Human needs theory has been used to ground the motivations that drive human behaviour. Understanding users' needs provides insight into the experiences they may wish to have. Needs are embodied in the assumption that all humans struggle to 'define the essential qualities of an experience' (Sheldon et al. 2001). Thus, if human needs can be defined, captured and designed for, the resulting experience will have a greater chance of being positively received by users whose goal is to optimise their satisfaction.

The most popular early example of a model of human needs was Maslow's hierarchy of needs, which was initially accepted by academics and industry. However, Maslow's model was criticised for lacking the necessary experiential basis to support its claims and verify its accuracy (Wahba and Bridwell 1976).

A more modern theory of human needs is Self-Determination Theory (SDT) (Ryan and Deci 2000). This theory assumes that there are three main human needs and that all of these needs must be met in order for a person to be happy. They are autonomy, competence and relatedness. Autonomy captures the need for humans to be free to make decisions and determine their own future. Competence describes the need for anyone to be useful and to be able to use their skills to meet the challenges that arise, while relatedness describes the need to be social and to stay connected to those dear to us.

User experience theory provides a number of valuable concepts for design and may be used as a focus mechanism to enable designers to find a design solution that satisfies one or more human needs. Max-Neef introduced the concept of need satisfaction (Ekins and Max-Neef 1992), which emphasises a particular way of satisfying a human need. For example, I can satisfy my need for competence by learning how to brew artisanal coffee, while others can satisfy this need by playing volleyball. These activities can all satisfy the need for competence, but they have nothing else in common with each other.

3. ***Donald Norman's Three Levels of Emotional Design***

Different theories about user experience focus on different aspects of the experience, and emotion is one of them. Norman introduced a framework for emotional design (Norman 2004). The framework describes three levels of experience, as shown in Fig. 6.1, which can be triggered by three levels of design.

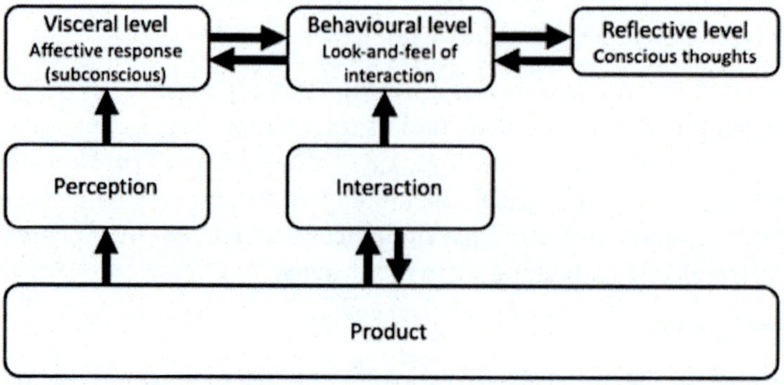

Fig. 6.1 Donald Norman's framework for emotional design

The first level concerns the emotional response, or the deep inner feeling (visceral), that occurs when a user interacts with a product. At this level people begin to construct their first impressions from the appearance of the product, and interactions with the product create immediate emotions that shape the user's experience. For example, a user's first interaction with a car they are considering buying will bring about impressions such as the smell of the car, the colour, the feel of the interior seats, and also the first interaction with the car's HMI will influence the user's opinion of the car. A good impression will add to the user's experience. This part of the requirement is easy to design, but the first impression is quickly replaced by the behavioural level of experience.

The second, behavioural level of experience concerns the look-and-feel of the interaction. The interaction with a product can build on the user's initial impressions and increase the user experience through its usability. In the case of a car, for example, the behavioural level of experience relates to driving the car and thus feeling the performance of the car and experiencing the interaction with the car's HMI. Is it easy to use the GPS function? Can users easily connect their smartphones? How does the Advanced Driver Assistance System (ADAS) behave in use? The answers to these types of questions and the way the product behaves in different usage scenarios provide the user with an experience of using the product. The behavioural aspect gives a more lasting feeling in the design than the first impression. Each time the product is used, the user experience is further shaped by the behaviour of the product in different situations.

The third level of the framework for emotional design concerns the reflective or cognitive level of design. At this level, the experience arises through reflection about our interactions with the product. Here, we reflect about whether the product enables us to achieve a meaningful experience in terms of our ideals, values and beliefs, thus forming a stronger and more lasting connection. The environmentally conscious car owner will have a preference for a car that is energy efficient or uses a lot of renewable materials in the manufacture of the car. Conversely, for accomplished business people, certain features of a certain type of car can symbolise success.

4. ***Product experience framework (Product experience)***

   Another theory that describes UX is Desmet and Hekkert's (Desmet and Hekkert 2007) product experience framework. This framework focuses on the experience of using a product, and the product experience is divided into three levels: Aesthetic experience, Attribution of Meaning and Emotional Response. "Products have the ability and the means to make us feel one or more pleasurable sensations", which leads to an aesthetic experience. In other words, this level of experience corresponds to the stimuli perceived by our senses, similar to the physiological pleasure in the Jordan's framework of the 4 pleasures. The first impression of an automotive HMI falls into this level of experience.

Attribution of meaning relates to Jordan's level of psychological pleasure and Norman's level of behavioural and reflective design. For example, if the car is where you and your girlfriend first bonded, then the association of this memory with the car gives the car a new meaning.

The emotional response includes the emotions elicited by the product itself or its use. The emotion results from the appraisal of an event in terms of whether it has relevance to oneself. The outcome of the appraisal can be positive or negative, giving rise to positive or negative emotions, respectively. For example, a car can be evaluated positively if the owner's interest is in having the freedom to travel, but negatively if it does not match the owner's sense of frugality.

5. ***Be-do Experience Model***

   Inspired by activity theory, Hassenzahl (2010) introduced an empirical model that divides the user's goals into three levels: a *motor* goal associated with the operational technique (*how* to interact with the system); a *do* goal associated with the user's activity (*what* to do with the system); and a *be* goal associated with the user's motivation, emotion, and meaning (*why* to interact with the system. For example, I drive a car for the purpose of going to work, which is *why* I do it, the *be* goal; I steer the car and speed up or slow down, which is *what* I do (the *do* goal); I use the steering wheel and pedals to operate the car, which is *how* I do it, the *motor* goal. "The distinction between these three levels is a conceptual tool to address the different levels of interaction with technology" (Hassenzahl 2010). The Be-Do experience model broadens the designer's scope by using activity theory within an empirical framework, urging designers to consider not only how users interact with technology, but also why they do so. For example, most people do not drive because they want to drive, but because they want to go somewhere. But it is easy for designers to forget these reasons, especially when questions about the *what* and *how* narrow the designer's attention on the details. According to Hassenzahl, the *why* can best be understood in terms of important psychological needs, for instance as proposed by Sheldon et al. (2001), such as Autonomy, Competence, Relatedness, Popularity, Stimulation, Self-respect, Luxury (note that the first three goals coincide with the goals in Ryan & Deci's Self-Determination Theory). In other words, users have positive experiences if the interaction with products satisfies one or more important psychological needs, and using this insight as a guide in design may enhance the design for positive experiences.

6. ***Threads of experience***

   McCarthy and Wright (2004) argue that experiences must be studied in their entirety and in specific application scenarios, rather than in isolation from the context and background of their use. The authors introduce four lines of experience as the focus, which are important components of UX, but this does not mean that the other components are ignored.

The first component is the Sensorial, which relates to how one feels, like Norman's (2004) innermost sensations and Jordan's (2002) Physio pleasure. The second component, the Emotional, emphasises the importance of emotion in shaping the experience, and McCarthy and Wright (2004) highlight the undeniable relationship between the emotional and sensory components, which can be illustrated by the experience of visiting a car dealer when interested in buying a new car, where for the first time entering a new car is strongly stimulated by the senses: the bright showroom lights, the smell of the new car and the modern interior and combination dashboard evoke emotions that constitute the first interactive experience with the car and set expectations for future experiences. The third component is the Compositional part of the experience. In this section, the authors place great importance on the way in which the experience develops and follows the interaction between its parts. In the same way that music played by an orchestra is not just the sum of the individual instruments involved, but also depends on the skilful collaboration between the various players, so too does the experience, and the relationship between its components contributes significantly to the interactive experience. Let's take the experience of driving a convertible through the woods as an example: the open road allows you to travel unhindered, the sun shines directly on the tall trees and the breeze transmits the smell of the forest to the driver, which further enhances the driving pleasure. At the same time, other experiences can influence the mood of the moment. For example, the car can retain any auxiliary information to minimise distractions and immerse the driver in the present moment. For these parts, the overall experience is inseparable, while the elements and the relationships between them form the experience of driving through the woodland paths. The fourth component is the Spatio-temporal aspect of the experience. This experience line relates to the impact of the user experience on the perception of time and space. "All contexts depend on the quality of the time and space in which they are produced" (McCarthy and Wright 2004) (p. 91), and these four lines frame the importance of the impact of the experience on time and things and the specific place in which this experience is produced. For example, time seems to pass slowly when one is waiting for something; or perhaps one remembers that a lecture seemed to last only a few minutes. Perhaps an exciting driving experience affects time and space, making time fly by and distances seem shorter than they actually are.

In sum, user experiences have a number of characteristics.

- They are situated in a spatio-temporal context, and may affect the perception of space and time
- They are compositional, comprising all levels of human existence: physiological (sensorial), psychological, social, ideological
- They are strongly linked to a person's desire to engage in meaningful activities, where meaningful activities are those that satisfy one or more person's psychological needs.

For instance, a person may enjoy driving, or more precisely a particular ride, enjoying the physical sensation of driving, feeling relaxed because of the beautiful scenery, and feeling good because it gives a feeling of competence and allows him to be courteous to other road users.

The de-compositional approaches to user experience have led to a number of new metrics and structural models. Hassenzahl (2004) distinguishes between two qualities: pragmatic and pleasurable. He argues that the pragmatic quality refers to the product being used to achieve a certain behavioural goal (i.e., usefulness, ease of use, etc.). In contrast, pleasurable quality refers to the user's selfhood; it is related to stimulation, i.e., the ability of the product itself to produce stimulation and facilitate personal growth, i.e., the product's demonstrated ability to satisfy the need for self-expression, Tractinsky and Zmiri (2006) suggest three different product quality attributes: Usability, Aesthetics and Symbolism.

Hassenzahl (2004) proposes two distinct judgements of the overall quality of a product: beauty and goodness. He finds that goodness is primarily considered in terms of utility (i.e., practicality and usability). In contrast, he finds beauty to be a social issue, heavily influenced by identity (i.e., the ability of the product to satisfy the need for self-expression). Similarly, Tractinsky and Zmiri (2006) make a distinction between a satisfying experience and a pleasurable one. They found that perceptions of usability were better predictors of satisfying experiences, while perceptions of product aesthetics were better predictors of pleasurable experiences. But will this relationship be stable over time? This is something that many scholars do not provide an answer to.

6.4 Diversity in UX

Not all users like the same design. While in other areas, the emphasis is on homogeneity (Cooper 1999), in the field of interaction design, diversity is one of the hallmarks. For example, in the physiological-psychological domain, the principle of homogeneity of perception states that different participants will more or less agree on the same perceptual judgements, for example how much frictional resistance or inertia may be included in a haptic operation. This assumption of perceptual homogeneity allows the data from the study to be presented in a statistical way. But in the field of design, the situation is different. Csikszentmihalyi and Rochberg-Halton (1981) conducted a study in which they asked respondents to choose personally relevant objects in their own house and to describe what made them special and why, where it was kept, how and when it was acquired, what it would mean to be without it, and so forth. The study found that the value of these objects did not lie in certain objectively defined qualities, such as aesthetic elements that people consistently appreciated, but rather in understanding the personal meanings that people attached to these objects and how these objects participated in their social lives and created self-identities. These results suggest that while we may all agree on perceptual judgements, such as the common judgement made about the beauty of a given product compared to its colour, higher level judgements may have more important effects.

Karapanos (2010) discusses four significant sources of diversity in UX, emphasising the unique individuality and personalisation of the experience: the individual, the product, time and the situation (see Fig. 6.2).

Individual factors describe the very significant influence of individual characteristics on the user experience of a product, such as differences in human values

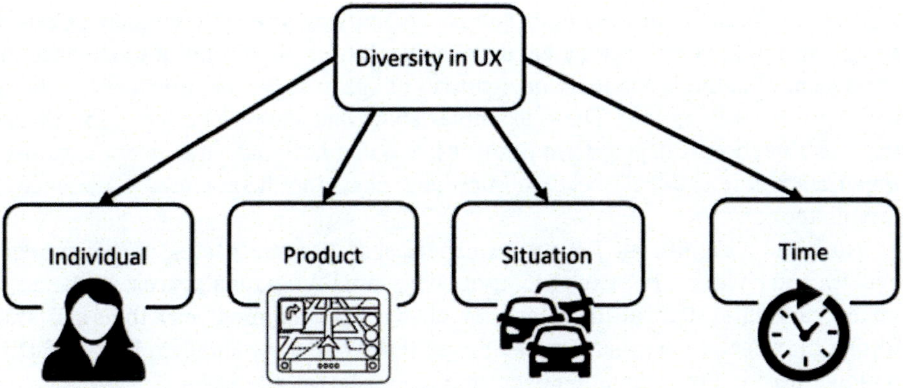

Fig. 6.2 Four different sources of diversity in user experience

(Schwartz 1992), which change the degree to which people value different qualities of interactive products (Karapanos and Martens 2007). Some people may prefer playful and exciting products, while others may value plain and conservative products. A person's upbringing and previous automotive experiences may lead that person to have a positive disposition towards future automotive experiences. A young woman who grew up in a family of car mechanics may feel affectionate and comfortable with cars, whereas a young person who has never seen a car engine will not feel this way. Further influences under this heading have been suggested by Hofstede (2001): in addition to judgements that are unique to the individual and significant to the individual due to their own relationships, there are others that may be shared with group members who may share common judgements that are significant to a particular society or culture, and yet others that are universal, at the innate or foundational level, judgements that relate to human commonalities.

The ***product attributes*** are another significant factor that influences the user experience (Jordan and Persson 2007). For example, while playful interaction is essential to the success of computer games, designs of this nature cannot be used in professional software. And the features of a car as well as the design of the exterior and interior can greatly influence the user experience of a car. A sporty car will make users expect and anticipate a sporty driving experience, while a luxury car will inspire expectations of smooth, comfortable driving.

The ***situation*** can also have a huge impact on the experience, similar to the impact of the context of use on usability. Even for the same product, the way in which individuals use it varies from situation to situation, which affects the importance they place on different attributes (Hassenzahl and Ullrich 2007); for example, driving on a congested motorway can dramatically change the experience of using a car, and assistance systems can help drivers to reduce traffic anxiety. Such features are becoming increasingly important to users. For example, will they be able to make phone calls, listen to music or stay connected while slowly moving through heavy traffic? Are the car's noise isolation and connectivity options up to standard? Considering a spring drive to a rural area, does the car have a huge sunroof to enhance the enjoyment of nature? Are there navigation options to make the user feel safe when looking for a destination? These examples show that such usage scenarios change the user's interior fitment system priorities, and that designs that may not have been important before suddenly become important, changing the potential experience of driving the car.

Time has a significant impact on experience. As time passes, the experience evolves and changes. For example, users who initially retain a distrust of the automated parking system in their car may change their opinion over time and after testing the system, thus changing their experience of the car and therefore gradually building trust in the automation over time (Gkouskos et al. 2015).

Thus, experiences may evolve over time.

1. Initial experience: These elements describe the first impressions and experiences that result from the first interaction with the product. The materials that make up the product, the UI elements that the product may have, etc. will greatly influence this initial experience.
2. Experience of use: These elements include the usability of the product and other experiences that arise from using the product.
3. Experiences gained through use: These elements include part of the experience gained through the use of the product, including the perception of the value and meaning of the product. For example, the freedom and independence that a car provides to its owner by allowing access to remote areas does not result from the experience of the car, but from the valuable experience gained through its use.

In the following section, we will look at the temporal nature of the user experience in more detail.

6.5 The Temporal Nature of the User Experience

As individuals use products, their perceptions of product quality change over time for a variety of reasons (Fenko et al. 2009; Karapanos et al. 2009; von Wilamowitz Moellendorff et al. 2006). For example, by using the product over a period of time, people get used to it and this changes their perception of its usability; at the same time, there is not the same excitement that there was at the beginning. More interestingly, they apparently assign different weights to different product properties at different stages of use. In their first interaction with the product, they may focus on the usability and excitement of the product. After a period of use, they may stop focusing on its usability, the novelty of its features, and focus on other aspects of the product. The expectation of approval from others also becomes more important. All these factors, including the individual, the product, the situation and the time of day, can change the judgement about the experience induced by an interactive product.

There are a number of framing theories that describe how user experiences are formed in different contexts, for example Forlizzi and Battarbee (2004) describe how experiences transcend from the unconscious into a cognitive state and eventually become 'experiences', memorable experiences that can also be conveyed in social interactions. Battarbee and Koskinen (2005) elaborate on the social mechanisms that enhance or diminish experiences when people engage in social interactions. McCarthy and Wright (2004) describe how experiences gradually evolve, moving from initial sensations and anticipation to processes such as reflection and retelling. Some of these frameworks work through a micro perspective, i.e., how experiences are formed, modified and stored, while others raise macro temporal issues. For example, how is the distribution or stability between unconscious and cognitive experiences maintained? Does the experience decrease over time as the user's familiarity increases? How do the underlying motivations for doing these things change over time?

Karapanos (2012) distinguishes three stages in the product use process (leaving anticipation aside): orientation, incorporation and identification. These stages reflect the different qualities of the product and give the experience of the product a dynamic temporal pattern, evolving from familiarity through functional dependence to emotional attachment. At each stage, the quality of the products appreciated varies. These forces motivate the transition between the three stages in the use of the product (Fig. 6.3).

Anticipation concerns the value of expectation before any actual experience of use, leading to behaviour that may shape the experience, such as the expectation that the product will give a good experience, or the fear that it will produce a bad experience.

Orientation is the initial experience of the user. This experience occurs mainly during the first week of exposure to the product. It is an experience of excitement and frustration when we encounter certain novel features, and the initial experience also depends on how well the product is designed to facilitate learning.

Incorporation refers to how products become meaningful in our everyday lives. Here, long-term usability becomes more important than initial ease of learning. The

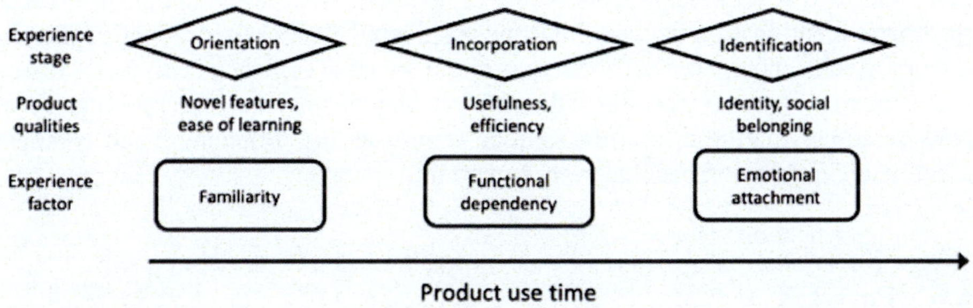

Fig. 6.3 The temporal nature of the experience, consisting of three main factors, increasing familiarity, functional dependency and emotional attachment

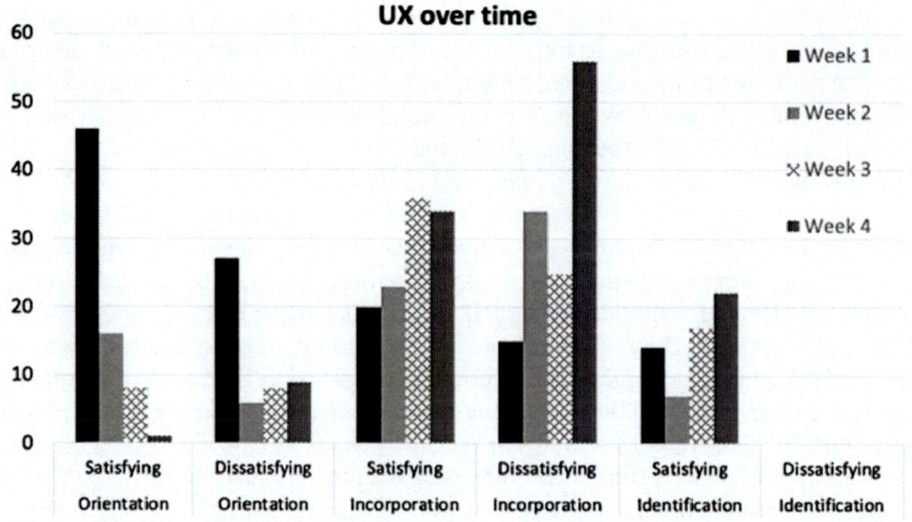

Fig. 6.4 Evolution of the user experience over time: number of satisfying and dissatisfying remarks relating to orientation, incorporation, and identification over time (Karapanos et al. 2008a, b)

usefulness of the product becomes the main factor influencing our overall assessment judgement.

Identification refers to the stage when we have accepted the product, and it participates in our social activities in our daily lives, conveying a part of our self-identity that distinguishes us from others or creates a sense of community belonging.

In a study of how the user experience evolves over time, Karapanos et al (2008a, b) collected experiences from six users over four weeks after having bought a particular brand of mobile phone. The users kept a record of their experiences at the end of each day. The remarks were classified in terms of the different UX factors (orientation, incorporation, identification). The results are shown in Fig. 6.4. As can be seen, users made fewer remarks relating to orientation (learnability, excitement about novel features) as time progressed, while they made more remarks relating to identification (usefulness, long-term usability) as time progressed. Also, the number of remarks relating to identification increased as time progressed. The peak for identification in the first week concerned mostly remarks about social acceptance. This study learns that excitement over novel features doesn't last, and that instead, with prolonged use, considerations relating to usefulness, long-term usability (efficiency) and personal identity and expression become more important.

6.6 Post-UX

In recent years, several limitations of the focus on UX have been pointed out.

1. UX focuses on the experience of the individual user.
2. In the UX approach, the user is a passive consumer.
3. UX focuses on the hedonic aspects of product/system use, at the exclusion of the social and societal aspects of product/system use.

One might argue that the work of Jordan and Hassenzahl provide handles to deal with some of these limitations. For example, one of the four pleasures that Jordan (2002) distinguishes is Ideo pleasure, leaving room for ideological considerations of consumers such as sustainability in deciding whether to buy and use products/systems. And Hassenzahl (2010) argues that UX should be understood in relation to users engaging in meaningful activities, where meaningful activities relate to people's psychological needs such as autonomy and competence. It appears that this goes beyond the idea of satisfying needs of pleasure and enjoyment that appears intrinsic to UX.

However, two developments have given rise to what may be called a Post UX perspective on product and system design (Brand and Rocchi 2011; Gardien et al. 2014). In the first place, the availability of platform technology has opened the way for consumers to be more actively involved in product/system design. Not only are users empowered by social media, enabling them to be much more active in providing reviews and comments on new products and systems, but also developments such as streaming services and app stores allow users to actively contribute playlists, music, videos and applications themselves. In the second place, the need for sustainable interaction with the world has resulted in an increased awareness of both industry and consumers that there are values beyond satisfying one's needs for individual pleasure, and that the ethical aspects of product/system use require different design considerations and production mechanisms.

This is not to say that UX is no longer important. It is still important for designers to aim for good user experiences. However, both designers and users should be aware of the need to broaden the scope of design and take into consideration moral values as well, to ensure a sustainable economy.

References

Battarbee K, Koskinen I (2005) Co-experience: user experience as interaction. Co-Design 1(1):5–18

Brand R, Rocchi S (2011) Rethinking value in a changing landscape. A model for strategic reflection and business transformation. A Philips design paper. https://www.researchgate.net/publication/272941478

Cooper A (1999) The inmates are running the asylum. Macmillan Publishing Co., Inc

Csikszentmihalyi M, Rochberg-Halton E (1981) The meaning of things: domestic symbols and the self. Cambridge University Press

Desmet P, Hekkert P (2007) Framework of product experience. Int J Des 1(1):57–66

Dewey J (1934) Art as experience. Minton, Balch & Company, New York

Ekins P, Max-Neef MR (eds) (1992) Development and human needs. Real-life economy understanding and wealth creation, London

Fenko A, Schifferstein H, Hekkert P (2009) Shifts in sensory dominance between various stages of user-product interactions. Appl Ergon 41(1):34–40

Forlizzi J, Battarbee K (2004) Understanding experience in interactive systems. In: Proceedings of the 2004 conference on designing interactive systems: processes, practices, methods, and techniques, pp 261–268

Gardien P, Djajadiningrat T, Hummels C, Brombacher A (2014) Changing your hammer: the implications of paradigmatic innovation for design practice. Int J Des 8(2)

Gkouskos D, Pettersson I, Karlsson M, Chen F (2015) Exploring user experience in the wild: facets of the modern car. In: Lecture notes in computer science (including subseries Lecture notes in artificial intelligence and Lecture notes in bioinformatics), vol 9188. https://doi.org/10.1007/978-3-319-20889-3_42

Hassenzahl M (2004) The interplay of beauty, goodness, and usability in interactive products. Human-Comput Interaction 19(4):319–349

Hassenzahl M (2010) Experience design: technology for all the right reasons. Synth Lect Human-Centered Inf 3(1):1–95

Hassenzahl M, Ullrich D (2007) To do or not to do: differences in user experience and retrospective judgements depending on the presence or absence of instrumental goals. Interact Comput 19:429–437

Hofstede G (2001) Culture's consequences: comparing values, behaviors, institutions, and organizations across nations. SAGE Publications

Jordan PW (2002) Designing pleasurable products: an introduction to the new human factors. CRC Press

Jordan PW, Persson S (2007) Exploring users' product constructs: how people think about different types of product. Int J Co-Creation Des Arts 3(1):97–106

Karapanos E (2010) Quantifying diversity in user experience. Unpublished doctoral dissertation, Eindhoven University of Technology. https://doi.org/10.6100/IR658635

Karapanos E (2012) Modeling users' experience with interactive systems. In: Studies in computational intelligence, vol 436. Springer

Karapanos E, Martens J-B (2007) Characterizing the diversity in users' perceptions. In: Baranauskas C, Abascal J, Barbosa SDJ (eds) INTERACT 2007, LNCS. Springer, Heidelberg, pp 515–518

Karapanos E, Hassenzahl M, Martens JB (2008a) User experience over time. In: CHI'08 extended abstracts on human factors in computing systems, pp 3561–3566

Karapanos E, Wensveen S, Friederichs B, Martens J-B (2008b) Do knobs have character? Exploring diversity in users' inferences. In: CHI 2008b extended abstracts on human factors in computing systems. ACM, Florence

Karapanos E, Zimmerman J, Forlizzi J, Martens J-B (2009) User experience over time: an initial framework. In: CHI 2009: Proceedings of the 27th international conference on human factors in computing systems. ACM, New York, pp 729–738

McCarthy J, Wright P (2004) Technology as experience. The MIT Press, UK London

Norman DA (2004) Emotional design: why we love (or hate) everyday things. Basic Civitas Books

Norman D, Miller J, Henderson A (1995) What you see, some of what's in the future, and how we go about doing it: HI at Apple Computer. In: Conference companion on human factors in computing systems. ACM, New York, NY, USA, p 155. https://doi.org/10.1145/223355.223477

Ryan R, Deci E (2000) Intrinsic and extrinsic motivations: classic definitions and new directions. Contemp Educ Psychol 25:54–67

Schwartz S (1992) Universals in the content and structure of values: theoretical advances and empirical tests in 20 countries. In: Advances in experimental psychology, vol 25. Academic Press, pp 1–65

Sheldon KM, Elliot AJ, Kim Y, Kasser T (2001) What is satisfying about satisfying events? Testing 10 candidate psychological needs. J Pers Soc Psychol 80(2):325–339

Tractinsky N, Zmiri D (2006) Exploring attributes of skins as potential antecedents of emotion in hci. In: Fishwick P (ed) Aesthetic computing. MIT Press

Von Wilamowitz Moellendorff M, Hassenzahl M, Platz A (2006) Dynamics of user experience: How the perceived quality of mobile phones changes over time. In: The 4th Nordic conference on human-computer interaction, workshop user experience—towards a unified view, pp 74–78
Wahba MA, Bridwell LG (1976) Maslow reconsidered: a review of research on the need hierarchy theory. Organ Behav Hum Perform 15(2):212–240
Wakkary R (2009) Experience in interaction design: five propositions. In: CHI '09 Workshop on building a unified framework for the practice of experience design
Wang C, Terken J, Hu J, Rauterberg M (2020) Enhancing social closeness between drivers by digital augmentation. Int J Human-Comput Interaction 36(5):477–494

Chapter 7
Driving Behaviour and Changing Behaviour

Abstract Automotive industry and policy makers have taken all kinds of measures to reduce the number of (fatal) accidents and make driving safer. However, drivers are a central factor in establishing safety. While safety measures are important in making driving safer, if drivers do not comply, efforts by industry and policy makers may not be effective. In this chapter, we look at driving behaviour, in particular at driving styles. We discuss the phenomenon of behavioural adaptation and look at efforts to apply technology to make drivers change their behaviour.

The occurrence of accidents has led to the development of measures to improve safety, both passive and active safety measures. Passive safety measures aim to mitigate the effects of an accident. Examples are safety belts, airbags and pedestrian protection systems. Active safety systems aim to prevent the occurrence of an accident. Examples are electronic stability control, anti-lock braking systems, automatic emergency braking, but also head-up displays and blind-spot warning systems.

Although it is not always easy to measure the effect of particular measures, still there is evidence that they may have drastic effects on safety. For example, Fig. 7.1 shows the influence of seatbelt compliance on the number of casualties in 46 high income countries.

There are a number of things to say about driving safety. In the first place, while safety measures aim to improve safety, the introduced measures may also have unexpected effects, in that people adapt their behaviour to the new situation resulting from the introduction of the measures, therewith reducing the effectiveness of the safety measure. This phenomenon is known as behavioural adaptation. In the second place, not all people drive in the same way. Instead, people have different driving styles, and some driving styles may be associated with higher risk of accidents. Both the phenomenon of behavioural adaptation and the existence of different driving styles have led to the insight that improving traffic safety may not just be a matter of introducing more and more passive and active safety systems, but that technology may also contribute to actively influencing people's attitudes and behaviour through so-called persuasive technology. In this chapter, we will go into these developments.

© China Machine Press 2023

F. Chen and J. Terken, *Automotive Interaction Design*, Springer Tracts in Mechanical Engineering, https://doi.org/10.1007/978-981-19-3448-3_7

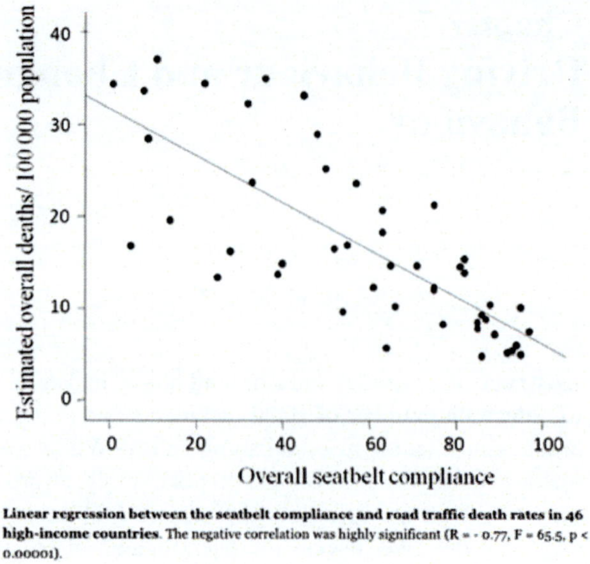

Fig. 7.1 Relation between seatbelt compliance and estimated deaths in 46 high-income countries. *Source* Abbas et al. (2011). Licensed through Creative Commons Attribution License

7.1 Safety Measures and Behavioural Adaptation

Kulmala (2010) lists a number of ways in which safety systems may affect safety. The most important ones are the following.

1. Certain measures may influence the modal choice or the route choice. For instance, speed limits and other measures to reduce the speed in residential areas may lead drivers to take another route, therewith increasing the safety in the residential area.
2. Intelligent injury reduction systems and crash reporting may mitigate accident consequences.
3. Systems that give information, advice and assistance, or that take over part of the task may influence the driving task. For example, a speed alert by an active acceleration pedal may lead the driver to reduce the average speed.

As already mentioned above, not all ways in which safety systems affect safety are positive. Examples are the following:

4. Active safety systems may lead to long-term modification of driving behaviour. For instance, users of Adaptive Cruise Control may rely on the system so much that their response time to hazards increases.
5. Safety systems may lead drivers to change their behaviour, which in turn is imitated by other drivers whose vehicle does not have such systems. For instance, Electronic Stability Control may lead drivers to drive faster through curves on wet roads, and this may make other drivers feel that they can also drive at higher speeds. Similarly, Connected Driving systems, by which vehicles receive information about deceleration and braking from the leading vehicle, may make

drivers drive at shorter distances, but this may lead drivers of non-connected vehicles to also drive at shorter distances.

6. Similarly, knowledge about the presence of safety systems may lead vulnerable road users (bicyclists, pedestrians) to change their behaviour. For instance, knowledge that automated vehicles are equipped with automatic emergency braking systems may lead pedestrians to take more risk in crossing the street.

In general, a safety measure targets an intended effect, which is to increase safety. This is called the "engineering effect". For example, the speed limits and additional measures under point 1 above may lead drivers to reduce their speed in a residential area. However, the same measure may also have other effects, and this is called "behavioural adaptation". These unintended effects may be positive. For instance, drivers may choose a different route, and this may have a positive effect on traffic safety in the residential area (although this may be at the expense of traffic safety at the alternative route). Or they may decide to choose a different transport modality (public transport) altogether, as a result of the introduction of the measure. However, more often the effects are negative, as mentioned under points 4–6 above, affecting traffic safety negatively.

Usually, the positive and negative effects may both occur at the same time, so that it becomes important to find out what the net result of a certain measure is. One way to do this is to measure the number of accidents before and after the introduction of a particular measure. However, at the same time other things may also change: other measures may have been introduced, the weather may have changed, there may be road construction works, and these confounding factors may influence traffic safety as well, so that obtaining an accurate estimate of the effect of a particular measure may be difficult. Obviously, however, this should not keep policy makers and engineers from thinking about and implementing measures and systems that aim to make traffic safer.

One thing to note in this context is that the occurrence of behavioural adaptation in influenced by a number of factors:

1. The size of the intended effect (the engineering effect) of the measure. An improvement of the car's lighting system will likely lead to larger behavioural adaptation by night than during daytime: the increase in speed that a driver exhibits when has car is equipped with such an improved lighting system may be larger at night than during daytime.
2. How easily the measure is detected. A collapsible steering column aiming to mitigate consequences of an accident may not even be noticed, and not lead drivers to change their behaviour.
3. Whether there are already behaviours that pave the way for a certain measure. Mandatory periodic inspection of the car aimed to check whether the car is still safe to drive may be more easily accepted because car owners bring their vehicle to the dealer for its annual check.
4. Whether the measure intends to prevent accidents or to mitigate the effect of accidents. Measures that intend to prevent accidents by making the car easier to control in adverse circumstances, such as Electronic Stability Control, are

more likely to lead to behavioural adaptation than measures that aim to reduce the severity of injuries in accidents, such as air bags.

5. Utility gain. Changes in behaviour that lead to utility gains such as a reduction in travel time are more likely than changes that do not lead to utility gains.

While behavioural adaptation with adverse consequences cannot be avoided, at least insight in factors contributing to its occurrence can help to anticipate the occurrence of behavioural adaptation and to think about ways to mitigate its effects.

7.2 Driving Styles

Everyone who is a regular driver knows from own experience that there are clear differences between people in the way they drive. Most people consider themselves to be good drivers, and most people dislike driving behind the proverbial 80-year old lady who drives slowly and overly cautious. Also, we all know the stories about young people going out on Saturday evening and exhibiting risky behaviour that we consider irresponsible and dangerous. Finally, there are also the stories reflecting our cultural stereotypes about driving behaviour in other countries. All these stories relate to the concept of driving styles. Generally speaking, driving styles are behaviours that people display typically, that is, under normal circumstances. This definition also implies that a particular person, who has a particular driving style, does not display this behaviour all the time. For instance, a newly married person who displays a more assertive driving style under normal circumstances, may show very different behaviour when picking up his parents in law from the railroad station. Furthermore, the connection between driving and risk-taking mentioned above has been the topic of much research, aiming to understand the factors determining the occurrence of risky driving and to propose and evaluate measures to further safe driving.

When conducting research into differences in the way people drive, there are a number of questions. The first question is how to measure driving styles. The second question is which are the actual different driving styles that have been inferred from these measurements. The third question is which factors govern the occurrence of particular driving styles. And the fourth question is how safe driving styles can be promoted and how driving styles that may have negative consequences can be countered.

There are typically two ways to measure driving styles. One is through self-report, the other is through measurements of the driving behaviour. Several self-report questionnaires have been proposed; commonly used ones are the Driver Behaviour Questionnaire (DBQ, Reason et al. 1990), the Driving Style Questionnaire (DSQ, Ishibashi et al. 2002), the Multi-dimensional Driving Style Inventory (MDSI, Taubman-Ben-Ari et al. 2004). Such questionnaires consist of a number of items describing relevant driving behaviours and ask respondents to respond by a scalar answer ranging from "not at all" to "very frequently". The data are then submitted to a factor analysis and the factors emerging from the analysis are then interpreted by mapping them onto dimensions of driving behaviours. The DBQ asks for the occurrence of different types of errors and maps the outcomes onto three types of errors: deliberate violations (e.g., "deliberately violating speed limits at late night"), dangerous errors (e.g., "failing to note a pedestrian crossing") and 'silly' errors (e.g., "forgetting where one's car is parked"). The DSQ maps the result onto six dimensions: speed (e.g., "Do you drive fast"), calmness (e.g., "Do you remain calm when things happen very quickly and there is little time to think"), social resistance (e.g., "Do you dislike people giving advice about your driving"), focus (e.g., "Do you drive cautious" and "Do you find it easy to ignore distractions while driving"), planning (e.g., "How often do you set out on an unfamiliar journey without first looking at a map"), and deviance (e.g., "Do you ever drive through a traffic light after it has turned red"). The original MDSI mapped the results onto eight dimensions: Risky driving (e.g., "like to take risks while driving"), Angry driving (e.g., "swear at other drivers"), Careful driving (e.g., "always ready to react to unexpected manoeuvres by other drivers"), Patient driving (e.g., "at an intersection where I have to give right-of-way to oncoming traffic, I wait patiently for cross-traffic to pass"), Anxious driving (e.g., "feel distressed while driving"), Dissociative driving (e.g., "intend to switch on the windscreen wipers, but switch on the lights instead"), High-Velocity driving (e.g., "when in a traffic jam and the lane next to me starts to move, I try to move into that lane as soon as possible") and Distress-reduction driving (e.g., "while driving, I try to relax myself"). There have been later evaluations of the MDSI in different countries, from which smaller numbers of reliable dimensions emerged. Importantly, the underlying idea of such questionnaires is that driving style is not a uni-dimensional concept but multi-dimensional: one is not either an aggressive or risky or calm driver, but for each driver a multi-dimensional profile arises from the questionnaire. On the other hand, certain correlations may be expected to exist, and Taubman-Ben-Ari and Skvirsky (2016) arrive at four main styles: Reckless and careless driving style, Anxious driving style, Angry and hostile driving style, and Careful and patient driving style.

Most studies measuring driving styles from observable behaviour are conducted using driving simulators, but studies of real-life driving behaviour have been conducted as well, collecting data from the CANbus or from dedicated in-vehicle data recorders. The disadvantage of driving simulator studies is that the behaviour of participants in such studies may not be representative of their normal driving behaviour, also because of limitations in the fidelity of the simulation, but the advantage of driving simulator studies is that the situations to be studied are under control of the researchers. In real life, the situations in which the relevant behaviour (such as "when in a traffic jam and the lane next to me starts to move, I try to move into that lane as soon as possible") can be measured may occur infrequently, so that only few measurements can be collected from which to determine the individual's driving style. As a result, studies aiming to derive driving styles from observable behaviour in real-life situations have focused on a few main dimensions, such as calm/patient versus aggressive/hostile, and the main indicators are acceleration behaviour, speed in turns, distance and the size of critical gaps.

A third way to infer driving styles from behavioural indicators is to look at the involvement in car crashes and the individual's history of traffic violations.

Studies have been conducted on the question of whether outcomes from self-report methods have external validity, that is, whether they correlate with the outcomes studies of observable behaviour. Some success has been achieved in this respect, although usually the amount of variance explained is rather modest, and again the evidence supports only a few main dimensions. This may also be due to the fact that behaviours such as "Do you dislike people giving advice about your driving" (DSQ) and "swear at other drivers" (MDSI) are difficult to observe, both in driving simulator studies and in real life. What matters most, from an interaction design perspective, is whether there is a relation between driving style and involvement in accidents, and indeed there is evidence that observable behaviours such as acceleration behaviour, speed in turns, distance and the size of critical gaps have predictive value for involvement in accidents. Thus, the ability to derive data concerning these behaviours from the CANbus or from dedicated in-vehicle data recorders offers opportunities for the design of applications that aim to counter adverse driving behaviour. In fact, such data have been used by insurance companies to influence driving behaviour by rewarding safe driving behaviour by lower insurance rates and in this way preventing involvement in accidents, which is to be preferred over the current practice where the insurance rate goes up *after* involvement in an accident.

The third question concerns the factors that govern the occurrence of driving styles. Research indicates that driving styles are conditioned by several demographic and personality variables (Taubman-Ben-Ari and Skvirsky 2016). Demographic variables are most notably age and gender. Men have a higher tendency to engage in risky and angry driving than women. Also, younger drivers have a higher tendency to engage in risky and angry driving than older drivers. As a result, younger male drivers show the highest incidence of risky driving. However, not all young male drivers exhibit risky driving, so the next question concerns which personality variables differentiate risky drivers from more calm and careful drivers. Both the reckless/careless (risky) and the aggressive/hostile driving styles have been found to correlate well with sensation-seeking, and to correlate inversely with the traits of agreeableness and conscientiousness as emerging from the Big Five personality inventory. The occurrence of an anxious driving style has been found to correlate well with neuroticism as measured by the Big Five personality inventory. And, as might be expected, the careful and patient driving style has been found to correlate with the traits of agreeableness and conscientiousness as emerging from the Big Five personality inventory.

Further contributions to the understanding of driving styles have been made by Wilde and Fuller. Wilde (1982) argues that, in order to predict and understand the effect of safety measures, we need to understand how driving behaviour arises. He proposes that one factor accounting for differences between drivers in the way they drive arises from the target level of accident risk they set for themselves. This is their accepted or preferred target risk. Among the factors that determine the target risk are person-related factors such as culture, gender, age, attitude, personality traits, but also transient states such as fatigue, being in a hurry and frustration. When driving, people compare the perceived accident risk with the target, and adjust their behaviour to reduce the discrepancy between perceived and target risk. If, for whatever reason, the perceived accident risk is lower than the accident risk they find acceptable, they

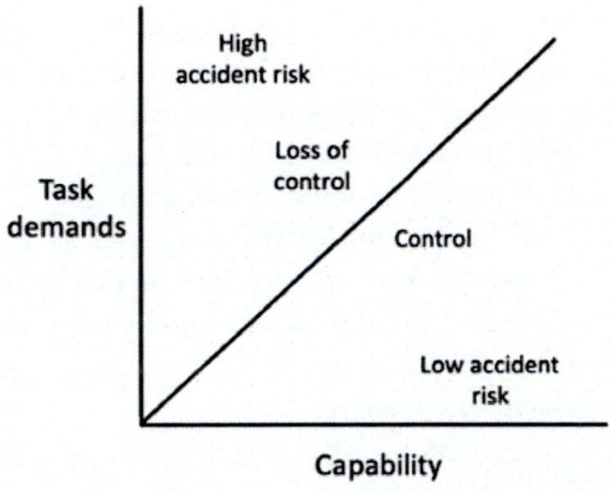

Fig. 7.2 Fuller's task capability interface model

will adjust their driving behaviour so that the perceived risk increases. In other words, they will drive more riskily. If the perceived accident risk is higher than the accident risk they find acceptable, they will adjust their driving behaviour so as to reduce the perceived risk. In other words, they will drive less riskily. Summarizing: people aim for homeostasis, aiming to maintain a particular state that they consider optimal, and adjust their behaviour in case of perceived discrepancy with this optimal state. From this it is clear that safety measures that affect people's perceived accident risk, will make them adjust their behaviour and drive more riskily to restore the optimal state. The extent to which they do this depends on the size of the effect of the measure on the perceived risk, as we have seen above.

Likewise, Fuller (2005) proposes that drivers select a preferred level of task difficulty and adjust their driving behaviour such that the difficulty of the driving task does not exceed their capability as judged by themselves (see Fig. 7.2). If the driving task becomes more difficult, e.g., because of adverse weather conditions, they will adjust their driving behaviour so that their driving skills again match the challenges posed by the task. In this view, differences between people arise from differences in their preferred level of task difficulty and differences in their level of skill as perceived by themselves. People who overestimate their own skill will be inclined to take more risk than people who have a more adequate estimate of their own skill. For Fuller, as for Wilde, measures that aim to improve traffic safety may not have the desired effect. For Fuller, if such measures lower the perceived task difficulty and therewith the demands posed by the driving task, people will adjust their behaviour and drive more riskily, so that again there is a balance between task demands and driving skills.

The important contribution of both Wilde and Fuller is to point out that, in order to devise effective measures to improve traffic safety, we need to anticipate the possibility of behavioural adaptation. Simple measures applying in the same way to everybody may not achieve the desired effect because of this behavioural adaptation. Of course, the positive (intended) effect may outweigh the negative (unintended) effect, so that the net result of a particular measure may still be positive. For instance, even if drivers increase their speed if dangerous curves in a highway are removed and some accidents result from this higher speed, the net result in terms of lower accident risk may still be positive. But both Wilde and Fuller argue that, in order to increase the likelihood that safety measures will have the desired effect, people's desire to be safe must be influenced, that is, they must change their target level of risk or the target level of task difficulty.

The discussion of the effects of safety measures has brought us to the fourth question, of how safe driving can be promoted and unsafe driving can be countered. In Sect. 7.3 we will look at some more recent approaches to this challenge, by means of smart technology enabling a more personalised approach towards attitudinal and behavioural change.

7.3 Changing Behaviour

The idea that design should not only aim to adjust technology to the needs of people or to create good user experiences, but may also be deployed to change behaviour is associated primarily with the work on Persuasive Technology by Fogg (2003). According to Fogg, technology may influence people in three different ways. Technology may provide people with *tools*, and as such shape the way people perform certain tasks, making it easier to perform the task or restructuring the way the task is performed. For instance, automatic transmission makes the gearing of the vehicle easier compared to manual transmission. Also, a navigation system may influence where and how people drive by offering navigation advice and speed limit information. Secondly, technology may function as *media* that convey a certain message. For instance, the Neon Drunk Driver Simulator was developed to simulate the experience of drunk driving for university students. A real automobile was prepared to respond unpredictably and sluggishly to the driver's inputs, and this vehicle was then taken to university campuses for students to experience drunk driving on a dedicated track on the campus. The message came across to some extent: It was found that the experience did not make students change their intention to drive when having drunk, but at least they said that they would think twice before getting into a car with a drunk fellow student as driver. Thirdly, technology may function as a *social actor*, and offer suggestions for change and rewards. For instance, a driver coach system may monitor the driver's driving behaviour and provide feedback and suggestions for change. Social actors may induce behavioural change in a number of ways, including praise ("Congratulations" or "Well done") and authority ("You should ...").

There have been a number of applications of technology as *social actors* in the context of driving, primarily for two purposes, sustainability and safety. Nowadays, truck manufacturers offer systems that provide feedback to truck drivers in order to make them adjust their braking and gearing behaviour and thus reduce fuel consumption. From a design perspective, it should be kept in mind that the savings do not only benefit the owner of the truck but that the drivers themselves also benefit from the savings. Another application, EcoDrive, was launched by Fiat in 2008 (Fig. 7.3 Left). They developed an application by which drivers could store data about their

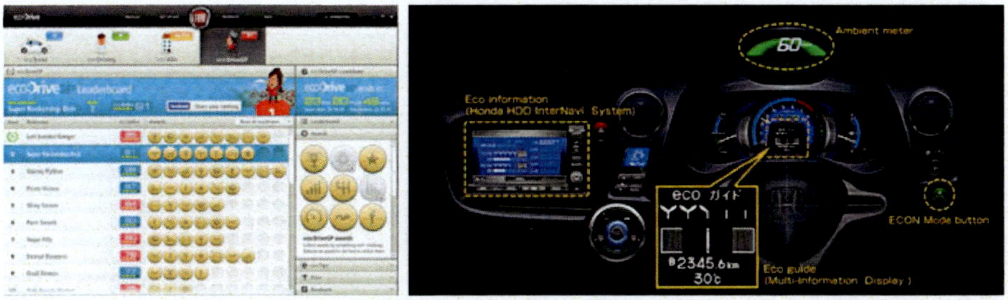

Fig. 7.3 Left: Fiat's EcoDrive system. *Source* https://www.baxtr.co/fiat-ecodrive. Reprinted with permission from AKQA/Fiat. Right: Honda's Eco Assist system. *Source* Honda Global, Reprinted with permission. https://global.honda/newsroom/news/2008/4081120aeng.html

driving behaviour on a USB stick and, after reaching their destination, upload the data in their personal computer and compare it to the behaviour of other drivers, thus turning sustainable driving into a contest and applying the *persuasive strategies* of comparison and competition. The application of such a *gamification* approach turned out to be very successful. Likewise, Honda offers the Eco Assist system that shows a growing tree, reflecting the success of the driver in achieving a sustainable driving manner (Fig. 7.3 Right). Nowadays, many companies offer coaching systems that provide suggestions and rewards for sustainable driving. Persuasive technology has also been applied to counter risky driving. An insurance company enabled young drivers to install a dedicated device in their car by which information could be collected about their driving behaviour. Fast acceleration and high speed in curves were seen as signals of a risky driving style. Through calm driving, young drivers could get a discount of their insurance fee.

However, not all drivers may be equally sensitive to the same persuasive strategy, as there is evidence that people may differ as to the persuasive strategy they are sensitive to. In fact, if a wrong persuasive strategy is applied in a system that people cannot turn off, the system may invoke resistance, or even the opposite behaviour (reactance). In the context of automotive, some research efforts have been made to develop such more personalised persuasion systems, but the results so far are inconclusive, for two reasons. In the first place, persuasive strategies that have been shown to be effective in other domains, for instance, the persuasive strategy of scarcity—(cf. "only two rooms left!" in hotel room booking systems) may not be easily translated to the automotive domain; possibly, persuasive strategies may have to be devised specifically for the automotive domain. In the second place, it is not yet clear how to determine which factors determine which persuasive strategy would work best for a particular person, and how to measure such factors.

Finally, gamification approaches usually imply rewards such as fun and excitement, which may not be long-lasting. Likewise, systems offering monetary rewards raise the question of what effective approaches to attitude and behavioural change are. While extrinsic rewards such as money are good for behavioural change, usually they are bad for attitude change. They are effective only as long as the extrinsic reward remains present, and once they are taken away, the driver may fall back into the old behaviour again. Even so, persuasive technology offers an interesting direction for attitude and behavioural change.

References

Abbas AK, Hefny AF, Abu-Zidan FM (2011) Seatbelts and road traffic collision injuries. World J Emerg Surg 6(1):1–6

Fogg BJ (2003) Persuasive technology: using computers to change what we think and do. Morgan Kaufmann, San Francisco

Fuller R (2005) Towards a general theory of driver behaviour. Accid Anal Prev 37(3):461–472

Ishibashi M, Okuwa M, Akamatsu M (2002) Development of driving style questionnaire and workload sensitivity questionnaire—their application to analysis of route choice preference. In: Proceedings of symposium on mobile interactions and navigation, pp 15–18

Kulmala R (2010) Ex-ante assessment of the safety effects of intelligent transport systems. Accid Anal Prev 42(4):1359–1369

Reason J, Manstead A, Stradling S, Baxter J, Campbell K (1990) Errors and violations on the roads: a real distinction? Ergonomics 33(10–11):1315–1332

Taubman-Ben-Ari O, Skvirsky V (2016) The multidimensional driving style inventory a decade later: review of the literature and re-evaluation of the scale. Accid Anal Prev 93:179–188

Taubman-Ben-Ari O, Mikulincer M, Gillath O (2004) The multidimensional driving style inventory—scale construct and validation. Accid Anal Prev 36(3):323–332

Wilde GJ (1982) The theory of risk homeostasis: implications for safety and health. Risk Anal 2(4):209–225

Part III
Interaction Design/UI Design

Chapter 8
Interaction Design Theory

Abstract Design is not just a matter of relying upon your own intuition. Instead, there is a wealth of theories and practical insights to guide design professionals. In this chapter, main theoretical frameworks in the area of interaction design are summarised: activity theory, distributed cognition, ecological interface design, hierarchical design theory and stages of action theory. Furthermore, automotive HMI design may be guided by user interface design principles and guidelines.

In addition to the psychological theories mentioned in the previous sections, there are many other theories that are good guides for interaction design. However, it is not uncommon for people who do interaction design to simply rely on their intuition and experience, ignoring or not understanding that there are theories that can be of great help to them in their design. Theory is useful in two ways: to explain phenomena and to predict the future. Of course, it is not impossible to generate designs that feel good but are not grounded in theory. In this case, interaction design is considered more like an art. Also, when people encounter a new design challenge, they look at other people's designs, but they don't understand how to evaluate others' designs, find the strengths and remove the weaknesses, because they have no methodological framework to support the evaluation. In order to make quality controllable and replicable, it is better to treat interaction design as a skill, where the expertise includes practical knowledge of the relevant design theories.

Design theories can be divided into many categories, and Table 8.1 gives a pathway to help designers find the right theory (Shneiderman et al. 2018).

Any theory that helps the designer to explain his design principles and to predict the outcome of the design is a useful theory. Theories about human perception and cognition provide insight into how people process information, make decisions and learn and may provide theoretical foundations of the how and why of design decisions. Relevant theories were treated in Part II. Descriptive, interpretive, prescriptive and predictive theories provide more specific knowledge supporting the understanding of interaction and guiding decisions when designing user interfaces. The following is a brief introduction to a few of the commonly used theories, but for those who wish to gain a deeper understanding of these theories, further reading of

© China Machine Press 2023

F. Chen and J. Terken, *Automotive Interaction Design*, Springer Tracts in Mechanical Engineering, https://doi.org/10.1007/978-981-19-3448-3_8

Table 8.1 Classification of design theories

Theoretical types	Description
Human movement capacity theory	Explanation and prediction of human muscle movement ability, similar to muscle reaction speed, clicking, line drawing, etc.
Human perception theory	Theory about the human ability to perceive information through the senses of sight, hearing, smell and touch
Human cognitive abilities theory	Theories about human skills and limitations concerning problem solving, long—short term memory, etc.
Descriptive theory	Describes user interfaces and their use with a common name and language
Interpretive theory	Systematic explanation of the causal relationship between events and causes
Prescriptive theory	Design guidelines for designers to help them make design decisions
Predictive theory	Facilitates comparisons, and give possible future scenarios
Methodological theory	Provides methodological frameworks for user research and evaluation

the relevant literature is required. Methodological theory provides theoretical frameworks for conducting user research and doing evaluations. This will be treated in Part IV.

8.1 Activity Theory

Activity Theory is based on anthropology/human consciousness. The basic components of activity are a subject (a person or a group), an object (a goal that motivates activity) and purposeful interaction between them (Activity) (Leont'ev 1978). Activity theory asserts that human activity is an activity that connects subject and object, and that the activity process maintains the identity of and influences the development of both subject and object. The theory focuses on the relationship between interacting people and products in relation to each other in the environment (Kaptelinin and Nardi 2006). Rather than looking at isolated individuals, activity theory seeks to analyse the structure and processes of activities in the use of products in order to understand individuals and the social entities that arise between them. Vygotsky's (1978) theory of the relationship between 'thought' and society suggests that the subject is inherently social and that 'thought' is embedded in the interaction between humans and the world. In other words, human behaviour in the social environment (Kaptelinin and Nardi 2006) generates consciousness. Activity theory provides a very general philosophical framework for understanding human culture, human work as a whole, and is a guide to ways of working, organisational development and design (Kuutti 1995).

The development of Activity theory has a long history. From the 1920s to the 1930s, recognition of activity theory can be traced back to the philosophical debates of Kant, Fichte and Hegel, Hegel's idealism and Feuerbach's materialism, as well as to the dialectic philosophy of Marxism and Engels, and Vygotsky's (1934) psychology of Soviet culture and history. Vygotsky's (1934) theory reflects an idea consistent with dialectic philosophy, the materialist view that 'social existence determines consciousness', i.e., human thought does not exist independently but is social in nature because our behaviour is constantly influenced by culture, environment, language or the world around us with which we interact. The subject and the object are the result of a cultural, environmental, linguistic and social context. Subjects and objects are developed in culture, and human entities are, by nature, social. Vygotsky suggested the need to understand the development of phenomena through a cultural and social lens, to understand the nature of consciousness and thought (Vygotsky 1962, 1978; Leont'ev 1981). An 'activity' is the smallest meaningful basic unit that supports the activity of an individual in a given context (Kuutti 1995). The hierarchy of activities provides multiple levels of behavioural interaction in the world (see Fig. 8.1).

The difference between one activity and another is rooted in the object or 'motive' of the activity (Leont'ev 1981). Human activities are directed towards objects that are relevant to various needs. For example, for safety reasons, drivers operate in a safe manner (activity, motivation) In activities characterised by different objects, transitions in the course of the activity may lead to increased conflict or contradiction (Nardi 1995). An activity will generally result from a combination of operational steps. People initiate actions to achieve goals, which directly guide actions, and any actions are carried out under specific conditions. For example, a driver wants to avoid a conflict with the vehicle behind him on the left when changing lanes on a motorway, so he needs to take the necessary steps to change his speed or the timing of the lane change. Each course of action can then be broken down into a series of lower-level steps, i.e., manoeuvres. For example, the driver first looks in his rear-view mirror and left-hand mirror to see if there are any cars in the left-hand lane and

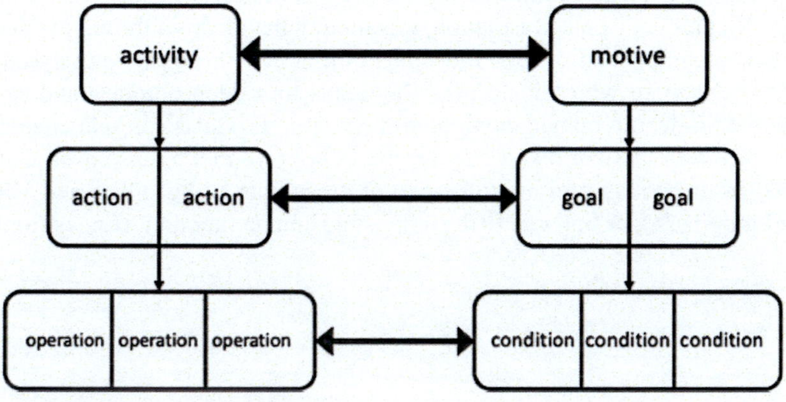

Fig. 8.1 Activity theory (After Kuutti 1995)

at what distance, and then decides whether to start the lane change. When deciding to change lanes, he will first hit his left turn signal, then start accelerating and turn the steering wheel to change lanes, so a meaningful action (changing lanes) leads to a series of manoeuvres. It is important to note here that the components of the activity are not fixed, they can change as the conditions change (Nardi 1995). That is, a change in context can reassign components of the activity. For example, in the case of the lane change above, if the driver is watching the left lane and notices that a car is accelerating to overtake him, he may wait for a moment until the car on his left has passed the car before starting the other series of operations.

Vygotsky's and Leontiev's research has developed a new perspective in which activities are not seen as a collection of linear movements, but should be analysed through a dynamic lens, in terms of cultural and historical development. These studies focus on the activities of individuals in their social environment within the framework of subject-object interaction. Their scope is further extended to collective activity through the triadic model (i.e., subject-object-community) (Engeström 1987, 1999), where 'community' and its concepts in relation to subject and object are used to illustrate the relationship between individuals and groups (e.g., social norms, culture, rules and practices) and between groups and organisations (e.g., division of labour). Often, research on collective activity further enriches the content and scope of the theory.

The application of activity theory to interaction design dates back to the 1990s. Prior to this, it was found that cognitive psychology, dominated by information processing theory (described in Chap. 3), was no longer adequate for the increasingly complex demands of interaction design. In analysing human behaviour, motivation and human-system interaction, the cognitive psychology of the individual cannot be separated from the wider social context. It was considered more appropriate to use activity theory to consider all aspects from motivation to operational activities in order to solve problems in organisational change and to complete system design, to use activity theory to understand operations, actions and activities, and to determine the way in which operations are carried out from a holistic perspective, considering the physical and social conditions of the activity and integrating them into the system design. No product exists in isolation, it is merely the medium through which some activity is accomplished, and the meaning of the activity needs to be analysed alongside the context in which it occurs. This leads to a deeper understanding of the role of the product in human–computer interaction. Products can influence the way we interact with the world and may represent how we make sense of it. Culturally specific products may turn out to be part of the human being (Flach and Voorhorst 2016; Pirsig 1974), or 'integral to the indivisible human function' (Engeström 1999).

8.2 Distributed Cognition

Distributed cognition (DCC) focuses on the relationship between people and the objects they use and the context in which they use them. This theory considers not only the cognitive characteristics of the individual people, but also the internal and external manifestations of the objects they use and the environment in which they work. The theory states that cognitive activity does not only take place in the human brain of the individual person, but is processed and transmitted through various media, including networked computers, display systems, paper and books, and that the cognitive activity is shared between people and artefacts. This is illustrated in Fig. 8.2.

While traditional cognitive psychology focuses on cognitive activity as occurring in the individual's brain, distributed cognitive psychology takes a more macroscopic view, considering that cognitive activity is shared between people, computers, all kinds of systems that can store information, including books, and that all these agents take part of the information processing work. The individual human brain no longer has to do all the work on its own. For example, we can store a lot of information in a computer and extract it when we need it. And we can share information with other people. There are many tasks that a single person cannot do on its own, so the processing of information takes place among the different people working together and the various devices they use. For example, when a pilot is flying, the main pilot, his second-in-command and the ground dispatcher need to be in constant dialogue in order to complete the mission.

When a task requires multiple people to work together, the solution to the problem is distributed between these people and the environment and systems with which they work. There is verbal and non-verbal communication between them, and there are certain rules for such communication. There are mechanisms for cooperation and

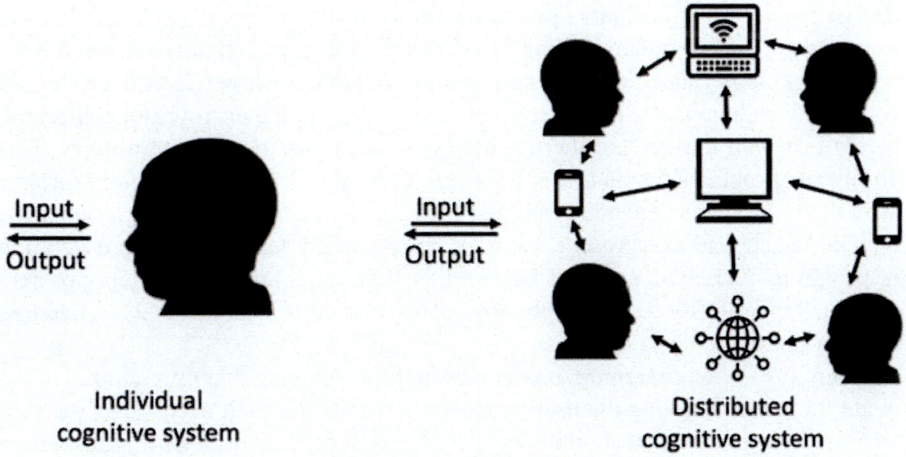

Fig. 8.2 Distributed cognition versus traditional cognitive psychology

communication patterns between them, and the knowledge to solve the problem is shared between them, and there are ways of accessing the knowledge.

With the development of computers and computing systems of all kinds, especially artificial intelligence, the cognitive activities that used to take place only in the human brain are extended to our computing systems, a typical example being the mobile phone. Our lives have become increasingly dependent on our mobile phones, which have become almost the other half of our brains. The phone numbers of our friends, our addresses, the records of our daily life, the search for all kinds of information, even the payment of accounts, etc. are all inseparable from these systems. Therefore, when designing systems, we need to consider how information is distributed in these human–machine-environments and how they communicate with each other in order to optimally and efficiently accomplish tasks.

8.3 Ecological Interface Design

Ecological interface design has emerged to answer limitations of a user-centred design approach that make User-centred Design less suited to deal with complex systems that are found for instance in power plants and aviation: (1) User-centred design focuses on the individual user and the 'user's' experience of the product. However, users may only know certain aspects of the system and may not have comprehensive knowledge; (2) User's may have limited knowledge of relevant technologies, for example, most drivers may not be aware of how the auxiliary safety system works; (3) The users may have unrealistic expectations of the product, due to lack of understanding capabilities of the product. This may result in expectations that are difficult to realise in design; (4) User-centred design emphasises user involvement in the design, but this may be difficult to achieve and provide little added value for projects as large and complex as in the field of automotive design; (5) User-centred design assumes that designers provide users with the products they need, but, due to the users' limited understanding of the complex domain, this may not be the best solution. This is related to the issue of how to distinguish between what the user wants and what he really needs. It is not always clear to the user what the difference is. (6) It is well known that there is a great deal of variation between users. Their ability and proficiency with technology varies greatly. Users also have very different attitudes to products depending on their lifestyles and educational backgrounds. So, how do we choose to represent 'users' in user-centred design? (7) Rapid iterations of design may create a sense of haste rather than thoughtfulness. Often, problems identified in usability testing might have been avoided if in-depth analysis had been done earlier in the design process.

Therefore, when designing, it is important to understand that user-centred design is not the same as giving the user whatever he wants, nor is it necessarily the same as providing whatever information he wants at each particular moment. The systems we design can neither help the user make decisions nor do everything for him. Car driving is a complex task, with complex car structures, complex driving environments, complex road conditions and complex road user behaviour, and it is difficult for users to really understand what their own needs are in the midst of all this complexity.

Human behaviour is inextricably linked to the environment and the technology used, and the system and environment have a strong influence on human cognitive psychology—this is the domain of ecological psychology, which emerged when psychology moved from the laboratory to the real world. The core of ecological interface design is to analyse problems from the perspective of the human–environment system and the context in which the action is generated. In interface design in the field of human–computer interaction, the ecological approach has had a major influence on the development of systems-oriented theory to support operators in complex socio-technical systems (Rasmussen 1986; Rasmussen and Vicente 1992) Design must think in terms of the limitations of the environment and technology, the limitations of human capabilities, etc.

Ecological interface design theory is rooted in Cognitive Work Analysis, which was developed to support the analysis of systems in complex environments. It incorporates the insight that, in complex environments, designers cannot anticipate all possible scenarios, and a traditional user-centred approach may result in tools that are poorly equipped to deal with unexpected situations. Therefore, design should analyse the complex environment and, only when such understanding is achieved, user needs can be understood and tools can be designed for users that enable them to solve problems in unforeseen situations. The constraint-based analysis identifies three types of constraints in complex environments: Constraints on action, Functional Constraints and Constraints on Information. It is not the average user who understands these three things best, but the expert who can see the bigger picture, the patterns and relations between things. Thus, the steps in analysing a problem become first analysing the environment and then analysing what people do in that state, how they do it, and what they know. Figure 8.3 gives a framework for the idea of eco-interface design.

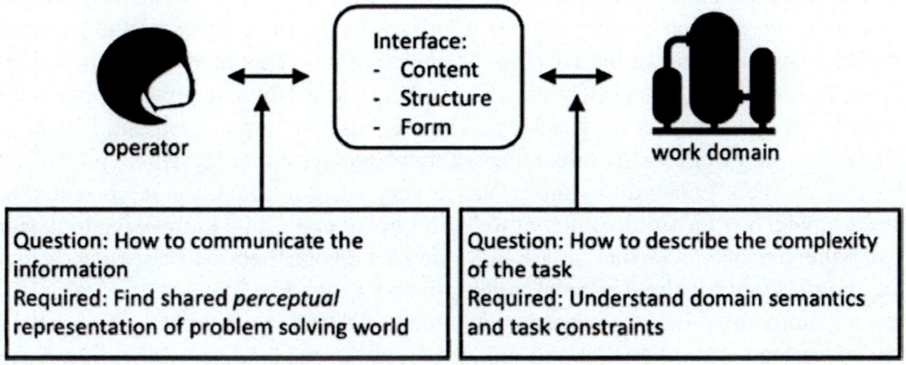

Fig. 8.3 Design framework for the eco-interface

The main task of an interactive interface is to help the user, or human operator, to complete a task. Between the interface and the task to be completed, the designer needs to think about how to describe the complex task. Here, the designer needs to identify the task constraints and domain semantics. The main question to be answered when designing the interaction between the interface and the operator is how to talk to the user. Here, the designer needs to find out how the domain semantics can be mapped onto perceptual operations, so that complex reasoning is replaced by perceptual operations and users are supported in constructing an appropriate mental model of how to solve the task and their mental load is reduced.

In order to do so, ecological interface design (EID) uses the concept of Abstraction Hierarchy (AH). This concept was introduced by Rasmussen (Rasmussen and Vicente 1992), who adopted a systematic approach to representing the domain constraints associated with goals at each level. These levels include.

- Functional purpose, i.e., the overall purpose of the system, the meaning of the system, i.e., why does the system exist? What is the main purpose of the system?
- Abstract function (abstract function), describes the causal structure of a process including mass, energy, information or value flows.
- General Function (general function), the systemic processes that structure the behaviour of a response, e.g., basic functions, the flow and storage of energy.
- Physical function (physical function), the states and properties associated with interactive components, the components and the relationships between them.
- The physical form, the position and appearance of the components, etc., and the relationship between these levels (Rasmussen and Vicente 1992).

The advantage of using AH is that it can provide goal-directed descriptions of the problem space, functional relationships and domain constraints, which allow for good analysis of complex information (Flach and Voorhorst 2016; Vicente 2002) and enrich problem solutions through a certain amount of mental correlation. The use of AH as a basis for ecological interface design can be very helpful in creating representations to help analyse problems and propose solutions.

The abstraction hierarchy analysis is generally difficult to perform and the analyst needs to be professionally trained. However, it is often referred to for complex systems, and Salmon et al. used this approach to do an analysis of the Victorian road transport system in the UK (Salmon et al. 2007). This analysis is appropriate for almost any road transport system. This is because the fundamental purpose of any road transport system is to be safe, efficient, accessible and convenient. Figure 8.4 gives the framework for the objectives of this large system analysis.

Due to space constraints, this is only a very basic introduction to eco-interface design, which is far from sufficient to learn how to apply this knowledge to design, but rather meant to open one's eyes. Eco-interface design theory is becoming increasingly important for the design of automotive systems. For instance, we can divide events on the road into three categories: familiar events; unfamiliar events that can be understood and predicted; and unfamiliar and unpredictable events. The design of the eco-interface is an attempt to help users cope with these three types of events.

Level of abstraction	Overall system (Road transport system)	Subsystem (Driver-vehicle-road system)	Component (Driver, vehicles etc.)
Functional purpose	Safe, efficient and acceptable mobility		
Abstract function	Reduced road user risk; Increased road user safety; Road user satisfaction		
Generalised function	Safe and efficient traffic flow	Compliant road user behaviour; Effective occupant protection	
Physical function			Driver capabilities and limitations; Vehicle capability
Physical form			Vehicle components

Fig. 8.4 CWA (Cognitive Work Analysis) analysis framework for transport systems. After Salmon et al. (2007)

8.4 Hierarchical Design Theory

Hierarchical design theory is a descriptive theory that was developed for the design of graphical user interfaces (Foley et al. 1995). It distinguishes four different layers and guides the design process from one level to the next.

Level 1, ***the conceptual layer***: where two mental models play a role, (1) a model of how the designer thinks the user will operate the system and (2) a model of what the user will actually do. Here the interface concept and the interaction framework are determined, including the pixels of the interface image and the computer program used.

Level 2, the ***semantic level***: here the semantics of the user input and the semantics of the system output are determined.

Level 3, the ***syntactic level***: the design conveys the semantics of user actions and how they are combined into computer sentences to perform certain tasks.

Level 4, the ***lexical level***: relies on the processing device to make precise calculations on user instructions.

This theory has its roots in computer graphics design and is a top-down design process that can be used in conjunction with software architectures to produce useful modular designs. It is important to emphasise here that decisions made at the first level affect operations at later levels. For example, if the output of a system is in a language and image that is familiar to the user, and the content that requires user input is also in a language that is familiar to the user, this can greatly reduce the learning costs and improve the usability of the system, as well as improving the user experience.

The key here is to decompose complex problems. Any complex system can be decomposed into subsystems, which can be further decomposed, but there are many different ways of decomposition, and it is the designer's job to choose a decomposition method that is understandable and memorable.

Similarly, the designer needs to break down the user's behaviour into smaller units of action. It is as if a building is constructed from a brick, a steel bar, a piece of glass and other such basic building materials. The designer's job is to be able to find these basic elements and then, through a series of actions, build complex systems. Simple systems are popular with users, so another challenge for designers is to make complex systems simple by simplifying the steps. So, the general guideline is to cast the output of a system in a language and image that is familiar to the user and cast the content that requires user input also in a language that is familiar to the user.

8.5 Stages of Action Theory

Stages of action theory was developed by Norman (Norman 2013). He divides human interaction with the world into seven stages (Fig. 8.5): (1) forming goals; (2) forming intentions (what options are available for reaching the goal); (3) identifying action sequences; (4) performing actions; (5) perceiving the state of the system; (6) understanding the state of the system; and (7) evaluating the state of the system. Interaction is successful, if each stage can be mapped onto the next stage without problems. But at each stage problems may arise. To illustrate the idea: One may have a clear goal, for instance, to get to an unfamiliar destination, but there may be a problem in choosing an option to reach the goal: how should I enter the destination into the navigation system, if there is no clear interface element visible in the interface that says: enter destination. Likewise, after having performed an action in the interface, one may perceive changes in the interface, but one may have trouble understanding what they mean and how to proceed.

The theory identifies four key points where users can go wrong.

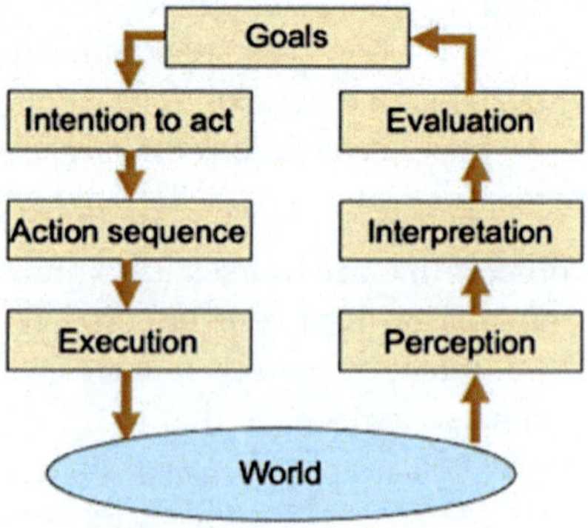

Fig. 8.5 Norman's seven stages of action. After Norman (2013)

- The potential for users to generate inappropriate action goals.
- The user interface may be designed using unfamiliar or inconsistent icons and text, resulting in the user misunderstanding the information in the interface and not finding the appropriate action from it.
- The user may not know how to identify his action and how to perform it.
- The user misinterprets the feedback due to design issues.

Based on the analysis of common problems, Norman derives four principles that a good design needs to satisfy.

- The state of the system and the various possible actions should be directly visible.
- The conceptual model and the system language/icon should be consistent.
- The user interface needs to be able to reflect well the relationships between the behavioural stages.
- Users need to be given continuous feedback.

The Stages of action theory is a commonly used theory in interaction design, not only as a basis for designers, but also as a basis for experts to evaluate the design of a system. Consistency is a very important principle in interaction design. Consistency can be seen in a variety of ways, such as colours, layout, icons, fonts, font sizes, button sizes and so on. For example, the word 'delete' can be called different things in the same system: delete, remove, destroy, erase, etc. Some people think that such variations create a more personalised design, but if a design does not meet the needs of usability, the user experience will be diminished. And consistency is an important part of usability.

There are various aspects of consistent design, including interaction styles in addition to the colours and layouts mentioned above, and consistency in this area allows the user to understand the interface better. Inconsistent design, such as the placement of virtual keys and the use of colours, can increase the user's reaction time by 5–10%, while inconsistencies in the use of functional terms can increase reaction time by 20–25%. This shows that consistency is a very important design point to develop users' operating habits.

8.6 User Interface Design Principles

Design principles refer to basic, general and widely applicable rules, while design guidelines are generally more practical summaries of rules that tend to apply to more specific needs. There is no generally agreed inventory of principles for user interface design, and many different inventories can be found. Also, specific maxims are labelled design principle in one inventory and guideline in another inventory. Therefore, rather than summarising different inventories or listing the commonly agreed upon principles, we refer the reader back to Sect. 2.6, where already design principles and guidelines were presented. Here, we consider a few principles guiding the process of user interface design.

1. ***Understand the user's level of proficiency***

   Everyone who designs is reminded about the importance of knowing the users. From what point of view do we understand the user? The first thing we need to know about our potential users is their basic information: age, gender, physical ability, cognitive ability, educational background, cultural background, training, motivation, purpose and personality. When designing products such as cars, designers try to target a specific group of people, but often many users do not fit the characteristics of this specific group. So, if possible, understanding the user should be multidimensional and a continuous process, as the user is constantly changing. One important difference concerns the users' level of proficiency. Successful designers are aware that users learn and know how to solve problems in different ways. In general, we divide users into three categories.

   - ***Newcomers***, or first-time ***users***. There is a difference between a novice and a first-time user. For example, new drivers who have just obtained their driving licence are not familiar with the car and the user interfaces in the cabin, whereas first-time users can be skilled drivers, as they may have owned another car before and are familiar with driving a car but are using the specific interfaces for the first time. To avoid any uncertainty, the designer should try to use a familiar operating design and a familiar dialogue style and simplify the steps as much as possible so that novice and first-time users can perform simple tasks without having to learn them, thus making them less anxious and building their confidence. For basic controls such as steering wheel, pedals and gear lever, which are standardised, this condition is satisfied, but for other controls, such as for HVAC, wipers and ADAS such as Adaptive Cruise Control, as well as the information and entertainment systems, standardisation has not yet been achieved, and good design is crucial. Feedforward and feedback from the systems should be direct and clear. And if the user has made a mistake, it needs to be clear how to improve it. This is true both for the design of cars and the in-car systems, especially for the design of features that are not commonly used. Even if the user is not a novice or first-time user of the car, if a particular function is being used for the first time or infrequently, the experience is like being a first-time user.
   - ***Users with some experience***. Many people are users at this level, for particular functions. They have a better knowledge of the task, some idea of the interface and have operated it a few times, but their biggest problem is that they cannot remember the structure of the menus and where a particular action is located and the order in which actions have to be performed. Therefore, the interaction should be facilitated by the structure of the menus, the consistency of the words and the use of a recognisable design, instead of requiring the user to remember the instructions. For this type of user, giving contextualised hints at the right time can be a great help.
   - ***Experienced people***. For these people, their requirements are fast operation and quick reaction. They do not need too much feedback and at the same time prefer to have shortcuts.

Some of the differences between users relate to user preferences and interests. We made a distinction between structural and functional mental models in Sect. 3.7. Some users may want to know the fine details of how the system works and develop a structural mental model, while the majority of users will be fine with developing a functional model that guides the interaction and helps them to determine how to achieve their goal. The system should not be designed such that it expects users to have a structural model of the system.

While it is relatively easy for designers to design for just one type of user, it is most often the case that the needs of multiple users need to be met, which requires multi-layered design. With the introduction of artificial intelligence, systems can learn the characteristics, interests, and operating styles of users, which opens up new design possibilities.

2. ***Understand the task***

Task analysis is a process that gives the designer an insight into the task that will be designed. A detailed explanation of how to do a task analysis is given in the methodology chapter (Chap. 11, Sect. 11.7). An important aim of task analysis is to find the basic elements that make up a task, and the choice of appropriate elements is very important. A large task can be decomposed into several subtasks and divided into multiple layers. For example, the task "play music" can be decomposed into several parallel subtasks, such as "play from saved list" or play through different players. If it is "Play from saved list", the next level of task is "Find the music you want to listen to", where there can be several parallel actions, such as "Enter song name" or "Enter the name of the singer" or "View from the list", etc. It is also clear from this that the results of the task analysis are directly linked to the current state of the art. At present, one has to select a different player, or source of music, but perhaps in the near future, when the system is more intelligent, this step can be removed and instead one might be able to go straight to "Select a track to play".

During the task analysis, it is analysed which tasks (functions) are frequently used and which are not. It is advisable to have dedicated buttons for frequently used functions or, if there are no physical buttons, virtual buttons that can be accessed with one click. For the less frequently used functions, the first level of the menu should be available, and for the less frequently used functions, a search line should be available.

There is no uniform format or standard for the design of what functions go in which level of the menu in a car, which require physical buttons and which are available as virtual buttons. Different car companies have their own considerations. Recently, a driver in Germany caused a crash while using a touch screen to adjust the speed of his windscreen wipers. He was held fully responsible. This case shows that poorly designed interactions can put drivers at unnecessary risk.

3. ***Understand the context***

Systems are used in many different contexts, and systems need to be responsive to variations in context, also to warrant safety. Simple context variations are daylight conditions, and navigation systems may automatically switch between daytime and night-time mode. More advanced systems may adjust to the traffic conditions and the state of the driver, such as variations in mental load or driver distraction. In connection with differences between users, the system may have to offer different interaction modes to different types of users, so that infrequent users can interact in different ways than frequent users. For instance, infrequent users may be interested in more basic functionality than experienced users. Making all functionality available at once may make the interface overly complex for infrequent users (cf. Carroll and Carrithers' (1984) Training Wheels approach).

8.7 Guidelines for User Interface Design

Based on 30 years of practice, Shneiderman proposed eight golden rules for user interface design that can be applied to the design of most user interfaces (Shneiderman et al. 2018). Similar guidelines have been proposed by Nielsen (1994) and Norman (1988) (see also Sect. 2.6).

- ***Strive for consistency***: A consistent sequence of actions should be performed in similar situations; the same terminology should be used in prompts, menus and help screens, and consistency in colour, layout, use of text, font size, typeface, etc. Exceptions such as asking for confirmation, deleting a command, or not displaying a password back should be well motivated, understandable and as few as possible.
- ***Seek versatility***: The user interface should be designed with a degree of plasticity so that it can facilitate content conversion. This is because there is a need to adapt to different user groups, whose backgrounds can vary considerably depending on gender, educational background, mastery of technology and familiarity with the system. For beginners, the system may need to provide more explanations so that users can understand why something happens and what action to carry out and how, while for skilled users, shortcuts, or concise operations are required.
- ***Provide feedback***: The system needs to give proper feedback to the user. If the interaction involves multiple steps, there should be feedback for each step of the operation. For common and small actions, feedback may be simple. For instance, for the action "Turn on radio" there is no need to provide feedback like "Turning on the radio". Instead, the sound of the radio may be sufficient feedback. For less common and large actions, there should be clear and explicit feedback. Visual feedback can clearly show the changes caused by the action.
- ***Design dialogs that produce a closing effect***: It should be clear to the user when the goal is achieved, instead of making the user wonder whether additional actions are still needed, or alternatively, requiring further actions without the user being aware of it.

- ***Prevent and correct errors***: Interfaces should be designed to prevent user errors, e.g., by blocking out inappropriate actions where possible, by greying out unavailable options, etc. If the user makes a mistake, the system should provide a reasonable, constructive and clear method of correcting the error. For example, if the user presses a wrong key during an operation, the system should not simply revert to the original state but allow a step back to correct the error and then continue the operation.
- ***Allow easy reversal of actions***: An "Undo" option allows the user to be able to reverse a previous action. This is important as it allows the user to use the interface without feeling anxious, especially when operating unfamiliar functions.
- ***Keep the user in control***: The experienced user wants to feel like being in control, instead of having the feeling that the automated system is in control and there is no possibility to influence the process. Also, experienced users do not want anything to surprise them, to not find the information they need, and to not achieve their purpose.
- ***Reduce the working memory load***: Humans have limits to their working memory capacity. They do not like pages that require them to remember certain information and then use that information on another page.

Here, we explain further about error prevention. This is very important in terms of design. User errors are inevitable, and the way to improve this is to design feedback messages about operational errors. The error feedback language should not be vague or even intimidating, nor should it be merely a status report, but it should tell the user what is wrong and what can be done to change the error. In Sect. 3.9.3, we mentioned the classification of errors and their causes. The best way to prevent errors is, of course, to not give the user the possibility to make mistakes, for example, by having a layout of input keys that is large enough and operates in a way that fits the user's mental model so that the user can operate them blindly without confusion. Providing as few options as possible also reduces the likelihood of errors.

It should be kept in mind that the guidelines for interface design were proposed at a time when user interfaces required explicit interaction, to be initiated by the user. With the emergence of intelligent systems that may take initiative themselves, the question is whether these guidelines still apply. In particular, the question is whether users still want to feel in control, and if so, how to ensure active human control in an automated system. When automated driving systems act autonomously to preserve safety, such as when using automated emergency braking to prevent an accident, users will likely not have a problem not feeling in control. Similarly, if the system adjusts parameters to satisfy pre-sets, such as when adjusting the HVAC settings, the user is usually fine with the system acting by itself (although it is debatable whether this should be considered an instance of 'intelligent systems'). The situation might be different, however, with non-time-critical, non-routine actions. A detailed analysis of representative concrete examples in different domains, taking different characteristics of intelligent systems into consideration, should be conducted to further our understanding of this issue (see for instance Amershi et al. 2019).

References

Amershi S, Weld D, Vorvoreanu M, Fourney A, Nushi B, Collisson P, ... Horvitz E (2019) Guidelines for human-AI interaction. In: Proceedings of the 2019 CHI conference on human factors in computing systems, pp 1–13

Carroll JM, Carrithers C (1984) Training wheels in a user interface. Commun ACM 27(8):800–806

Engeström Y (1987) Learning by expanding: an activity-theoretical approach to developmental research. Orienta-Konsultit Oy, Helsinki

Engeström Y (1999) Activity theory and individual and social transformation. Cambridge UP, New York

Flach JM, Voorhorst F (2016) What matters: putting common sense to work, Dayton, USA

Foley JD, van Dam A, Feiner SK, Hughes JF (1995) Computer graphics: principles and practice in C, 2nd edn. Addison-Wesley, Reading, MA

Kaptelinin V, Nardi BA (2006) Acting with technology: activity theory and interaction design. MIT Press, Cambridge

Kuutti K (1995) Activity theory as a potential framework for human-computer interaction research. In: Bonnie AN (ed) Context and consciousness: activity theory and human-computer interaction. Massachusetts Institute of Technology, pp 17–44

Leont'ev AN (1978) Activity consciousness and personality. Prentice-Hall Inc, Engelwood Cliffs NJ

Leont'ev AN (1981) Problems of the development of the mind. Progress, Moscow

Nardi BA (1995) Studying context: a comparison of activity theory, situated action models, and distributed cognition. In: Bonnie AN (ed) Context and consciousness. Massachusetts Institute of Technology, pp 69–102

Nielsen J (1994) Heuristic evaluation. In: Nielsen J, Mack RL (eds) Usability inspection. John Wiley & Sons, Inc, New York

Norman DA (1988) The design of everyday things. Basic Books, New York

Norman DA (2013) Design of everyday things (Revised edition). Basic Books, New York

Pirsig RM (1974) Zen and the Art of Motorcycle Maintenance. William Morrow and Company, NY, USA

Rasmussen J (1986) Information processing and human-machine interaction: an approach to cognitive engineering. North Holland, New York

Rasmussen J, Vicente KJ (1992) Ecological interface design: theoretical foundations. IEEE Trans Syst Man Cybern Part A 22(4):589–606. Retrieved from http://ieeexplore.ieee.org/xpls/abs_all.jsp?arnumber=156574

Salmon PM, Regan M, Lenné MG, Stanton NA, Young K (2007) Work domain analysis and intelligent transport systems: implications for vehicle design. Int J Veh Des 45(3):426–448

Shneiderman B, Plaisant C, Cohen M, Jacobs S, Elmqvist N, Diakopoulos N (2018) Designing the user interface—strategies for effective human–computer interaction. Pearson Education Limited

Vicente KJ (2002) Ecological Interface Design: Progress and Challenges 44(1):62–78

Vygotsky L (1934) Thinking and speaking (re-print 1). The M.I.T. Press

Vygotsky L (1962) Thought and language. MIT Press, Cambridge, MA

Vygotsky L (1978) Mind in society: the development of higher psychological processes. Harvard University Press, Cambridge, Massachusetts

Chapter 9
User Interface Design

Abstract In this chapter, practical knowledge for interaction design is summarised. Interaction technologies such as gesture interaction, speech perception, and multimodal interaction, and interaction styles such as direct manipulation and menus are discussed. Also, we go briefly into the design of alarms and warnings. Finally, we take a look at the design of the in-car infotainment system.

A user interface is a tool designed by the designer to help the user talk to the system to complete a task. In this chapter, we look at user interface design in more detail.

9.1 User Interface

There are currently many names used to describe user interfaces, such as graphical, command, voice, multimedia, invisible, environmental, emotional, mobile, intelligent, adaptive, smart, tangible, non-contact and natural, to name a few. Some descriptions are functionally focused, while others consider interaction styles more than anything else. The devices used for input and output also vary relative to the different interaction designs. There are many books on the subject, so we will not go into detail here, but will just select a few interfaces that are commonly used in cars.

The most common is the graphical user interface (GUI) and there are many books and articles on the design of this interface, so we won't repeat them. The key component is the WIMP (Windows, Icons, Menus, Pointer), but in terms of graphics, there are more complex 2D and even 3D motion graphics, which may be accompanied by sound effects. Here we focus on the design of menus and icons, as these are inevitably used extensively in the current increasingly complex design of in-car infotainment systems.

Menu Screen

The menu interface is available in a variety of styles: flat lists, drop-down lists, pop-up lists, contextual lists and extended lists such as scrolling and cascading. Expandable menus (Fig. 9.1) allow more options to be displayed on a single screen

© China Machine Press 2023

F. Chen and J. Terken, *Automotive Interaction Design*, Springer Tracts in Mechanical Engineering, https://doi.org/10.1007/978-981-19-3448-3_9

Fig. 9.1 Expandable menus commonly used in vehicles

than a single flat menu, making searching more flexible and allowing multiple options to be selected in the same window, and therewith making cascading menu interfaces the most popular. The disadvantage is that they require precise mouse, or click control, which may result in clicking or selecting the wrong option. Furthermore, by far the biggest issue is what is the best name/label/phrase so that the user can find the function he is looking for. Finally, the design of the position of the different contents in the list is crucial.

Icons

Icon design is widely used in cars, and unfortunately there are nearly 80 icons related to cars and driving that can be displayed in cars, and all of them are industry, or international, standard (Fig. 9.2 shows some of them). The reason for the widespread use of icons is based on the accepted assumption that they are easier to learn and remember than commands. Applying them also allows for a more compact design and transformable placement on the screen. The use of icons now permeates every interface. But in practice, even automotive design experts are not able to recognise all the in-car icons.

The mapping between the representation of an icon and the underlying reference it represents can be multifaceted, such as similarity (e.g., a picture of a file representing the target file); analogy (e.g., a picture of scissors used to represent a "cut"); arbitrary (e.g., the use of X to represent "delete"). The most effective of these icons is the similarity-based icon. Unfortunately, there are many car functions that are difficult to represent with a similarity-based icon. Therefore, a combination of text and icons will be less confusing for the user.

Fig. 9.2 Some of the on-board icons. *Source* https://thenounproject.com Creative Commons License

Direct-Manipulation Interfaces

Direct-manipulation interfaces are interfaces where all actions are visible, can be performed quickly and can be reversed (corrected) at any time, using only clicks or similar input devices such as joysticks, and no command input. This type of direct-manipulation interface is commonly found in gaming-based interfaces, virtual reality and augmented reality interfaces. The driving of a car proceeds typically by direct manipulation. The control of the steering wheel is directly linked to the direction of travel of the car. The force of the accelerator and brake is directly related to the speed. For the direct handling interface, the following 3 principles are generally followed.

- Objects of interest are visible, and continuous action is accompanied by continuous movement.
- There are direct physical or virtual keys that have a sense of substance, avoiding the need to give commands following a certain syntax
- Fast, continuous, reversible movements, the effects of which can be demonstrated immediately.

Direct manipulation interfaces have the following benefits.

- Newcomers can learn the basics very quickly, sometimes only needing to be shown once by someone with experience.

- Professionals can quickly master the operation of the vast majority of functions and can even define new ones.
- Knowledgeable users can retain the concept of operation even if it is not used regularly.
- No operational error messages required.
- The user can immediately see whether their action will achieve its purpose, and if not, they will simply change the direction of the action.
- Users generally do not feel anxious because the interface is intuitive and easy to understand. Also, the operation can be corrected at any time.
- Users feel confident in the operation of the interface because they feel in control and can also anticipate the consequences of each action.

Direct interaction interfaces may deploy metaphors that are familiar to users in their everyday lives (e.g., moving an interface element representing a file to an interface element resembling a waste bin to delete a file, literally, "throwing the file into the waste bin"). This makes it less costly for users to learn and remember.

Of course, there are some difficulties with the design of direct manipulation interfaces. The first is their ability to handle images, the changing visual effect of continuous images and the relatively large screen area it takes up. At the same time, the choice of a familiar metaphorical representation is also a challenge, and if icons are used, this again implies a problem of understanding them.

9.2 Symbol Recognition

It is often said that a picture is worth more than a thousand words. Therefore, in our lives we often see icons used to represent an object, a concept or even a function. The prerequisite for this is that we are able to recognise the meaning of the icon at a glance, drawing on our previous experience and knowledge. Each icon is given its own meaning, as shown in Fig. 9.3. These icons are basically recognisable to everyone, even without textual descriptions. An important factor in the design of an icon is that it is clear what it represents and that it is not similar to other icons, let alone ambiguous. Unfortunately, this is not the case with many icons, especially in-car icons, as shown in Fig. 9.4. Currently, there are more and more in-car icons, but many icons cannot be recognised by drivers. Since icons are used to replace text and allow the user to capture the information represented more quickly, the user should not be expected having to learn to remember them. There are therefore two key factors in the design of icons, one being the semantic distance, i.e., the similarity between the icon used and the meaning it represents. For example, the icons for toilets for men and women and the icon for escalator, in Fig. 9.3. The other is familiarity, that is, whether the icon is recognisable to everyone, as in the case of the no smoking icon in Fig. 9.3.

In-car icons are difficult to recognise because they represent meanings that are not visible and recognisable to the average person in everyday life. With the development

Fig. 9.3 Commonly used icons in everyday life. *Source* https://www.webappers.com/. Creative Commons License

Fig. 9.4 Icons for the interior of the vehicle

of safety assistance systems and autonomous driving technology, more and more icons will be introduced. In order to help the users understand the meaning of the icons without adding additional working memory load, one or a few words are often placed underneath the icons for clarification. In this way, people who are familiar with the icons will be able to recognise them quickly, and those who are not familiar with them can be reminded of them with the help of text.

9.3 Sound Icons

Just as words can be represented by means of sound—language—so icons have their counterparts in the form of sound icons, of which the various warning and alarm sounds are typical examples. There are two types of sound, the earcon and the auditory icon. Earcons are music-like synthesised sounds that represent relatively abstract meanings and are acquired through learning. With the development of electronic technology, this kind of sound image is becoming more and more important. Different mobile phone brands, for example, design their own sound for switching on and alerting people of incoming calls, so that when they hear the sound, they know what brand of phone it is and whether it is a call or a new text message that has come in. In car design, too, there is an increasing focus on such sounds, such as the sound signalling that the car door is still open and the sound of the in-car system turning on. Of course, it is also possible to give a particular sound a specific meaning. This meaning, then, needs to be acquired through learning. As with text icons, factors guiding the design are semantic distance and similarity. But for earcons minimising semantic distance is often hard to achieve, and most effort may be put into maximising the similarity. Auditory icons are everyday sounds that can be recognised without special learning, such as the sound of raindrops or the sound of running water. In our daily lives, these sounds form the basis of our awareness of our surroundings. In design, the sound of falling raindrops may be used for instance to warn the driver about upcoming rain showers.

9.4 Alarms and Warnings

The design of alarms and warnings needs to address two dimensions: what meaning needs to be conveyed, and how can the meaning be mapped onto one or more modalities. With respect to meaning, a distinction can be made between alarms, which require immediate action in order to avoid safety hazards or damage to the engine, and warnings, which inform the driver about matters that require attention but are not critical. Since alarms require immediate action, the main function of the alarm is that it draws immediate attention to the situation requiring action, and it is less important that the alarm signals the nature of the hazard. In the driving context, the driver's visual attention is usually focused on the road, and it cannot be assumed that the driver will perceive a visual signal in the dashboard or mid console, unless it is excessively bright and has sharp transients (flickering); and by bright daytime light even an excessively bright signal may not be noticed. Therefore, the auditory modality is a better choice. Using spatial audio, the driver's attention can be drawn to the location where the hazardous situation arises. E.g., if there is an immanent engine failure due to oil shortage, the audio may come from the dashboard. There, more information may be offered using graphical messages. For warnings, the visual modality may be used, either warning lights or text messages, or the auditory modality may be used, in the form of spoken messages. The advantage of text and spoken messages is that they have a larger bandwidth ('bandwidth' denotes the amount of information that can be transmitted through a channel) than non-speech audio or warning signals, so that they can code information about the nature of the problem.

Auditory signals are usually considered more obtrusive than visual signals, so that drivers may develop a dislike of auditory signals. For instance, a truck manufacturer developed a so-called "virtual rumble strip" system, by which truck drivers were warned about lane deviations. If one of the wheels crossed the lane marking, the system would play a sound that resembled the sound of the tyres when driving over real rumble strips. However, truck drivers disliked the warning and looked for ways to turn it off. It was found that truck drivers preferred haptic warnings for crossing the lane markings over auditory warnings. Nevertheless, auditory alarms may be effective in situations where urgent action is required. In that case, immediate response is more important than a pleasant user experience.

Furthermore, the identification of situations where a warning or alarm needs to be emitted is not always straightforward. For instance, in the case of a system that warns the driver that s/he is getting tired and on the verge of falling asleep, the system needs to identify the state of the driver on the basis of one or more indicators such as eyelid closure (PERCLOS, see Sect. 4.9). This is typically a signal detection and classification problem, which, in addition to hits (the correct identification of a state of drowsiness) may result in misses (situations where a state of drowsiness is not detected) and false alarms (situations where a state of drowsiness is erroneously detected). Inherent to such classification problems, the number of false alarms and misses is related. If the number of misses needs to be minimised, the number of false alarms will increase excessively, and the other way around. Obviously, in case of alarming situations such as when the driver is close to falling asleep, one would like to minimise the number of misses, but this can be done only at the expense of a large number of false alarms. In such cases, auditory warning signals may be considered annoying by drivers.

If we include the temporal dimension, the situation may evolve from a warning into an alarm: in first instance, the driver may be warned about a particular situation, such as a vehicle in the left lane on the highway that prevents the driver from overtaking. The warning may consist of a light signal in the left mirror, based on the assumption that the driver will check the left mirror when intending to overtake and will see the warning light. However, if the driver misses the warning light and the overtaking vehicle and still intends to overtake, the situation evolves into a hazardous situation requiring an alarm. In that case a directional auditory alarm may sound that comes from the direction of the left mirror or window, drawing the driver's attention. Alternatively, a haptic signal may be emitted through the steering wheel.

9.5 Gesture Interaction

Gesture interaction refers to the interaction between a user and a machine using direct hand movements, and generally requires the support of computer technologies such as gesture recognition, motion tracking, body gesture recognition, and facial expression recognition (Pang et al. 2014). Hand posture, or gesture, refers to an action that is performed using only the hands. Human body movement, including gestures, is a form of non-verbal communication that originates from natural human interaction (Kortum 2008). Human body language can be divided into innate and acquired gestures (Microsoft 2013), with innate gestures having meaning based on one's daily activities and knowledge of the world, and acquired gestures that the user must learn before s/he can interact meaningfully with the system.

There are a number of basic gestures, including closed palm gesture, open palm gesture, five fingers gesture, pointing index finger gesture and L-gesture (Fig. 9.5). These are the more natural gestures, but they are also being developed by a number of organisations, and their meaning is often redefined by the developer according to the task situation, resulting in their own gesture language. When used in practice, these gesture languages require guidance and instructions that the user must learn and remember.

The biggest challenge in gesture design is to enable the user to learn gesture-function connections quickly and to minimise the user's memory load. Therefore, a prerequisite for gestural interaction to be widely used may be to standardise the gestural language and to find a gesture language that is easy to remember. The user's experience, including perception of physical laws, experience with existing human–machine interface models, socio-cultural practices and feedback methods, will all influence the use and acquisition of individual gestures.

Body movements are fleeting and do not leave any visible traces. Therefore, the user needs to be provided with the necessary feedback as to whether the physical

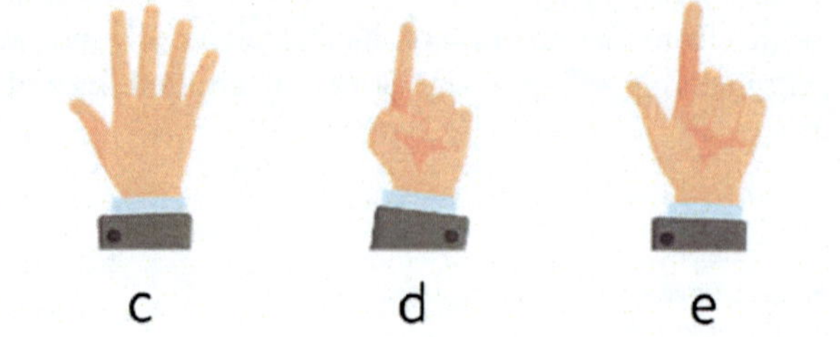

Fig. 9.5 Some examples of basic gestures. **a** Open palm gesture; **b** Close palm gesture; **c** Five fingers gesture; **d** Pointing index finger gesture; **e** L-gesture

input has been correctly entered and successfully recognised. Currently, this feedback comes mainly from the visual and auditory channels, especially the visual channel. The introduction of gestural interaction is intended to reduce the visual load on the driver, which is the biggest problem for the driver, so feedback design is a major challenge if gestural interaction is to be introduced into the vehicle. The reliability of the gesture interaction is another issue. Unlike traditional interaction methods, in somatosensory interaction the user may inadvertently perform an action that may trigger a function that was not intended. This can result in the user not being able to understand the current state of the system or the connection between the action and the result, resulting in a reduced sense of control or even loss of control over the system.

9.6 Speech Perception

We will not go into the details of phonemes, syllables and single words in phonological perception, which may be very different between different languages such as Chinese and English. Speech is spoken by a speaker, travels through the air, and is processed by the auditory mechanism and the nervous system of the listener. Broadly speaking, this whole process is the process of speech production and perception. Here, we focus on speech perception and understanding. Speech understanding is a complex skill that people master at a young age, and language is a primary way for humans to interact with each other. Since the invention of the computer, researchers and engineers have attempted to enable machines to "hear" human speech, to understand its meaning and to respond correctly. Speech technology aims to enable machines to process and understand human speech. It is a rapidly developing discipline and is an important component of automotive voice interaction. It involves three main sub-disciplines, namely automatic speech recognition (ASR), natural language processing (NLP) and speech synthesis (SS).

In speech understanding, as in written language understanding, humans combine a bottom-up approach (trying to recognise the sounds and works spoken) and a top-down approach, using their knowledge about the language, the context and the world to 'guess' probable interpretations and prune impossible and improbable interpretations. In the context of automatic speech understanding, the context typically exists of an application scenario, which limits the possible interpretations of the spoken message. For instance, when the driver uses voice interaction to control the car radio, a speech recogniser that is specialised for the car radio favours interpretations that match commands that apply to the radio. Typically, such recognisers apply a "command and control" type of interaction, requiring the user to know the commands by which the radio can be controlled. On the other hand, search engines typically accept natural language, leaving more freedom to the user to convey messages. In recent years, natural language processing has achieved high performance, due to the enormous amount of data that have become available for training through internet technology. However, the high performance does not apply to all languages equally.

Also, the performance may differ between speakers. For instance, the performance may be worse for non-native speakers, in particular if their pronunciation of the language deviates much from the canonical pronunciation by native speakers.

In the process of human communication, there are other factors that may play a role in speech understanding, in addition to the spoken message conveyed by the speech itself. (1) The movements of the mouth when speech is produced. For people with hearing problems, lip reading plays a major role, but for normal people it also plays a major role in the understanding of speech, enriching the language we hear. In particular, sounds that may be difficult to understand because of ambient noise, may be disambiguated because their visual representation is not affected by the ambient noise. Processing visual language has become a part of our 'listening' through the constant practice of face-to-face communication in our lives. Imagine watching a TV series and the voice of the voice actor does not match the actor's diction; this may make you feel uncomfortable and it may even interfere with your understanding of what you hear. (2) Non-verbal cues, including the way a message is pronounced, but also gestures, body posture, facial expressions, etc. Paralinguistic cues may provide information about whether a speaker is in a neutral mood, is stressed or irritated. (3) Feedback from the listener: The listener's face may indicate whether she understands an agrees what was said or may show a confused and questioning expression when two or more people are communicating. Speech technologists have attempted to include these sources of information into the process of speech understanding and voice interaction. However, given the massive amounts of data needed to train automatic speech recognition programs to achieve high recognition performance, including these additional sources of information is usually not considered to be cost-effective. Instead, developers of voice interaction systems put effort into the development of effective dialogue strategies to identify and deal with problems in the interaction.

9.7 Anthropomorphism

In the context of speech-based interaction, it is easy to get caught up in a form of Anthropomorphism, where human-like qualities are attributed to inanimate objects (e.g., cars, computers). This phenomenon is often used in advertising, such as dancing butter, drinks, breakfast cereals, etc. This type of design is increasingly being used in human–computer interaction, where it intends to make the user experience more enjoyable and motivating, and to make people feel at ease and reduce anxiety. A typical example in automobiles is the Nomi design of the Nio ES8, as shown in Fig. 9.6.

In voice interaction, this design is even more prevalent. Let's compare the two welcome message designs: "Hi Ming, welcome back, where were we last? Oh yes, I'm looking up a Sichuan restaurant" and "I've found a Sichuan restaurant". The former is very anthropomorphic, while the latter tells you exactly what you're talking about with a system. Here, we encounter an interesting question: do users prefer the

Fig. 9.6 Anthropomorphic design on a car. *Source* https://www.nio.com

anthropomorphic expressions and dialogues of the system, or the more straightforward, unemotional responses? The answer to this question is particularly important for the design of voice interaction. An overly anthropomorphic design can give the user the false impression that the system's intelligence is comparable to that of a human being, especially with the guesses of age and gender implied in the voice. There is evidence, however, that the use of voice interaction as such, whether more social or more factual, already suffices to induce expectations about intelligence and conversational competence (Nass and Brave 2005), although it must be noted that most studies only concern the initial experience. In conjunction with the issue of the temporal evolution of the user experience mentioned in Sect. 6.5, the question remains how long this anthropomorphic design will be satisfactory for the user? As time progresses, users may well prefer the more factual communication style, as it takes less time and hence is more efficient.

9.8 Multimodal Interaction

Multimodal interaction has become an active area of research in human–computer interaction with the increase in computing power of computers and the development of artificial intelligence. This combination of different forms of input (e.g., voice, gesture, touch, gaze, etc.) is known as multimodal interaction, and its goal is to provide the user with multiple options for interacting with the computer in order to support the user's natural choices. In contrast to traditional single-modality interfaces, involving primarily pressing buttons and keys, controlling a cursor on the screen through a mouse or similar input device and looking at a screen, multimodal interfaces can be defined as combinations of multiple input and output modalities, doing justice to the fact that people's interaction with the world is richer and more expressive, spanning multiple modalities.

In the context of interaction design, modality refers to the different sensory modalities (auditory, visual, haptic and so forth). Multimodal interaction means that a system supports multiple modalities both for input and output. For input, multimodal interaction typically involves input by voice (speech) and touch (gestures), but also facial expression and eye gaze information. For output, multimodal interaction typically involves sound (both speech and non-speech) and graphics, but also haptic information. Here new questions arise: how to design such multimodal interactions? What information is suitable for which modality? What operations are suitable for what modality?

Independently of whether tasks involve visual or auditory resources, information may be encoded in different ways. There are two main types of code, an analogue/spatial code and a categorical/symbolic code. The analogue code is best represented by the classic disc watch or the traditional disc speedometer and by spatial audio, while the symbolic code is best represented by language and text (either spoken or written). This distinction is important, because messages with spatial content, such as communicating the position of an object, or an action in space, such as the opening and closing of a window, the control of the steering wheel, etc., are best expressed in the spatial domain, e.g., by showing the position of the object, by showing the relevant button to open the window and by showing how to control the steering wheel. At the same time, in the case of a symbolic message, such as a text message, it is easier to say or type it (or listen to it or read it).

If we have to do different things at the same time (time sharing), it is more efficient to combine different modalities than to code all information in a single modality. For example, when driving, the visual requirements are high. The driver's vision should be focused on the road, but if there is an email that the driver needs to read, and both driving and reading require his visual attention, it is hard for him to ensure driving safety, because both tasks are taking up his visual resources. However, if this email is read to him in speech, then the interference with the driving performance is relatively modest.

To summarise, Table 9.1 presents different interaction modes and their advantages and disadvantages.

In the context of driving, speech-based natural language and command language have the advantage of being less distractive than the other interaction modes. Menu selection and form filling may also be conducted using speech, but will probably require visual interaction as well and are therefore more distracting. And direct manipulation requires eye-hand coordination and is distracting by definition.

9.9 Infotainment System Design

In the early days of automotive development, there was little information available to the driver other than speed and remaining fuel, and the design of this information was relatively simple. However, with advances in electrical/electronic engineering and information technology, more and more systems were added. The exchange of information between the driver and the system has also become increasingly complex. In addition, advances in technology have given more opportunities to make

Table 9.1 Comparison of the advantages and disadvantages of different interaction modes (Shneiderman et al. 2018)

Advantages	Disadvantages
Direct manipulation • Visual representation • Easy to learn • Easy to remember • Avoiding mistakes • Encourage exploration • User satisfaction	• Somewhat difficult to program • Need for visual attention
Menu selection • Reduced learning time • Reduced keystrokes • Structured decision-making processes • Allowing the use of dialogue management tools • Errors are easily corrected	• May produce too many menus • Slow operation for frequent users • Take up screen space • Request a quick response
Form filling • Simplifying data entry • Available for easy management • Available with form management tools	• Takes up screen space
Command Language • Suitable for highly competent users • Easy scripting and history retention	• Requires learning and memorising • Error-prone
Gestures • Do not require eye-hand coordination • If already existing gestures from everyday life (innate gestures) are used, easy to learn and remember	• When using innate gestures, only a few different gestures can be used, so only suitable for limited functionality (e.g., move to next track; increase volume etc.) • If also using acquired gestures, it requires extensive learning • Issues with feedback: the user doesn't know whether the correct gesture has been used and whether the gesture has been realised correctly
Natural Language • Does not require eye-hand coordination • Connects to people's natural expressive abilities • Reduces the burden of learning grammar	• Needs clarification dialog • Context may not be displayed • In case of written language, many keystrokes may be required • Unpredictable

the driving experience more enjoyable by providing information and entertainment media. The collection of information and entertainment systems is often referred to simply as IVIS—In-Vehicle Infotainment System. Interaction with the IVIS usually occurs while the driver is engaged in driving tasks, and the design of the user interface (UI) of the IVIS is no longer simple. In this section, we will consider basic IVIS interaction design.

In the early 1990s, thanks to technological advances and the creativity of engineers, a typical car radio could contain over 90 functions. A car radio manufacturer began to wonder how many of these functions were actually used by the driver, so he came up with the brilliant idea of just asking the customer. The outcome was that most customers only used between five and ten of the available functions. This does not only mean that the engineers' creativity produced features that are not used, but also that customers are paying for features they never use. In addition, too many features make the UI unduly complex and given the limited display space on a car radio and the limited interaction time available to the driver, the driver will only use those features that are important and relatively easy to learn and remember. This mismatch between design supply and customer needs can be avoided through a user-centred design approach. While the intention of IVIS is to make the driving experience more enjoyable, this mismatch can lead to a negative user experience. A user-centred design process should therefore be holistic and should not just focus on making interactions easier, but should aim to build a positive and satisfying user experience.

When designing an in-car infotainment system, there are two main issues: (1) where to place the display and controls, and (2) how do people interact with the system. There are two principles that can guide design decisions. The decision on location may be guided by the principle of reduced line-of-sight distance. The design of the interaction may be guided by the principle of reduced time. The principle of reduced line-of-sight distance holds that features critical to the driving task should be placed as close as possible to the main line of sight (representing the direction of the driver's gaze when monitoring traffic in front of the vehicle).

The principle of reduced distance leads to a distribution of functions such that the controls/displays that are related to DRT (driving-related tasks) systems are basically arranged on the dashboard or on the steering wheel, while other controls/displays are located on the centre console and centre screen (Fig. 9.7). However, due to the rapid increase in the number of DRT systems and the advent of ADAS, there is an increasing tendency for more and more information to be placed on the dashboard, leading to

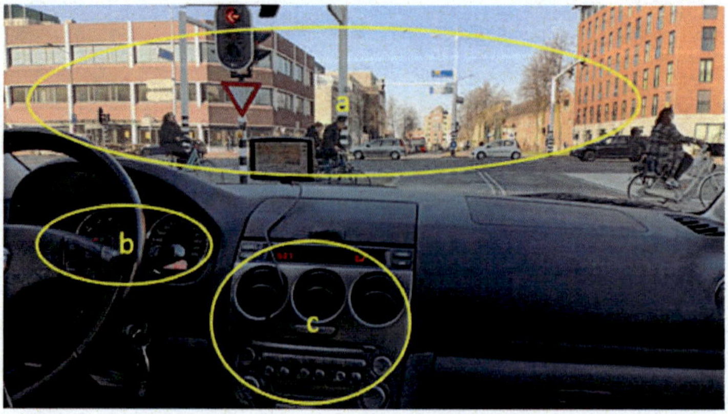

Fig. 9.7 Areas available for distribution of UI functions. **a** Primary line of sight; **b** Dash panel; **c** Mid panel/console

Fig. 9.8 Harman's "dashboard of the future". *Source* https://www.motor1.com/news/226657/harman-dashboard-of-future-ces/

lack of space and overcrowding. Furthermore, concepts have been proposed where the whole dashboard is a screen (see Fig. 9.8) and can be configured at will. The risk is that it may lead to further crowding of the cockpit with information displays and interfaces, but the advantage is that it allows screen access to a passenger.

The principle of reduced time states that interactions should be designed so that the driver's gaze is taken off the road for as short a time as possible. In order to reduce the potential distractions (driving distractions) that can occur when interacting with in-vehicle systems, especially for NDRT (non-driving-related tasks) systems, the main challenge when applying traditional interface techniques, including visual displays and (virtual) buttons and keypads, is to design the displays and action sequences so that the number of required actions is minimised without making the displays too complex. The design process requires a task analysis first.

New interaction technologies have been introduced, enabling designers to take different approaches to the principles of distance and time reduction. Head-Up Displays (HUD) offer the possibility of further reducing the distance from the main line of sight. Haptic interfaces provide haptic feedback to the driver, accelerating interaction and reducing the visual pressure on the driver who must coordinate movements through the eyes. BMW's iDrive rotary button is an example of this. A further development in this direction are shape-changing interfaces, in which the shape of the interface can be dynamically changed to suit the interaction. The basic idea behind shape-changing interfaces is that the driver can recognize different shapes by hand, providing information about the state of the system and thus enabling interaction without the need for visual support. Gesture-based interaction allows touchless interaction through mid-air gestures. Voice-based interaction avoids the need for eye-hand coordination, which makes it ideally suited for reducing off-road glance time.

One challenge with most novel interaction technologies is that they can easily lead to developers overestimating the driver's capabilities, thus overloading the driver's resources. For example, with HUDs, manufacturers may put too much information on the display, which can lead to driver distraction. With haptic feedback, too many different feed forward/feedback modes may be difficult for the driver to learn and differentiate. The same applies to gesture-based interaction: too many different gestures can be difficult for the driver to learn, and since the driver does not have feed forward/feedback when gesturing, it can be difficult to determine the accuracy required to achieve gestural interaction. Regarding speech-based interaction, ideally the driver could use natural language and have a natural conversation with the system. However, natural language takes time and users tend to reduce the length of the discourse in favour of command-based speech, which requires the driver to balance between commands that are deemed convenient and those that the system can understand. In addition, speech interfaces require good dialogue design to deal with natural language phenomena and possible misunderstandings.

Another way to achieve the principle of time reduction is to create predictive interfaces, which aim to cut down on the number of interactions by trying to predict what the user wants to achieve at each moment.[1] This allows the system to reduce the number of options, thus making it easier for the driver to select one. Obviously, this requires good predictive skills, otherwise the options the driver is looking for may not be among the predicted possibilities, and the interaction will only become complex and frustrating. Artificial intelligence/machine learning techniques are a necessary part of the toolkit for designing such interfaces.

As mentioned earlier, the principle of reduced distance leads to a distribution of functions such as driving-related (primary) functions being provided in the dash panel, inside the steering wheel or behind the steering wheel, while other (secondary) functions are provided in the centre console. Yet another use-related consideration for the positioning of controls is the frequency of use. Firstly, even for those functions that are not directly related to the driving task, those that are used frequently should be designed according to the principle of reduced distance in order to reduce the time spent with the eyes off the road. Secondly, considering Rasmussen's 'skill-rule-knowledge' (SBB-RBB-KBB) framework (see Sect. 3.8), it can be inferred that over time, drivers can blindly access frequently used functions.

Developments in software and display technology have created enormous opportunities and scope for design flexibility. While traditional in-vehicle user interfaces include hard buttons the shape and function of which remain fixed throughout the lifetime of the vehicle, touchscreens include soft buttons the function of which is dependent on the context. With the development of smart web technologies for cars, new software versions can be updated and downloaded, and the layout and functionality can be modified and expanded in each new version over the lifetime of the car. In addition, these new designs allow the user to customise the functions. This development is often referred to as the Digital Cockpit.

The flexibility of the design clearly offers advantages. Furthermore, as more and more ADAS/ADS functions are introduced into vehicles, the display space can easily be expanded to accommodate new functions. However, from a user-centred design

[1] https://www.mitsubishielectric.com/news/2014/0210.html.

perspective, the digital cockpit offers opportunities as well as challenges. Firstly, it may be difficult for manufacturers to resist the temptation of featurism, i.e., to include features that do not reflect the needs of the user. As the technical possibilities create opportunities for manufacturers and suppliers to implement features that can be sold as unique selling points, design from a user-centred perspective should consider which features create real value for the user. Secondly, rigorous consideration needs to be given to the use of touchscreens, as interaction with them often requires hand–eye coordination, which may run counter to the principles of reduced distance and reduced time. This undesirable consequence can be mitigated by introducing predictive interfaces and sensing technologies that increase the size of icons and soft buttons on touchscreens when fingers are close to the screen, but the eyes still need to be off the road. In short, more than ever, a user-centred approach to design is needed to discipline design so that technology translates into real value for users.

References

Kortum P (2008) HCI beyond the GUI: design for haptic, speech, olfactory, and other nontraditional interfaces. Morgan Kaufmann

Microsoft (2013) https://www.go.microsoft.com/fwlink/?LinkID=247735

Nass CI, Brave S (2005) Wired for speech: how voice activates and advances the human-computer relationship. MIT press, Cambridge, MA, p 9

Pang SY, Guo RJ, Yao NP, Yu JL, Yu SH, Gao ZF (2014) A review and outlook of human factors research on somatosensory interaction. Chin J Apply Psychol 20(3):243–251

Shneiderman B, Plaisant C, Cohen M, Jacobs S, Elmqvist N, Diakopoulos N (2018) Designing the user interface—strategies for effective human–computer interaction. Pearson Education Limited

Part IV
Design Process and Research Methodology

Chapter 10
Design Process

Abstract In this chapter, we take a closer look at the design process. We summarise design processes from different fields of engineering and conclude that the human-centred design process provides the best structure to ensure that knowledge about and feedback from prospective users is integrated in the design process. Finally, we consider more recent developments, leading to the insight that design is not just a matter of trying to satisfy user needs, but that considerations regarding users and a design vision should be integrated in a balanced process.

Design is a complex activity that sits between science and craft. The aim of this book is to establish an experience-centred design process and support it through tools and design methods that will help designers establish a positive user experience. Despite the established scientific tradition of design research, few researchers have described the design process. This can be attributed to the fact that design is a non-linear activity and that it must be reconfigured to accommodate the various steps that make up the design problem. This is probably a major headache for many interaction designers, who would like to have a fixed method, a fixed process to follow, but unfortunately, this is difficult to do.

The design problem has been described as a 'wicked problem' (Rittel and Webber 1973), which is the main distinguishing factor between design research and other sciences, and the main reason for the scarcity of design process models. 'Wicked problem' refers to 'a problem that cannot be solved because of incompleteness, contradictions and changing needs'. Any design is a creative process, and no matter how many theories and principles there are, there is always a certain amount of uncertainty. A successful designer must be able to understand his users and at the same time have full knowledge of the various technical possibilities and a good sense of aesthetics. Design is not a state but a process, so it is dynamic rather than static; the design process is not hierarchical, so it will not just be bottom-up or top-down, it can be a variety of combinations. The process of design will also constantly discover new goals.

© China Machine Press 2023

F. Chen and J. Terken, *Automotive Interaction Design*, Springer Tracts in Mechanical Engineering, https://doi.org/10.1007/978-981-19-3448-3_10

10.1 Overview of the Design Process

Jones proposed a model of the design process (Jones 1992) that is intentionally abstract in order to subsume design activities of various kinds, and allowing for an iterative approach to enable designers to solve various design problems. Jones' design process includes stages of divergence, transformation and convergence (Fig. 10.1). The process model is provocative and can be adapted to suit the unique circumstances of each design problem.

Divergence "refers to expanding the boundaries of a design scenario so that there is a large enough and fruitful enough space to find a solution" (Jones 1992). The main objective of this stage of the process is to enable designers to develop a deeper understanding of the design problem and to translate this understanding into requirements that can be used later in the process. Typically, various methods of user research are used in this phase, such as ethnographic observations (Crabtree et al. 2012) and interviews. Jones notes that dispersal work requires legwork rather than armchair guesswork. That is, the designer must get out of the office and into the field, trying to understand people and their behavioural environment through observation, and from this understand the fears, desires and motivations of the user, rather than sitting in the office and guessing at them.

Transformation is the idea generation phase of the design process. Designers translate potentially complex requirements into designs by using their design skills and a range of design methods to 'decide what to emphasise and what to ignore'. Through ideation, possible design solutions are created, iterated and turned into prototypes.

Convergence is the stage of the design process in which various possible solutions are reduced to a final design result through a rigorous evaluation process. The prototypes and ideas generated in the conversion phase are tested against the requirements of the divergence phase as well as the designer's intuition and sensitivity. Evaluation methods such as checklists, rankings and weights help designers to evaluate and make decisions.

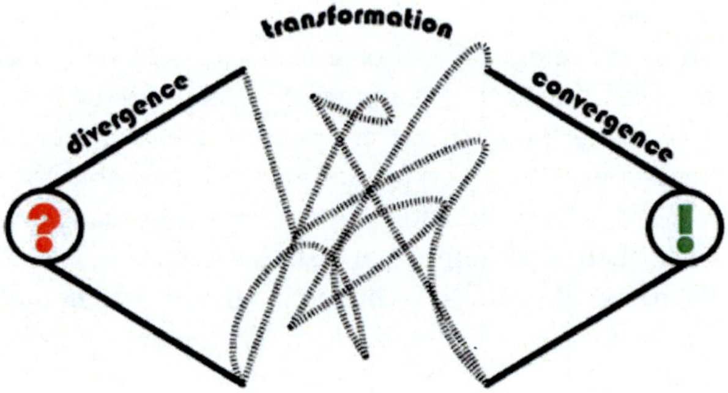

Fig. 10.1 The design process as composed of diverging and converging activities. *Source* https://kylewilliamsdesign.com/theories-and-concepts. (C) Tess Colavecchio. Reprinted with permission

Table 10.1 Interaction design process

Analysis	Synthesis	Evaluation
Defining the problem space of the design Collecting user requirements	Generating design concepts Prototyping	Evaluation using requirements Selecting the final design concept

A similar model of the design process can be found in several other academic works. Benyon (2010) describes the design sequence in his book Designing Interactive Systems in the section 'Techniques for designing interactive systems' as follows: understanding, which includes user research methods such as interviews and observation; envision and design, which includes design methods and evaluation. Sharp et al. (2019) outline the interaction design process as one that involves four basic activities: (1) establishing requirements, (2) designing multiple alternatives, (3) prototyping and (4) evaluation. Jarvis et al. (2012) give an annotation of the design process with clear steps to gather requirements and then conceptualise and evaluate them in iterations with increasing attention to detail. Taken together, these can be summarised in a three-step model of analysis, synthesis, and evaluation, as shown in Table 10.1.

Analysis is the initial stage of the design process. In this phase, the design team must usually define the design problem by gathering design requirements. A good Analysis phase summarises the main requirements that will guide the subsequent phases of synthesis and evaluation, while balancing the interests of the different stakeholders. The specific requirements can be used as a guide for the synthesis phase, where designers can use the requirements as inspiration or even as material in different conceptual approaches, and for evaluation, where the requirements can be used as a benchmark so that the design that best meets the requirements can be selected.

Synthesis is where ideas are generated, where requirements and other user data are transformed into design ideas through a creative process. The conceptual approach helps designers to generate ideas by realising ideas in the creative process of design.

Evaluation is where potential design solutions are evaluated against each other using the requirements from the analysis phase, and the final concept is selected.

Later stages are dependent on the outputs of the previous stages. This means that Analysis feeds requirements into Synthesis and Evaluation, Synthesis feeds design into Evaluation, and Evaluation may highlight the need for further iterations of Synthesis or even Analysis. These three stages are interlinked and not strictly defined, but rather imply the overall process.

In the next sections we consider different models of the design process that have been proposed in various fields.

10.2 Linear Design Process

The linear design flow, also known as the waterfall model, is a design flow often used in the mechanical and software engineering disciplines in the past, as shown in Fig. 10.2.

The issues to be addressed at the different stages of this process are as follows.

- Problem definition and requirements analysis phase: Defines the requirements, including the functions, hardware and software parameters that the system must perform. The requirements usually also involve validation criteria. User requirements are considered in this phase. Requirements may come from the customer/project owner, from previous projects (as complex systems are not always designed from scratch) and from the marketing department.
- During the feasibility stage, an analysis is carried out to find out if the available technology can be used to determine the project requirements at the appropriate time and within the set cost limits.
- During the conceptualisation phase, potential solutions are generated. A typical method used in this phase is to brainstorm.
- In the high-level design phase, the most promising ideas, which represent what the system will do and how it will be implemented, are further elaborated, Often, structured morphological charts are used. Finally, decisions are made and solutions are selected for the next stage.
- In the "detailed design" phase, the solutions derived in the previous phase are further detailed through detailed descriptions, technical drawings, models, etc., which provide the basis for manufacturing.

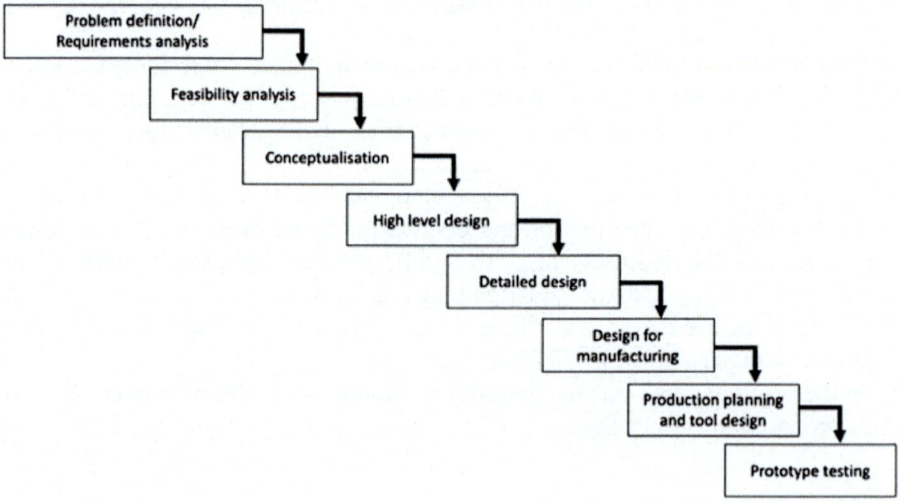

Fig. 10.2 Linear design flow

- The purpose of the design for manufacturability phase is to gather information to ensure that the system can be manufactured within set constraints (time, cost, reliability, etc.).
- In the production planning and tool design phase, the production process is selected and the tools required for the production of the system are designed.
- During the prototype testing phase, the prototype is tested to assess whether the system meets the requirements defined through the different phases of the project. Once the prototype meets the requirements, the production process can begin.

A limitation of the linear approach is that shortcomings in the design are detected only at a late stage, when it is costly to make substantial changes.

10.3 V-shaped Process

Given that vehicles are complex systems consisting of several subsystems, the V-model or V-cycle is sometimes applied by car manufacturers, as shown in Fig. 10.3. The first version of the V-model was proposed in Germany in 1992. Modified versions were proposed later on in Germany and in the US.

The V-cycle model focuses more explicitly on the integration of design and test/verification activities throughout the design process. Despite this, the V-cycle takes a linear approach to the design process, as testing is done after requirements gathering and design is complete. Testing/verification of subsystems is done at the unit level, followed by testing/verification of the system. In addition, most testing is technical: the test is designed to verify that the system meets the internal requirements and constraints. Validation, i.e., testing whether the system meets the customer's requirements, is done through final acceptance testing. Again, at this stage it is too

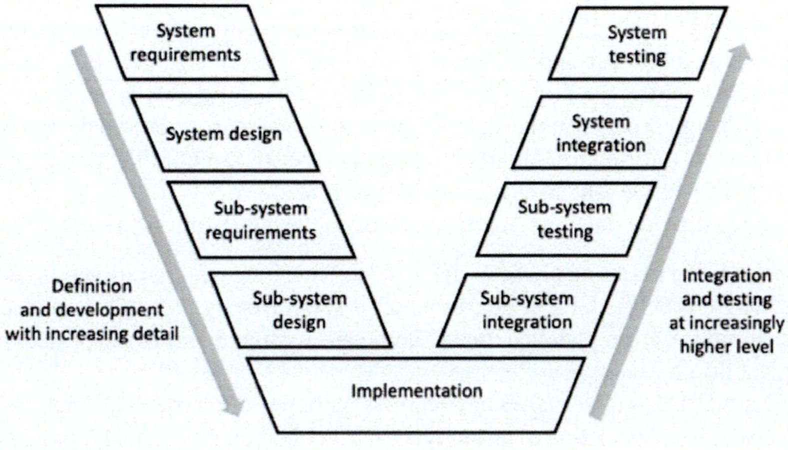

Fig. 10.3 V-cycle model

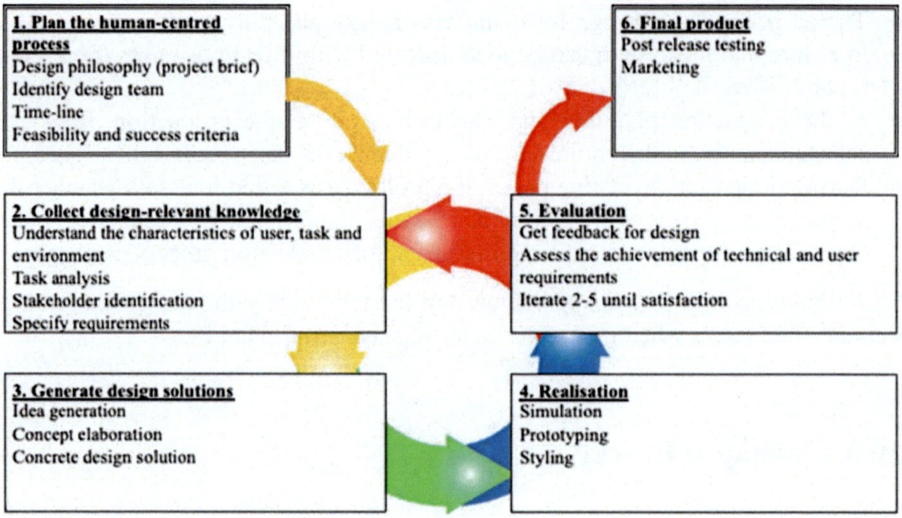

Fig. 10.4 User-centred design process

late to make substantial changes based on the results of the acceptance tests. Through developments in the field of simulation and model-based development, it has become possible to deviate from the strict linearity characteristic, which makes it possible to verify high-level designs before elaborating component subsystems, leading to a triple V-model of three interwoven V-models.[1] But again, most of the critical tests are purely for technical purposes.

10.4 User-Centred Design

The user-centred design approach was established in the 1990s to deal with the limitations of the linear approaches to design and ensure a more explicit focus on user feedback during the design process. Rather than testing only at the final stage of the linear design process for acceptance, it established a user-centred design process to obtain more information about and from the user throughout the design process. Figure 10.4 illustrates the user-centred design process.

Here, we provide a detailed description of each stage of the work.

- Phase 1 is the planning phase of the project, where the design rationale is defined, the budget is defined, the time-to-market is set, the design team is composed, and the communication plan is defined. These are written down in the project plan or project brief.
- In Phase 2, Design-relevant knowledge is gathered. This involves analysis of the context of use, consisting of an analysis of user characteristics, tasks, and physical,

[1] See https://insights.sei.cmu.edu/sei_blog/2013/11/using-v-models-for-testing.html.

social and organisational context. This may include gathering information requirements from users of the current system and potential users of the future system and gathering information about how users interact with the existing system in order to identify usability bottlenecks, for example by identifying typical errors. Also, at this stage additional stakeholders are identified that may contribute their own requirements and hence have to be consulted when design decisions need to be made. The outcomes of this process are translated into requirements, both functional requirements (what the system should do) and non-functional requirements (performance and usability criteria). If applicable, requirements emerging from previous projects (legacy) are added, and the functional requirements are checked for consistency with the legacy.

- In Phase 3, design concepts are generated and elaborated.
- In Phase 4, realisations are created of the design concepts through prototyping and simulations. The realisation also involves the styling ("look-and-feel").
- In Phase 5, the design solutions are evaluated. This involves technical evaluation, evaluation against the requirements, and evaluation with users: The design proposals are exposed to potential users, and they are invited to give their opinion.

  Phases 2–5 are iterated until the requirements are met. In later iterations, users are asked to interact with the system and perform typical tasks, so that possible operational bottlenecks can be identified.
- In Phase 6, the product is launched. After sales service is provided. Collecting user feedback may continue by means of user panels and/or web reviews.

While the numbering from 1 to 6 suggests a linear process, as stated in 5, the design process is iterative rather than linear and therefore feedback from users can be taken into account in later iterations. The iterative process will continue until the pre-defined quality standards (requirements) are met. Furthermore, it should be noted that the requirements gathering phase is also part of the iterative process, as shown in Fig. 10.4, so that requirements need not be fixed once and for all before the design process begins, but can be extended and revised later in the process. Finally, it should be noted that the user research does not necessarily precede the design activities. Instead, design teams may already start a short design round to come up with ideas and reflect on those ideas to organise their understanding of the domain. This may help to give direction to the user research, by sensitising the team to the domain and helping to refine the questions for the user research.

Typical methods applied in the different phases are as follows.

The analysis phase aims to gather information relevant to the design to support subsequent design decisions. From a system perspective, the context in which the system is used covers all the elements that influence the interaction between the user and the product, i.e., the users as well as the physical, organisational, social and technical context. As this phase is about gathering information about people, activities and the physical, organisational, social and technological contexts, it is also referred to by Benyon (2010) as PACT analysis (People Activities Context Technology analysis). How to do a requirements analysis is discussed in detail in Sect. 11.1.

Table 10.2 Different concept representations at different stages for different feedback purposes

Design representations	Desired user feedback
Paper prototype	Do users like the idea? Do they understand how to interact with the system?
Primary Prototype	Do users understand how to interact with the system?
Driving simulator test	How does the system enable the user to perform the intended task in the usage scenario?
Real car tests	How does the system enable the user to perform the intended task in the usage scenario?

The design and realisation phases concern the design activities. In each iteration, consideration is given to designing representations that express user needs and that provide the basis for presenting ideas to potential users to gather feedback. In the early iterations, the representations are global and rough (paper prototypes). As the project progresses, the representations become more realistic, moving from models to simulations to working systems with operational functionality. The general rule is always to spend appropriate effort to design a representation that is commensurate with the feedback sought at that iteration. Table 10.2 summarises the goals for user feedback at different stages of a practical design project.

The evaluation phase involves all activities intended to collect feedback about the proposed design solutions and about whether they meet the requirements. Under ideal conditions, feedback can be gathered from potential users outside the company. However, for reasons of confidentiality, it is often not allowed to consult external people at the early stages of the design process. Some companies therefore use so-called user representatives, i.e., company employees who are expected to represent potential users, because they perform similar activities (driving) in their daily lives, or because they have regular contact with actual users and can therefore be expected to have empathy with potential end users. Such user representatives can provide feedback at an early stage of the design process, rather than using external potential end users. Finally, workshops can be organised to conduct design reviews with panels of user representatives, and to collect feedback from user research teams or management. However, it needs to be clearly understood that these people do not represent the real users.

The nature of the feedback varies as the design project progresses. In the early stages, feedback is usually subjective and qualitative, for example, by asking test participants to think out loud during operation and conducting post-test interviews. In later stages, although qualitative subjective feedback can still be collected, the feedback may also become more objective and quantitative, for example, by collecting performance indicators (e.g., "How quickly does the system enable the driver to regain full control?") and quantitative subjective feedback, having test users fill in usability questionnaires, etc.

As mentioned above, initially the user-centred design approach aimed to ensure good usability of the system. Later, as the field grew to focus on the broader user experience, the user-centred design approach was expanded to include usefulness and pleasure as design goals. For usefulness, methods were developed to gather feedback from potential users on the potential usefulness of the system (e.g., "Does this concept make sense to people?"). As it is already difficult for designers themselves to imagine whether new concepts will be meaningful to people, methods have been developed to help people think about the future by making relevant elements of the context physical and relevant to their current experiences. The underlying assumption is that, by making current and future situations more tangible, participants can more easily be brought into future scenarios and think about the future. Furthermore, initial designs such as prototypes and models do not only aim to find out whether potential users can understand the logic of the interface, but also to help potential users understand the concept and consider whether it makes sense to them. In other words, the design activity focuses on a global view of the system, in particular answering the "What" and "Why" in the early iterations, and on the more detailed aspects of the interface (understanding the "How") in the later iterations.

To facilitate the gathering of information from potential users about the usefulness of a concept design, the involvement of potential end users can be restricted to so-called 'innovators' and 'early adopters', i.e., people who are open to new technologies. Rogers' Innovation Diffusion Theory (Rogers 1962) introduces the concepts of 'innovators' and 'early adopters'. The central tenet of this concept is that user groups are not homogeneous, but that users who are open to new technologies will be the first to acquire them as new users, introducing them to society and spreading the experience of using them through the media and personal communication. This can persuade the early majority to also consider buying and using these new technologies, and subsequently the majority will also consider using them. In the case of automated driving systems and electric vehicles, the early owners of Tesla can be seen as innovators and early adopters. Currently, automated driving systems and electric vehicles are in their 'early' stages. It makes more sense to discuss new concepts with innovators and early adopters than with the general public.

Various questionnaires have been developed to assess the overall user experience of a product, not just its usability. However, the question of whether a particular system is useful to people in their daily lives can only be answered by studying its use over a longer period of time. It is difficult to carry out such longitudinal studies before the system is actually introduced into the market. In some areas, the so-called "technology probe" method has been used. Technology probes are fully functional prototypes that can be used by selected test subjects/information providers for a week or two. Obviously, this generates more relevant feedback than a short user test of half an hour. However, in the context of automated driving systems, the production of such technical probes is challenging and expensive. Nevertheless, the use of a small number of prototypes given to selected users for a limited period of time to drive on different roads and in different scenarios can still provide very useful information for developers of autonomous driving systems. For instance, for some time Volvo has announced the Drive Me program with real families.

In summary, the concept of user-centred design has evolved from a focus on usability to a broader focus on the user experience. The aim of the user-centred design process is to ensure that information about and from users has a role in guiding design decisions at all stages of the design process. Of course, this does not relieve designers from the responsibility to use their own intuition and develop their own visions, as it is impossible to validate all detailed design decisions throughout the design process. However, the aim of the user-centred design process is to find the right balance between the designer's vision and the opinions of potential users. The fundamental goal of the user-centred design concept is therefore to develop a user-oriented attitude so that the designer feels the need to empathise with the potential user throughout the design process and to bring in the awareness of the user beyond his or her own preconceptions. The design solution is looked at from the perspective of the potential user with fresh eyes.

10.5 Agile Development Methods

Whereas in linear design methodologies and user-centred design methodology much attention is given to requirements specification and software documentation, in more recent approaches in software engineering the focus has shifted to actual software production for flexibility/agility and higher speed of software production. Typically, a larger project is divided into smaller projects, each executed in a short period of time ('sprints') to completion. Each sprint/iteration involves planning, analysis, design, testing and documentation and results in a usable outcome (feature or prototype). Sprints are executed by small, co-located teams of software developers, product managers, UX specialists and business analysts, usually with fewer than ten team members. Being in the same location allows for intense communication between team members. In agile methods, user requirements are replaced by so-called user stories, which specify what the user is assumed to need or want, and what value will be generated if the user's needs are met. The user stories are taken as starting point for the design team to create the functionality meeting the demands. Finally, the user stories are prioritised so as to ensure that the user needs with higher priority are addressed first. The created functionality is then taken by the UX specialists for testing.

Agile development differs from the conventional user-centred design approach, where extensive user requirements research is carried out in advance of product development; also, the process of agile development is a simultaneous development and integration of interaction design and system design. User requirements research and user testing are conducted in conjunction with technology development, and the product is continuously improved during the design process, thus ensuring that the product is designed to meet user requirements to a great extent. It should be noted, however, that later versions of the user-centred design process have included user research in the iterative process, therewith allowing user research to contribute and affect user requirements in later iterations.

Agile methods are particularly well suited for web and app development. Therefore, they can be applied well to the development of in-car infotainment systems.

10.6 Organisational Design Process

As mentioned above, a user-centred design process requires a user-centred attitude. Since the goal of a user-centred design process is to ensure that the design process results in meaningful products that are easy to use and satisfying to experience, user orientation cannot be limited to the human–machine interface (HMI) design department. Rather, a user-centred design process requires the entire company to work together towards the goal of improving the user experience of the product. To guide companies in organising their design processes so that they can best organise the work of the people involved to achieve this goal, a usability/user experience maturity model (ISO 18,529, 15,504) has been proposed, inspired by the software engineering maturity model. Table 10.3 summarises the company-based usability/user experience maturity model.

At level X, there is a lack of awareness throughout the company of the need to focus on the user in the design process. At level A, there is awareness, but the adoption of

Table 10.3 Company-based usability capability maturity model

Level X	Unconscious	No recognition of user-oriented needs. Issues in user satisfaction have not been brought to the attention of the company internally
Level A	Awareness	Problems with user satisfaction have led to concern within the company and recognition of user-oriented needs. This has led to system practices aimed at gathering user requirements and including them in the design process, but not routinely
Level B	When considering	People are employed who are considered to be trained in human-centred approaches and HCI and who are aware that human-centredness is not only about the interface but also about the whole system
Level C	Implementation	A human-centred process has been implemented. Staff with training in human-centred approaches are involved in all stages of the design process
Level D	Integration	The Human Factors department or HMI department successfully interacts with other departments in the company. Evaluations are conducted at all stages in a timely manner. An iterative design process is established so that feedback from the evaluation may later influence design decisions
Level E	Institutionalisation	The key role of human-centred skills in integrating the design process is recognised. Setting up the design and manufacturing process to transform the company into a learning organisation: implementing quality assurance methods in the organisation to improve its own processes

user-centred design methods and processes is still ad hoc, and it is not yet well established that user needs should be collected and considered in the design process. At level B, there may be a department that cares about user requirements (mainly about usability requirements) and human–machine interfaces, but the rest of the company does not consider that department to be essential. At level C, human factors experts may be involved as consultants in all stages of the design process, not only regarding usability and HMI design, but also practicality. At level D, other departments are convinced of the value of the HMI department and the design process is organised as an iterative process so that feedback can influence further design processes. At Level E, the human-centred design process is institutionalised and quality control mechanisms are in place to provide continuous feedback and to guide the organisational process. In summary, the Usability/User Experience Maturity Model expresses the view that a user-oriented orientation is not only relevant to the HMI sector, but that a human-centred attitude should be adopted by all those involved in the design process.

10.7 Rational Design and HCD

In this context, 'rational design' means that when a new epoch-making technology is created through technological change, or when different combinations and changes are made to an existing technology, this breakthrough may offer the possibility of new functions and applications, and the new technology brings about a new design. Such designs are not necessarily demand-driven. For example, we commonly use televisions, refrigerators, washing machines, etc. in our homes. In our current homes, these three items are still there. But new technologies have enabled changes in design and functionality, making the products fundamentally different from those of their counterparts 10 or 20 years ago. The constant change in technology brings with it the possibility of new functions and new user needs.

The introduction of new technologies gradually changes the meaning of the product. Twenty years ago, the car was still a means of transportation, and in most cases, it is still mainly a manually driven car, but it has given the driver many different possibilities and experiences. The task is easier than before. Onboard navigation systems have also become smarter. These changes have come about through the creation of in-car sensors and chips against reduced costs, and the refinement of various algorithms. These technological developments will also play a large part in the changes in traffic conditions now and in the future. In turn, such changes cannot be achieved without the development of smart network technologies, which are not only reflected in the automotive industry, but also in the lives of everyone, and in the management of society and the transformation of social culture. As a result, cars are no longer just a means of transport, but also a space to engage in all kinds of activities, preparing for work, staying in touch with people elsewhere and so forth. Smart cities, smart homes and smart mobility are among the biggest products of this. As such, it in turn drives the development of automotive technology and design.

What is the relationship between this kind of design from 'reason' and Human Centred Design (HCD)? The initial designs are generally rational, i.e., they are based on technological innovation, or they are new designs that combine different technologies in new ways. But this design does not generally achieve the desired quality or user experience. In order to provide quality and user experience, it is necessary to introduce the HCD mechanism and reach a higher level of quality through user-centred design. In order to further develop the product, technological changes are established and even the nature of the product may be fundamentally altered. Then, the HCD design process is introduced again to meet the desired quality and user experience.

It is clear from this that the relationship between innovation triggered by technological change and the improvement of product quality through HCD is not one of contradiction but of mutual enhancement.

10.8 Post-UCD

Traditional User-centred Design, as developed in the 90 s, has been criticised for allocating a rather passive role to designers. The focus of the user research was on identifying needs of the users, in terms of problems that users encountered, and the role of designers was to solve these problems through design solutions. Somewhat mockingly, traditional UCD was characterised as "Tell me what your problem is, and I'll solve it for you".

The view of UCD as problem solving was already nuanced by the UX wave (most notably Jordan 2002, and Hassenzahl 2010): According to Jordan, designers should not just solve problems but aim to design pleasurable products. And according to Hassenzahl, user needs were not just problems that users reported or that could be identified by observing users, but were to be interpreted in terms of deep psychological needs such as autonomy, competence and relatedness, so that the scope of user research broadened from identifying user problems and focusing on cognitive abilities and limitations to trying to understand how people pursue personal values.

Further criticisms of traditional UCD were brought forward by Verganti (2009), who makes a distinction between incremental and radical innovation. Incremental innovation is indeed aimed towards solving problems that users report or that can be identified by observing users, either with existing products/systems or with prototypes in user tests. On the other hand, radical innovation is primarily driven by new technologies, which provide designers with opportunities for radically different ways of performing everyday activities and radically new types of applications and product eco-systems (cf. the Technological Innovation phases of Rational Design as discussed in Sect. 10.7). Verganti argues that, in order to be successful, companies should spend a substantial part of their work on radical innovation (cf. Apple and Tesla). Furthermore, Verganti argues that, in relation to radical innovation, user research in the sense of finding out what users need or want runs into limitations, because users are unable to think into the future and formulate needs or wants for radically new applications. Instead, Verganti argues that radical innovation is design-driven rather than user-driven, or, otherwise said, opportunity-driven rather than need-driven. In order for companies to be innovative, designers should develop visions and translate these visions into product-service eco-systems. Asking users to provide feedback on such radically new applications would be beside the point, as they would not be able to provide meaningful feedback. Instead, getting the visions accepted and adopted by society may well take several years.

One might argue that this view is a bit too harsh, and that it might be preferable to help users to think into the future, as for instance is done by Pettersson (2017) and Buskermolen and Terken (2012). However, the point remains that a view according to which designers are problem-solvers does not do justice to the qualities of designers, and under-illuminates the role of a design vision.

Thus, in post-UCD approaches, the design process is seen as comprising at least four different sorts of activities: analysing (mostly involving user research), envisioning (involving the development of design visions), ideating (involving design activities) and validating/evaluating (providing arguments to support, modify and refute design visions and the products/systems/services emerging from them—again, this involves user research). Furthermore, there is no strict order in which these activities are performed. Instead, the different types of activities may be conducted in any order, continuing throughout the iterative design process. In this way, the design vision is developed throughout the design process, guided by feedback from user research, and resulting in an outcome for which the value for the user can be supported by convincing arguments. This also means that the development of a solid design vision requires thorough understanding of the user.

References

Benyon D (2010) Designing interactive systems: a comprehensive guide to HCI and interaction design, 2nd edn. Pearson Education

Buskermolen DO, Terken J (2012) Co-constructing stories: a participatory design technique to elicit in-depth user feedback and suggestions about design concepts. In: Proceedings of the 12th participatory design conference: exploratory papers, workshop descriptions, industry cases, vol 2, pp 33–36

Crabtree A, Rouncefield M, Tolmie P (2012) Doing design ethnography. Springer, London

Hassenzahl M (2010) Experience design: technology for all the right reasons. Synth Lect Hum Cent Inform 3(1):1–95

Jarvis N, Cameron D, Boucher A (2012) Attention to detail: annotations of a Design Process. In: NordiCHI'12, p 11

Jones JC (1992) Design methods. Van Nostrand Reinhold, London

Jordan PW (2002) Designing pleasurable products: an introduction to the new human factors. CRC Press

Pettersson I (2017) Travelling from fascination to new meanings: understanding user expectations through a case study of autonomous cars. Int J Des 11(2):1

Rittel HWJ, Webber MM (1973) Dilemmas in a general theory of planning. Policy Sci 4(2):155–169

Rogers EM (1962) Diffusion of innovations, 1st edn. Free Press of Glencoe, New York

Sharp H, Rogers Y, Preece J (2019) Interaction design—beyond human-computer interaction, 5th edn. John Wily & Sons, Inc

Verganti R (2009) Design driven innovation: changing the rules of competition by radically innovating what things mean. Harvard Business Press

Chapter 11
Analytic Methods

Abstract The design process involves different activities: analytical/research activities, design activities and validation activities. In this chapter, we go into methods to support the analytical activities, and summarise methods for doing user research and for analysing qualitative data. The outcomes of the user research may be captured by means of user profiles/personas, requirements and use cases and scenarios. Finally, we go briefly into methods for conducting task analysis.

11.1 The Five Key Elements of User Research

In interaction design, many people focus on how to design cooler graphics, thinking that this is the way to improve the user experience. But in fact, the form always serves the content. User research, therefore, is one of the most important aspects of interaction design methodology. Unfortunately, there are many interaction designers who never have any contact with real users and do not really understand their needs, but only determine their own interaction design solutions through indirect means, research reports done by other organisations, or rough research and ambiguous statements by suppliers, and by imitating the designs of benchmark cars. There are some designers who have never even driven a car themselves. Interaction design is an art with a strong scientific background, and one's eagerness to capture the real needs of the user and understand them is the key to good interaction design.

To get a true picture of what users want, you need a good methodology to do user research. There are five key elements to consider before choosing which method to use.

1. ***Establishing the purpose of the research***. User research serves many different purposes, and it is only when the purpose of the research is established that it is possible to select and design an appropriate research method. The purpose of the research can be varied. For example, do you want to find out which parts of the interaction of a car that is already on the market are preferred and used by users, and which parts are disliked and not used at all? Maybe you want to find out whether the user prefers visual or auditory information when using navigation? Do they prefer 2D maps or 3D maps? In what situations do

© China Machine Press 2023

F. Chen and J. Terken, *Automotive Interaction Design*, Springer Tracts in Mechanical Engineering, https://doi.org/10.1007/978-981-19-3448-3_11

users prefer to use the voice system? It could even be about specific interaction modes, or where and how information is presented. It could also be research into which new technologies drivers would very much like to use while driving. Different research methods are used for different purposes, different people are researched, and different data are collected.

2. ***Establish whom to research***. Once you have identified the problem you are researching, you can then identify whom you want to research. Sometimes the target group is very specific, for example, you need to find out how many years of driving experience people have with a particular car. Sometimes this is a relatively small group of users, but sometimes it is a very large group. In any case, no researcher can study every single person in a defined population, so he has to make a selection, and this selected group is the "sample". The selection of this sample will have a direct impact on the results of the research. Therefore, in order for the results to be free of any particular bias, a 'random sample' is the best approach in the strictest sense of the word. In theory, random sampling means that everyone in the user population has an equal chance of being part of the sample. In practice, however, this is very difficult to do, so in most cases the sample is obtained by 'convenience' or 'voluntary' methods. In practice, you put the message that you need a sample of people, and the conditions of the group, in a place where most people are likely to see it, and those who are suitable will sign up voluntarily to become part of your sample. On the other hand, as a researcher, you cannot choose your sample based on your own personal preferences. Random sampling is an important prerequisite for making the statistical analysis of your research data valid.
3. ***Relationship with the researched***. It is important to maintain a clear professional relationship between the researcher and the researched. There should be no personal feelings or other agendas or factors involved. In this way it is possible to ensure that the data are impartial. How this can be done varies from country to country; in the USA and the UK researchers often sign an agreement with the researched person (Informed Consent Form). In Scandinavia, no such written agreement is required, but there are verbal instructions. In any case, the researcher needs to be clear to the respondent about the purpose of the research and how the data will be used, and also to ensure that the respondent has complete freedom to interrupt the research process at any time. It is important that there is an understanding before the research begins that the researcher needs to know that the data he is collecting can be used where it is needed, and that the respondent needs to know that the data he is providing is not being used in a way he does not want it to be used. This mutual respect is very important to the research.
4. ***Multifaceted research*** (Triangulation). Triangulation means that there is no single approach to user research. There are four ways of being multi-faceted: (a) multi-faceted data sources, which means that data should not come from just one source, but should be searched for from different sources, collected at different points in time, in different locations and from different people; (b) multiple methods are used to collect the data; (c) different researchers are involved in

the collection of data; and (d) multiple theories are used to guide the collection and analysis of data. This multifaceted research is used to verify whether results from different sources, methods, researchers are consistent, which will enhance the reliability of the results.

5. ***Pre-experimentation***. After the above points have been confirmed, it is also necessary to try the data collection exercise on a small scale in order to check if there are any missing areas or problems that may arise during the practical exercise. The staff are also given training in the methodology.

The tools used for data collection will not be explained in detail here. However, when using audio or video recording equipment, it is important to seek the consent of the respondent and to promise that the data will not be used for purposes other than those promised, let alone for any purpose that may adversely affect the respondent (e.g., posting videos with identifiable participants on social media).

11.2 User Research Methods

Through the last decades we have accumulated a number of methods that can be used in interaction design. The key question is how to choose the right methods and how to use them. Methods can be classified in different ways. There are research methods for obtaining user requirements, design evaluation methods for validating designs, laboratory research methods, methods for operating in real car situations, static methods, and dynamic methods. There are also many books on methods, so we will only describe some of the commonly used methods here.

It is important to stress here that any method has its strengths and weaknesses. For example, insights can often be gathered through user research, focus group interviews and experiments with driving simulators before a new technology is introduced into society. On the one hand, this allows researchers and developers to gather user insights before the technology is launched and to use these insights to guide the development of the technology. On the other hand, there are disadvantages to this research method. In particular, surveys are usually conducted with people who have no experience with the technology, so their feedback may not be very useful. When experiments are conducted in driving simulators, the scenarios are usually artificial, safety is not an issue and the duration of the experimental phase is usually short, so they do not provide information about how the technology would be used, mis-used or abandoned in real life. The safety of the technology in question for use in vehicles requires a much longer study. Ideally, longer experiments should be conducted in the field, not by trained testers, but by 'ordinary users' (usually volunteers). It is therefore reassuring to see that many OEMs have started to conduct larger field tests to collect data, and that these studies are not only about technical performance but also about user experience. However, such tests can only be carried out if the technology is already sufficiently advanced. Because of the desire to gather insights into future user experiences early in the design process from a user-centred design perspective, which can be used to guide technology development, compromises need to be made. This compromise should be based on insights into the strengths and weaknesses of different approaches. Table 11.1 shows the advantages and disadvantages of some of the different approaches, which can be used as guidelines for choosing an approach.

The statements in Table 11.1 are very general in that they are abstracted from specific research questions. For example, if one is interested in understanding the public opinion on automated driving before it has been implemented, without having to understand the nature of the technology, then conducting questionnaire research is an appropriate approach. However, once automated driving technology has been introduced into society, people may question whether the public opinion that emerged from this research is still relevant. In previous research, many people were found to be concerned about the trustworthiness of automated driving systems. However, once self-driving technology is introduced into society, people are likely to receive more information about its performance from early adopters and the media, which will influence their opinions. Also, the survey may provide more trustworthy information if it is targeted at early adopters rather than the general public. In conclusion, it is

Table 11.1 Advantages and disadvantages of the different methods

Methods	Advantages	Disadvantages	Note
Literature reading	Highly efficient, a lot of information in a relatively small amount of time	The information obtained may not be directly applicable to the design project	This is an essential step at the start of any project
Questionnaire research	Understanding the wishes of the public allows for a large amount of data to be obtained in a limited time frame	Limited external validity as the general public may not understand the nature of the technology	The results obtained may not be significant if the people asked have no experience of automated driving systems
Spotlight interviews	A study to identify the possible advantages and disadvantages of the application of the technology	The results obtained can only be seen as a hypothesis and need to be further verified by other methods	Subjects need to be helped to imagine the future
Driving simulator experiments	Research on the different levels of automated driving systems (1–3) is very important	It's a designed scene, so it can be a bit unrealistic	Simulator driving time should ideally not exceed 0.5 h. The fidelity of the simulator has an effect on the results
Live testing at the test track	A great way to collect data on the driving experience	Driving scenarios still differ from real road conditions	Need to simulate road scenes with different materials
Practical road tests	Very good method for accumulating long-term usage data	More technically demanding	Obtain permission and be safe

recommended to verify on which technical information the respondents base their answers to the questions. If the information is only very concise, there may be good reasons to be cautious about the validity of the results.

As said before, the different methods have their own strengths and weaknesses. Methods are tools, and how they are used depends on the questions and conditions that need to be answered, and the experience of the operator. In this chapter, we will consider the methods listed in Table 11.1 and a few additional ones in more detail, with the exception of driving simulator experiments, which are described in great detail in Chap. 14.

11.2.1 User interviews

Interviews are a common user research method, often used in the requirements research and analysis phase of interaction design. Interviews are used in many user experience studies. The design team can gain valuable user experience narratives from them. For the same reason, there are different types of interviews, the main differentiating factor being the predetermined plan and structure introduced during the interview process. (1) Unstructured interviews, where the questions of the interviewer are inspired by the moment and the preceding part of the interview, are not pre-framed much before the interview starts, the only guiding factor being the overall topic of the interview. The interviewer is then free to direct the interviewee towards meaningful topics through the questions asked. Unstructured interviews can produce very different results based on individual responses, so it can be more difficult to draw some sort of generalised conclusions or patterned answers in data analysis. (2) A structured interview has the most scope specification, works with a pre-defined list of questions, asks many specific questions and cannot change the questioning in any way once the plan has been developed. (3) Semi-structured interviews are a good compromise between 1 and 2. They add more structure to the interview process by listing topics of interest and specific questions to be asked, but the interviewer can still deviate from the order of questions and the questions themselves and pursue potentially interesting topics into more depth. Semi-structured interviews have the potential merit of good coverage of the subject while feeling like a conversation.

User interviews are one of the most common user research methods and are widely used. In their article, Fontana and Frey call interviewing the art of science (Fontana and Frey 1998), but many people use this method without really understanding it and using it properly. For, no matter how carefully we organise our language, language itself contains a lot of uncertainty and incompleteness. The interview itself is therefore not just a tool, but a social art, which requires the interviewer and the interviewee to be on an equal footing. This involves two issues: (1) how the interviewer asks questions, how the interviewee is guided, and how the two interact with each other. (2) How the interviewee understands the questions asked by the interviewer, and how the interviewer understands the interviewee's responses. During the interview, there is not only spoken language between the two, but also body language, tone of voice, facial expressions, which are attached to the spoken language and may give it a different understanding. There are therefore a number of considerations when conducting an interview.

- Do not provide excessively long explanations of your questions; it is best to prepare relatively standard explanatory text.
- If the interviewee does not understand your question, you can reformulate your question.
- Try not to deviate from the dominant content of the question and to change the order in which the questions are asked, unless there are good reasons to do so.
- Do not allow others to interrupt the interview process, add other content themes and do not allow others to annotate the interpretation of the questions asked.
- Do not suggest an answer and do not give your opinion on whether the answer is right or wrong, and whether you agree or disagree, but remain as neutral as possible.
- During the interview, it is best for you to remain gentle and it is best not to play with your phone in between, otherwise the interviewee will think that you don't actually care about his point of view.
- Nodding and humming like in a normal conversation show the interviewee that you are attentive and may help to encourage the interviewee to talk. Summarising answers shows whether you have understood the answer, and usually also encourages the user to expand his answer.

  There are also a number of caveats when conducting interviews that are particularly important when working with interviews.

- Setting up and accessibility of the interview environment: not all interviews need to be done in a laboratory or office; some need to be conducted in the familiar surroundings of the interviewee, such as a driver who is driving. Getting the driver to allow you to sit in the passenger seat while accompanying him in the car and interviewing him then requires good preparation and acceptance by the interviewee.
- Understand the culture and language of the interviewee: this is sensitive and difficult; if you are asking questions related to technology but the interviewee does not understand the vocabulary, you need to say it in a way that is common to the interviewee, which will also give them a sense of closeness and identification. Be particularly careful about what can and cannot be said and not to touch their sensitive bottom line.
- How you introduce yourself: how you introduce yourself amounts to creating an image of you and your relationship with the interviewee, which will have an incalculable impact on the outcome of your interview. Do you introduce yourself as a student in a university? A leader of a company? Or a woman facing another woman, or do you put yourself in the position of having come to receive an education?
- Gaining trust: It is not necessary to gain the trust of the interviewee in every interview, but in a significant number of interviews trust is a very important cornerstone. With trust, the interviewee will come forward and say what he or she really thinks.

There are 3 main ways of asking questions in the interviews.

1. The questions are closed-loop, meaning that the user only has to answer "yes" or "no". For example, you ask "Is system X useful to you?"
2. Open-ended questions, which are asked to understand the reasons behind, for example, "What do you think about the usability of system X?"
3. Prompting questions, which usually follow the previous question, such as "Why do you think system X is not easy to use?"

The main thing at the beginning of the interview is to get both parties to relax and be natural. A good way of making the interviewee feel at ease is not to start with the interview right away, but do some small-talk before. Once the interview starts, it is best to start with a more general topic and introduce it clearly and naturally. Transitions to the next topic should also be clearly marked.

Interviews do not necessarily have to be one-to-one, but can also be conducted in small groups. Most group interviews are unstructured or semi-structured, giving the group members some room to play and flexibility, from which more ideas and issues can be identified. However, there are also disadvantages to group interviews, as the different personalities and social status of the group members can result in one or a few people dominating the group discussion, leaving others' ideas unheeded or not even given the opportunity to raise them. It is also easy to stray from the topic and be taken in directions of interest to certain individual interviewees. The organisational guidance of the interviewer is therefore important.

Advantages of using user interviews are.

- The possibility of obtaining more information about the user.
- The ability to feel relaxed and flexible.
- The interviewer has control over the pace and direction of the interview.
- The data can be statistically valuable if prepared well in advance (in particular in the case of structured interviews).

Disadvantages of user interviews.

- Methods are more time consuming and therefore difficult to achieve large amounts of data.
- The credibility and validity of the data may be difficult to demonstrate.
- Data may be entrained with the personal preferences of the interviewer and the interviewee.
- The data analysis process is also laborious and time-consuming.
- Sufficient interviewer skill is required, and the skill of the interviewer determines the quality of the data.

11.2.2 Focus Groups

Focus groups are a special form of user interviews. They consist of bringing together a small group of targeted users (typically six to eight), and setting up a conversation and/or discussion of the topic under discussion. The topic under discussion may be the targeted activity or technology (for instance "automated driving"), or a collection of design proposals, and the goal of the focus group is to identify main issues/concerns and majority and minority opinions. Preferably, if there are differences of opinion between the participants, the discussion should lead to consensus, but this is not necessary, as it is already valuable to learn what differences of opinion there are. Although several focus group sessions (up to three or four) may be organised, the total number of participants remains small, and the goal is not to achieve representativeness against the population.

The task of the facilitator(s) is to moderate the session and facilitate the discussion, guiding the discussion through questions, making sure that the discussion is not dominated by one or two participants, to ensure transitions to the next phase in order to prevent discussions from carrying on with little added value, and to summarise the outcomes of the discussion.

The session typically involves the following stages:

- introduction/welcome, including informed consent, goals and rules,
- introduction of topic/design concept(s),
- discussion (guided by questions),
- steering towards consensus (optional),
- wrap up, thanks and follow-up.

A focus group session typically lasts about an hour.

An advantage of the focus group method is that it is time-efficient, as opinions from several users are obtained in a relatively short time span, and that the discussion towards consensus may help to sharpen the opinions and motivations of users and give them more depth. A disadvantage is that, since only one participant can speak at a time, most of the time participants are actually not speaking, making the session less efficient in terms of eliciting information from the individual participants.

11.2.3 Questionnaires

Questionnaires are a well-established technique for gathering demographic data and user opinions. They are similar to user interviews and can also have closed or open questions. Every effort and skill is needed to ensure that the questions to be asked are clearly articulated and that the data collected can be analysed effectively. Questionnaires can be used on their own or in combination with other methods to clarify or deepen understanding. The method used and the questions to be asked will depend on the purpose of the research, the context, and the background of the user. The questions asked in a questionnaire can be similar to those used in a structured interview. One advantage of questionnaires is that they can be distributed to a large number of people, so that a large amount of data can be obtained on general views.

The design of the questionnaire will generally have a set format, for example, the questions will begin by asking for background information such as gender, age, family status, educational background, cultural background and details of the user experience such as the age of the car, skill level of the driver, experience of using different technologies etc. The following are general recommendations for designing the questionnaire.

- Make the presentation of the questions clear and unambiguous.
- If possible, ask closed questions and provide a range of answers for the respondent to choose from.
- Consider including the "no comment" option.
- Consider the ordering of the questions. The answers to questions may be influenced by the order in which they are asked. General questions should come before specific questions.
- Avoid complex multiple questions in one paragraph. It is best to ask only one question in a sentence.
- If using numbers to quantify answers, make sure the ranges are relevant to the subject, not more specific than needed, and do not overlap. E.g., if precise age is not needed, consider asking users to indicate to which age group they belong. Also, do not specify ranges as "0–10 years/10–20 years" etc., but "0–10 years/11–20 years" etcetera
- Make sure that the order of number scales is intuitive and consistent, and be careful about using negative numbers. For example, when asking for the user's opinion on a scale of 1 for low and 5 for high, all questions should be consistent from beginning to end. It is also preferable to have all questions be 1 for low and 5 for high. However, it is also possible to deliberately alternate the labelling of the numbers to avoid users not looking at the question carefully and identify response biases.
- When most questions are phrased in an affirmative statement tone and a few are phrased in a negative statement, there is a risk of users misinterpreting the question. However, some researchers have argued that changing the direction of the questions helps to check the user's intentions.
- Avoid the use of jargon.
- Consideration may be given to the need to use different versions of the questionnaire for different groups of people.
- Clear instructions are to be provided on how to complete the questionnaire.
- The questionnaire can be carefully worded and well laid out to make the information clear.
- A balance must be found between the use of white space and keeping the questionnaire as compact as possible.
- Longer questionnaires take more time to answer, which can lead to many people refusing to answer, or not answering the questions that follow.

A general model for designing questionnaires is given in Table 11.2 (Stanton et al. 2005).

There are two important issues to consider in advance when using questionnaires. Firstly, how many questionnaires do you need to know to send out to achieve a sufficient sample size? For large scale research, if the results need to be representative

Table 11.2 General models and examples of questionnaires

Type of problem	Case studies
Multiple Choice Questions	How long do you rate the average weekly driving time? 1–2 h, 3–4 h, 5–6 h, > 6 h
Measuring by numbers	I think the system is too complicated Strongly agree (5), agree (3), disagree (1)
Double comparison	Which did you find more mentally taxing, Task A or Task B?
Continuous measurement	What do you think of the usability of this system? (Rate any) 1 (very bad), … 10 (excellent)
Open-ended questions	What are your thoughts on the usability of this system?
Closed questions	Which of the following types of errors do you often encounter? 1. Can't find the position; 2. Misunderstood; 3. Press the wrong key
Filtered questions	Have you encountered any difficulties in operating the system? Yes, no (If yes, continue with question 11)

of the population, sampling techniques need to be used to select potential respondents. Usually, if project budget allows, a professional agency is hired to decide on the sample size, and to select a representative sample of sufficient size. The second is what is a reasonable response rate? Ensuring return rates is a well known issue in questionnaire research. Generally, surveys accept a return rate of 40%, but in reality the return rate is usually much lower. Some ways to encourage users to answer the questionnaire include: designing the questionnaire carefully so that participants do not get bored and give up. Provide a short overview section and tell respondents that if they don't have time to answer the whole questionnaire, it is fine to fill in just the short version. This will ensure that you get something useful in return. If mailing a printed questionnaire, include a self-addressed, stamped envelope for return as well. Explain why the questionnaire needs to be completed and ensure anonymity. Follow up with the respondent by letter, phone or email. Offer incentives, such as payment incentives, or small gifts.

Online questionnaires are becoming increasingly common because they can quickly and easily attract the attention of large numbers of people. There are two types: email and web-based. The main advantage of email is that you can target specific users. However, email questionnaires are usually limited to text, whereas online questionnaires are more flexible and can include checkboxes, drop-down and pop-up menus, secondary screens and even graphics. Online questionnaires can also provide real-time data validation. The advantages of questionnaire research are:

- Access to data from a large number of users in a relatively flexible way.
- If the questionnaire is well designed, data analysis can be very quick.
- No complex resources are required, and once the questionnaire has been designed, it can be delivered through multiple channels.
- Many questionnaire designs are readily available, such as QUIS, SUMI, SUS, etc., and these ready-made questionnaire forms can be very helpful for cross-sectional and longitudinal comparisons.

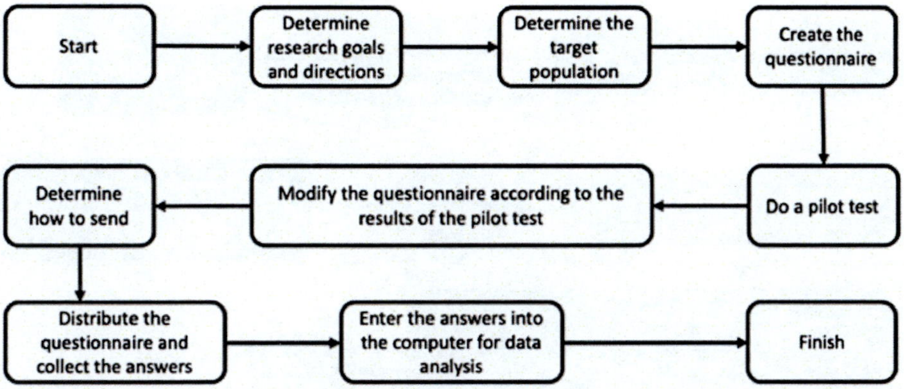

Fig. 11.1 Operational flow of questionnaire research

- Easy to manage and operate.

However, there are many limitations to questionnaire research, such as

- Questionnaire design and doing small tests to validate the questionnaire design is one of the more difficult steps.
- The reliability and validity of the data remain difficult to verify.
- The response rate to the questionnaire is generally low (usually around 10%).
- There is no guarantee that the person answering the question is thinking hard and answering the question.
- The data available are limited and there is also the potential for bias.

Figure 11.1 summarises the operational process of the questionnaire research (Stanton et al. 2005).

11.2.4 User Observation

The first thing to determine when preparing a user observation study is what the purpose of the study is. As with designing user interviews and questionnaire forms, only if the purpose is clear can we further define what needs to be observed. For example, before a new car model is designed, we need to understand the strengths and weaknesses of the benchmark car and how the user operates it. Also, we need to understand the user experience of our other models to understand which strengths need to be inherited and which weaknesses need to be improved. User observation is essential at this stage. The user observation method is also used if the use of different prototypes is evaluated in the course of product development. The aim may be to see if the user needs are met and if the product usability is good enough, so that the purpose of the observation is very different from that during requirements research.

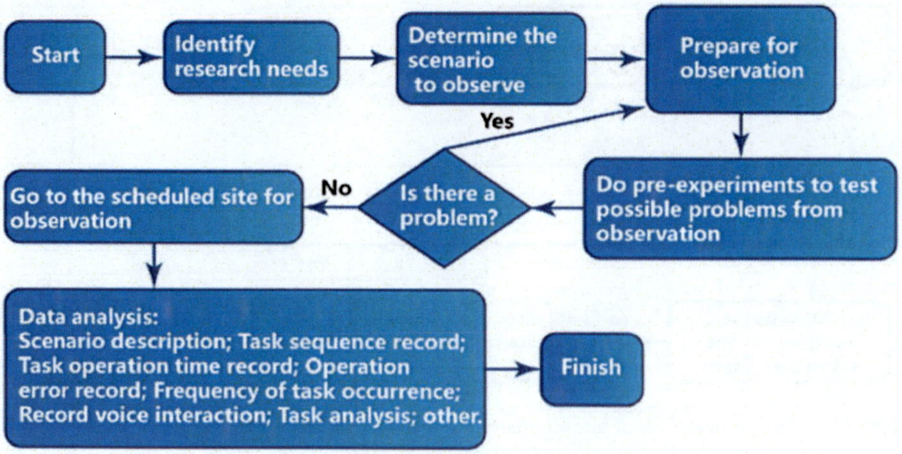

Fig. 11.2 User observation operation flow

User observation can take place at any time and in any place, sometimes without the need for detailed planning and programming. This method is also called the "quick and dirty" method. In this case, you have a pen and paper at hand to look at and ask questions at the same time. However, most formal user observation research takes place in two places: in the laboratory or at the application site.

For automotive design, laboratory observations typically occur in experiments using driving simulators, and the use of driving simulators to study user experience is described in detail in Chaps. 14 and 15. The main focus here is on observational studies in application sites. For automotive design, this observation usually takes place in the cockpit, with the observer sitting in the passenger seat and usually with a camera recording the driver's actions and road conditions, while the driver drives on a real road and the observer observes and asks various questions. Alternatively, the observer only maintains the conversation and the recordings are made by an assistant in the backseat (who may also control how the system works).

Audio and video recordings are indispensable devices when observing users. The user may feel uncomfortable with these devices at first, but it will not be long before they forget that they are there. Of course, the researcher needs to reassure the user that these recordings and videos will not be used for purposes other than the research itself.

Here we summarise the issues that need to be considered when conducting user observation studies.

1. Users: Who are the users we need to observe? What are their characteristics? What role do they play in the use of the product?
2. Time: What time of day do they usually use the device? Is there a connection between the use of this device and other devices?
3. Occasion: In what context do they use the equipment? What are the physical characteristics of this occasion? What impact does it have on the use?

4. Goal: what do they do with this device? What happens in this? What will they say? What do they do?

One common approach to the study of car driving is to install multiple cameras inside and outside the car to automatically record the driver's activity over a relatively long period of time, and then analyse it in detail afterwards.

Figure 11.2 gives an example of a common user observation process (Stanton et al. 2005).

11.2.5 The Repertory Grid Technique

The Repertory Grid Technique (RGT) is one of the oldest and most popular attribute-inspired techniques. It is derived from Kelly's Personal Construct Theory (PCT) (Kelly 1955, 1969), according to which each person has his or her own unique view of the world, which is determined by inherent dissimilarities between entities in the world. RGT has been widely used since its inception. It can be used to study the real thoughts of the human mind and can also be used to make product comparisons. RGT is an interview-based analysis used to compare users' thoughts or perceptions of similar products, systems or devices. RGT can be used early in the design cycle to gain insight into how users perceive problems in a product and to indicate user requirements and design preferences, or to evaluate existing product designs in light of user attitudes. RGT works by showing people three instances of something and then asking participants to divide the three instances into two groups, one group of two most similar instances and the other group of one instance that is most different form the others (Fransella et al. 2004). Participants are then asked to explain why (in which respect) these two instances are similar, as well why the third instance is dissimilar. The method is versatile and the output of RGT can be either qualitative or quantitative data.

For example, we compare a number of cars, and the participant is presented with three different cars A, B and C at a time. She may then put car A and car B together and put C apart from the A-B pair, explaining that A and B have a high-tech feel that distinguishes them from car C. Once this concept has been developed, the researcher may explore the participant's thoughts further in order to structure and elaborate on product attributes (see Fransella et al. 2004), in order to understand more in-depth what motivates participants to give a particular statement; it requires the participant to abstract from the concrete stimuli (Gutman 1982). For example, we would ask why they thought that cars A and B had a high-tech feel but car C did not.

This process is repeated for all possible product combinations until no new attributes emerge and then a specific list of attributes is produced. Each attribute is then expressed as a bipolar construct, either binary or scalar, and the participant is then asked to rate each product on the list of attributes. Further analysis may then be conducted using exploratory techniques such as Principal Components Analysis (PCA) or Multi-Dimensional Scaling (MDS) (Osgood et al. 1957).

With the recent increase in interest in user experience (Hassenzahl and Tractinsky 2006), RGT has become popular in the HCI field. Hassenzahl et al. (2000) and Fallman and Waterworth (2005) used this method to evaluate the user experience with mobile technology devices. Hertzum et al (2007) used the RGT method to examine the differences between designers and users' perceptions of three different cultural contexts. A more elaborate description of the RGT is provided in (Stanton et al. 2005).

11.3 Data Analysis

The user research methods described above provide mostly qualitative data (quantitative data are provided by questionnaire methods such as the System Usability Scale and are usually collected in later iterations of the design process, see Sect. 13.5). Qualitative data consist primarily of text (statements by users and/or notes from researchers). Such data may be analysed in different ways. In the first place, they may provide answers to concrete questions, such as in the case of structured interviews. In this case, the analysis consists simply of summarising the answers by means of descriptive statistics, showing the distribution of the different answers. More often, however, qualitative data are used to shed light on motivations and concerns of users with respect to a particular topic, such as the introduction of particular ADAS or, more generally, automated driving systems. In this case, the interview data are usually recorded using audio and video recording, supplemented by the interviewer's notes at the time. The recorded data may be transcribed, or summaries may be made.

There are three primary ways to analyse the qualitative data: Thematic analysis, Card sorting and Grounded Theory. Grounded Theory is a more formalised version of Thematic analysis.

11.3.1 Thematic Analysis

The goal of thematic analysis is to classify the data and identify main themes and patterns in the data. Here, we assume that the data consist of text, either transcriptions of things the participants/respondents/interviewees said, or things they wrote, or notes by the researcher. If audio/video recordings were made, either verbal transcriptions can be made, or the researcher may listen to the recordings and make notes or summaries. The analysis proceeds usually as follows.

1. Reading the text materials to get a good impression of the data.
2. Grouping the data in a meaningful way, either by question or topic that was used to guide the conversation, or by participant group (e.g., males and females, or older and younger drivers).

3. Coding the data to arrive at themes. The impression of the data resulting from step 1 already may give a good suggestion about the set of themes, so these may be used in coding the data. However, additional codes may have to be added while coding the data. In this phase, it is important to stay close to the data, and one or a few iterations may be needed to arrive at a final set of themes.

   A relevant question here is what the unit of analysis is. If answers to questions are short (one sentence), then this can be taken as the unit of analysis. If the data result from unstructured or semi-structured interviews, the contributions of the interviewees may be longer, where sentences in spoken language are usually not clearly marked as in written language. In this case, one may assign a single code to a span of text that is judged to be about a single topic and assign another code only if the speaker starts talking about a new topic; often, but not always, clear markers are used by speakers to mark such transitions. Thus, the question of what the unit of analysis is, is answered in a bottom-up way.
4. Identifying the main themes, relations between themes, and patterns of results. This step consists in essence of building a "theory" from the data. The main themes indicate what the central opinions, issues and concerns of the targeted users are, and how they differ between groups. Relations between themes can be derived if participants mention one theme in relation to another.

   If only a few participants have been interviewed, one should be cautious about making strong statements about differences between groups. Usually, if it appears from the data that there are differences between groups, it is advisable to collect additional materials to have an acceptable number of participants for each group.
5. Arranging a session with colleagues to discuss the analysis and arrive at consensus. Alternatively, the whole analysis may be teamwork, but even then it is useful to discuss the analysis afterwards and arrive at consensus.
6. Reflecting on the results/reporting the results. The key results are identified and representative quotes are selected. Quotes give liveliness to presentations and reports and give the design team the feeling of getting close to the user.

An easy and practical introduction to thematic analysis is Taylor-Powell and Renner (2003).

11.3.2 Card Sorting

In the card sorting technique, the qualitative data are written or printed on separate snippets of paper (see point 3 above about the unit of analysis), and then the following procedure is applied. Often, this is done by a team of two or more people.

1. The first snippet is laid apart to form a pile.
2. Then, the next snippet is taken. If this snippet is about the same topic as the first snippet, it is put on the same pile. If it is about a different topic, then a new pile is formed.

3. The same procedure is applied with snippet 3. If it is about the same topic as snippet 1, it is put on pile 1; if it is about the same topic as pile 2, it is put on pile 2; else, it is used to start a new pile.
4. This procedure is repeated for all snippets. An optimal number of piles is 6–8. If there are too many piles, piles that contain closely related content may be put together. If there are too few piles, the sorting may have been too coarse, and a more fine-grained sorting may be desirable.
5. After the final set of piles has been produced, the different piles are labelled, giving the themes.
6. Reflect on the results/report the results. This step is similar to step 6 of thematic analysis.

As can be seen, the main difference with thematic analysis is that in thematic analysis the labelling is done for each individual element, while in card sorting the labelling is done towards the end. Thus, card sorting is somewhat more efficient, but thematic analysis may give a somewhat more refined understanding of the data.

11.3.3 Grounded Theory

Grounded theory (Corbin and Strauss 2008; Patton 2002) is a more formalised version of thematic analysis, and the aim is to arrive at a theory consisting of a set of hypotheses that emerge from the data. Grounded theory typically proceeds as follows.

1. Open-coding the data, proceeding line by line, and identifying useful concepts. Codes are generated applying the constant comparative method: Each new unit is compared to the existing data and codes are created to connect units. Data or codes may contradict, expand or support each other. If new data or a new code contradicts existing codes, the code may have to be adjusted, and the contradiction needs to be explained. If new data expand the existing codes, this is a sign that new information is obtained. If new data support the existing data, this is a sign that saturation may have been reached and that no new insights will be obtained by further coding.
2. Memoing and theorising. The researcher writes memos (running notes about the concepts and insights that emerge from the data).
3. Sorting the memos.
4. Integrating and constructing the theoretical model. The categories emerging from the previous steps are linked together, and a core category is identified to which all the other categories are related, giving rise to a theoretical model.

Because much of the analysis of qualitative data relies on subjective judgement, subjectivity can be a potential threat to the quality of the analysis. However, for qualitative data, complete objectivity is not possible and not necessary. Therefore, this type of analysis will rely on the analyst's a priori knowledge and his sensitivity, which can also provide useful information rather than just guiding the analysis (Dey 1999). Sensitivity, as opposed to objectivity, shows the ability to pick up nuances and clues in the data (Corbin and Strauss 2008). It can help analysts to develop an understanding of what is really going on in the data.

11.4 User Profiling

User profiles are now commonly used in many car companies to illustrate the characteristics of the user population. As user research often indicates that user groups are not homogeneous and that there are differences within groups, user profiling provides a means of capturing the main differences between different segments within a group. Most of this information is collected by the marketing department or an external agency. If an exhaustive user study has been carried out, information about the users may be provided to the design team in the form of detailed documentation, but designers are not fond of extensive reports, and prefer a more condensed and inspiring format. Here we focus on a particular type of user profiles, called personas (Cooper 1999; Cooper et al. 2003), and explain what elements should be included in a persona for interaction design and how, so that it can help design.

Initially, a persona is a brief description of a fictive user, with a name and even a portrait photo. It may include a short story about a typical day in the life of the person, often in the form of a narrative that tells the important characteristics of the person and their purpose for using the product to be designed. It may also include short summaries of their motivations, interest, preferences and pain points. The purpose of a persona is not to provide accurate information, but more to resonate with the user. By constructing several user profiles (two to four different user profiles are considered a good number to guide the further design process), it is possible to be sensitive enough to characteristics, motivations, preferences and attitudes and the way these vary within the population. One point to emphasise is that the different personas may represent characteristics of different categories of users within the population, and therefore may result in different design solutions to meet the needs of the different categories during conceptual design. Ideally, personas are based on extensive user research, but if there is a lack of resources to conduct such extensive research, the user profiles may be based on data from limited user research, or even inspired by some real people. Most importantly, the value of a user profile is in helping the design team to remain focused on the user throughout the design process. Figure 11.3 shows an example of a persona, to illustrate what it may typically contain.

Personas provide a way to communicate relevant user characteristics between the team conducting the user research and the design team. In addition, the process of constructing personas itself may become a tool for reflection. Workshops may

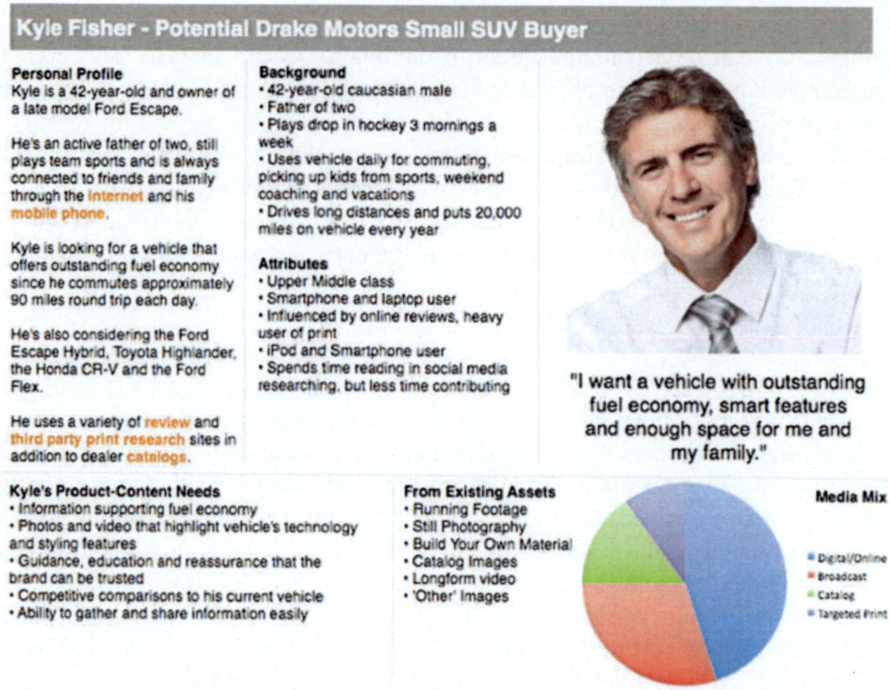

Fig. 11.3 What a persona typically entails. *Source* coolerinsights.com/2016/03/abcs-building-customer-profile/ Courtesy D. Eizans. Reprinted with permission

be organised, in which the user research team and the design team will exchange information about the users and then the teams will build the personas together. If the personas are prepared by the user research team and delivered to the design team, the opportunity for designer reflection will be missed.

By printing the personas as posters that can be displayed on the walls of the design studio, it is possible to ensure that the personas are always present, thus helping the design team to remain focused on the user. In addition, at key moments designers may ask the personas questions ("What would you think…, "What would you do in this situation?") and then identify with the personas to try and answer the questions themselves. In this way, personas provide a tool for reflection during the design process.

11.5 Establishing Requirements

Establishing design requirements is a key step in interaction design. Requirements research is about discovering the space for the problem that needs to be solved and identifying what needs to be designed. For interaction design, requirements research

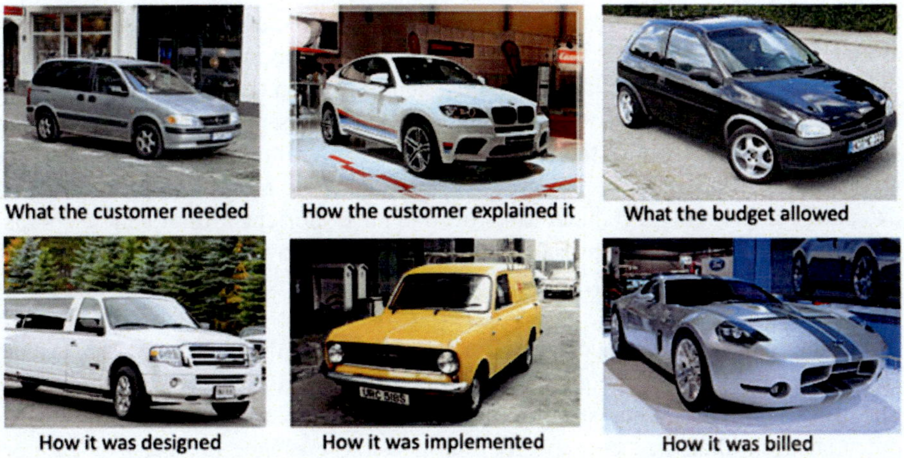

Fig. 11.4 The lack of clarity of requirements by all parties involved can result in a final product that is nothing like what the user wants. After an idea by Sharp et al. 2019. All pictures licenced under Creative Commons License

involves understanding the users of the product, their capabilities, how the product to be designed can help them, what tasks they need to perform, what goals they want to achieve, what kind of environment they operate in, what constraints there are, etc. The establishment of requirements is a guarantee of the correctness of the designed product.

The importance of requirements research can be illustrated in Fig. 11.4. If the requirements are not studied in depth and agreement from all parties is not ensured, the final product may be the opposite of what the user wants.

Requirements are a description of the specific properties of the product to be designed, what it is intended to do and how the user will use it. For example, a car navigation system is designed to help drivers choose their route prior to or while driving. So, we need to understand what the driver cares about when driving. He needs to know the distance to his destination, the remaining time, the road conditions, the possibility of choosing his own route, a timely reminder to change lanes, or enter a turn. If there is a traffic jam, is it possible to redo the route planning? These needs are reflected in the design of the navigation system. At the same time, as in-vehicle technology evolves, the needs of users will change. For example, when entering L3 level automated driving, users may want the navigation map to show which roads are eligible for L3+ automated driving. Or when the driver sets up automated driving, should the map be displayed differently to manual driving?

The design requirements of a product involve many dimensions, such as the users who use the product, the user interface which is used to interact with the product, the actions taken by the users when using the product, the relevant data contained in the product, the control system, the environment in which it is used, etc. An important aspect of the requirements study is the knowledge of the users themselves. This includes not only the direct users, but also all relevant interest groups, the "stakeholder" groups. Who are they? Their nationalities, educational backgrounds, attitudes to new technologies, etc. The use of any system includes different groups of skilled people, from novices to experts, and how often they use the product—do they use it only occasionally, or frequently? For novices, the system may have the responsibility of educating the user, including giving hints, clear constraints, clear feedback, etc. Expert users may want the product to be more flexible and to have extended functionality. Frequent users need to have shortcuts. Infrequent users may need clearer guidance or menu paths. At the same time, novices may become experts. And with the same system, some features will be used frequently, while others may only be used occasionally.

Different methods are used to gather design-relevant information about user characteristics to support subsequent design decisions. This relates to human perceptual and cognitive abilities and limitations, for example, the technical expertise of potential users that may be expected, the possible presence of visual or auditory impairments in the user group, their learning and memory abilities and their physical abilities and limitations (including dexterity), strength and range of physical movement. As many of these characteristics show correlations with age, it is also necessary to determine the typical age distribution of the user group as well as the gender distribution. It should be noted, however, that these correlations may have limitations. For example, while it is often assumed that older people may have a limited understanding of modern technology, this notion may not be correct. Studies of user abilities include physical abilities, human movement, and cognitive abilities. Examples include anthropometrics, the workload the human body is capable of carrying, etc. Human perception of speed, sound, colour, etc. may need to be understood. The detailed content of the requirements research should be linked to the specific design.

It is also necessary to identify the typical tasks or activities that the system should support. Such information may come from an analysis of existing systems or a conceptual description of a future system. The answers to these questions introduce the existence of other stakeholders, hence the stakeholder analysis. The stakeholder group comprises all those affected by the system.

The physical, organisational, social and technological environments in which technology is used are also areas that need to be studied. For example, it may be relevant to gather information about when and where the system is used. Should there be separate modes for day and night? Is there organisational support? Is there a service system that intervenes to help solve problems when users are unable to make the system work? Are there other people around (colleagues, family) who can influence the user's opinion? And what are the peripheral technologies with which the system needs to interact?

As mentioned above, the requirements gathering phase is part of the iterative design process and is an important part of what makes a project successful. In the initial user requirements study, user profiles/personas and storyboards are used to visualise the acquired user requirements so that the different members of the interaction design team have a common goal of understanding and communication (storyboards are described in Sects. 12.3). Information about the user represented by the

Table 11.3 Types of requirements and what they contain

Limitations of the design project	Pre-defined constraints Communication terms and specialist vocabulary Relevant facts and assumptions
Functional requirements	Scope of work Business data models and databases Range of products Detailed requirements for functions
Non-functional requirements	Visual and haptic needs Usability requirements Operational requirements Work environment requirements Maintenance and care needs Confidentiality requirements Cultural needs Compliance needs
Project-related requirements	Start-up projects Off-the-shelf solutions New issues Mission Migration to new products Risk analysis Budget and expenses User manuals and training Solutions

user profile, the activity represented by the scenes or episodes in the storyboard and aspects of the context can be refined or modified as more information becomes available during the design process.

We can divide the requirements we need to collect into three categories: functional requirements (including data requirements), non-functional requirements and user experience requirements. Functional requirements are, as the name suggests, the functions that the system needs to satisfy ("WHAT the user is able to do using the system"). These requirements are easy to understand and intuitive. Non-functional requirements are the overall requirements that govern the operation of an interactive system ("HOW the user is able to work with the system"). User experience requirements refer specifically to the non-functional requirements relating to the user interface and the interactive system. Requirements are grouped into multiple dimensions and a general overview is given in Table 11.3 (Sharp et al. 2019).

It is important to emphasise here that the target audience for requirements research is not just the end user of the product, but the entire stakeholder group. For example, in a project where we worked on the design for the centre console of taxis, the stakeholder team we researched for the design requirements included:

1. The driver: they are the ultimate target of the product. The functional requirements of the driver, their working environment, the state in which they perform their tasks, the impact of their operation on safety, and the requirements for usability and user experience are all elements that must be studied to establish the design requirements.
2. Service providers: currently in-car infotainment systems are provided by various service providers via the Internet. The characteristics of these service providers, the content of their services, their user interfaces, technical features, usage ecology, etc., also constitute elements that need to be studied in detail before design.
3. Interaction technology providers: they provide the technologies that enable multimodal interaction; the feasibility, functional characteristics and technical limitations of these technologies form part of the design requirements.
4. Software platform designers: the technical possibilities and limitations of the software platform affect the user experience of the driver.
5. In-vehicle active safety technicians: the methods and limitations of the use of these technologies also influence the design of infotainment systems
6. Stylists: The styling design, which determines the size of the central control screen, the position of the various controls, the stylistic features of the car inside and out, the design of the central control screen, also need to be considered in terms of stylistic consistency.
7. Company executives: many non-functional requirements are involved in the design of the centre panel, which may impose design constraints.

This long list may not be exhaustive and not all of the above groups need to be taken into account in the design of every screen feature. How much consideration depends on the content of the design. One thing is certain, however, and that is that the design requirements research should focus on the needs of all stakeholders, and multiple representatives of each stakeholder group involved need to be researched. Multiple data collection techniques should be used, not just one. It is important to note that the design needs of the stakeholder groups will vary and may even be conflicting, so reconciling their needs until a consensus is reached is an important part of the requirements research.

In addition to the several conventional methods mentioned here, props such as low-fidelity prototypes and task descriptions can also be used to support the data collection process. The task analysis approach allows for a deeper understanding of how a task will be operated. By using low-fidelity prototypes to communicate with users, it is possible to gain a deeper understanding of user needs, operational characteristics and usability design points at an early stage of the design process, and iterations on low-fidelity prototypes can shorten the overall design time, reduce errors and lower design costs. Once you reach the high fidelity prototype stage, or the sample generation stage, most of the design development costs have been spent and it is not easy to make changes.

When formulating requirements, it is important to formulate them in such a way that it possible later on to verify whether the requirements have been satisfied. While this is obvious for functional requirements, this also applies to non-functional requirements. This is important in particular in communication with the client. For instance, a requirement that says "The system should be easy to use" is too vague, and if you tell the client that the requirement has been satisfied, she will probably want to know what it means and how you established that the requirement is satisfied. Therefore, it is recommended to make the formulation of the requirement precise enough so that it is clear what it means that the requirement is satisfied; sometimes this is called the fit criterion. For instance, if the requirement is that "the system is easy to use", the fit criterion might be that it will be verified by having a sample of target users fill the System Usability Scale and that the score should be at least 75 (the maximum score being 100).

11.6 Use Cases and Scenarios of Use

The concepts of Use case and Scenario are often mixed together. Currently, many car companies are trying to build up a library of scenarios, but if these two concepts are not clearly separated, then the library of scenarios can be somewhat confusing. Both use cases and scenarios serve as a basis for requirements gathering.

Use case: The term was originally taken from the book Object-Oriented Software Engineering (Jacobson et al. 1992). After it was introduced, it acquired several different meanings. In the context of project definitions, the term Use case is often used in the sense of critical cases that the project should be able to demonstrate at key moments in the project. Elsewhere, the term Use case is more loosely used as applications of a certain technology. For instance, we may read about "use cases showcasing the use of artificial intelligence".[1] Staying closer to the original meaning, in the context of software engineering and user-centred design, Use cases are a more or less formal way to define the user's interaction with the system, serving as a basis for deriving functional requirements. A use case focuses on what information is exchanged between user and system, not on how this information is exchanged. Let us take an example to illustrate this. The example is taken from 5GAA (2020).

Short description:

"The use case concerns entertainment content delivery to the passengers of a moving or stationary vehicle. It is applicable to both automated and non-automated vehicles, where in the latter the driver is restricted in the content he or she is allowed to consume. For cars, up to four occupants can consume high-definition and immersive entertainment media content while the vehicle is stationary or moving. For buses and transporters up to 30 passengers can consume the same content under similar conditions. Each occupant may be interested in different content which may include video, gaming, virtual reality (VR), office work, online education, advertisement, etc.

[1] https://unfoldlabs.medium.com/ai-automotive-8-disruptive-use-cases-fd079926aea9. Visited Jan 12, 2022.

Contextual information can be embedded in the entertainment media depending on the location of the HV.

Goal: Access in-vehicle entertainment flow.

Use case (Main event flow).

1. *The system establishes a communication link between the HV and wireless service provider. The new communication link for the HD content does not disrupt the communication link for other use cases involving safety and other mission-critical services*
2. *The system establishes a business relationship with a HD content service provider.*

   (Note: this does not preclude wireless and HD content service providers from being the same entity)
3. *Each passenger individually chooses which HD content he/she is interested in before or after entering the car*
4. *Individual passengers request access to the chosen HD content each time they enter the car*
5. *Service providers identify each passenger's individual choices and the HV's location*
6. *Service providers check if the content is available and/or has permission to be accessed in the region where the HV is located and if the HV is authorised to receive the HD content*
7. *Service providers make the HD content available to individual passengers*
8. *Each passenger individually accesses and plays the HD content at his or her own convenience*
9. *Each passenger stops or pauses the HD content at his or her own convenience"*

From the use case, the requirements can be derived about what functionality the system needs to have.

A key point in the analysis of use cases is that the analysis should not be based on the existence of a technology and the way it restricts operation, but should be independent of the technology itself, so that the use cases analysed do not become exclusive to a technology.

Cockburn (1997) suggests that the use cases should be a series of goals and sub-goals in a tree structure based on the human purpose (goal). The main goal is the basis of each Use case. It does not contain any possible user interface or technical decisions. The basic use cases focus on user intent, or requirements, rather than interaction details, and on simplification rather than elaboration. The user interface designer can then use these basic use cases as input to create the user interface, without being bound by any hidden decisions. The analysis of these Goal and sub-goals is somewhat similar to task analysis. We will elaborate on task analysis in Sect. 11.7.

Scenario: Scenarios are narratives or short stories (Carroll 2000). A scenario describes the activities or tasks in the form of a short story, allowing exploration and discussion of the context, the requirements, and the people in the requirements story. Similar to use cases, a scenario does not explicitly describe the software or other technical support for completing the task. In other words, the scenario describes the context of a person's needs and processes in accomplishing a task but is not directly linked to a specific technology. The advantage of scenarios over use cases is that short stories are good for imagining the situation and creating empathy with the user, facilitating the understanding of what is to be designed and what it is for and for whom, by all members of the stakeholder group.

The level of detail of the information presented in scenarios varies and there is no specific guidance on how much or how detailed they should contain. Typically, scenarios are generated in workshops or interview sessions to help explain or discuss certain aspects of the user's goals. They can be used to imagine potential uses for the equipment as well as to capture existing behaviour. They are not intended to capture the full range of requirements, but rather offer a personalised description that provides little more than a perspective.

To better understand the difference, the Use case from above might read as follows.

Anna wants to access entertainment content when driving her car, using the in-vehicle entertainment system. After having entered the car, she checks whether the connection with the content service provider has been established. She chooses the content that she is interested in, and requires access to the content. The system checks whether the content is available, and, if so, makes it available. Anna plays the content, pausing or stopping it at her convenience.

This scenario focuses on one actor, Anna, the driver. Other scenarios might be formulated for passengers. That is, there may be more scenarios for a single use case. The purpose is the same, however: The scenarios serve as a basis for deriving user requirements. Whether one prefers use cases or scenarios may be a matter of taste. Use cases are more formal and technical, and if formulated using a modelling language such as UML, lend themselves for conversion into software specifications. Scenarios invoke a more vivid picture of the users and may be more suitable in design teams.

Analysis of use cases/scenarios: From the driver's point of view, we can perhaps divide scenarios into two categories, those that are related to driving and those that are not directly related to driving itself. In use cases/scenarios related to driving, the driver is required to complete tasks that are related to driving, often referred to as "primary tasks". In use cases/scenarios that are not directly related to driving itself, the driver is required to perform tasks that are not directly related to driving, such as listening to music, turning on the air conditioning, etc. These are often called "secondary tasks".

For the analysis of use cases related to car driving, we can take a use case such as illustrated in Fig. 11.5, where we take "setting up automated driving" as a goal. From this main objective, we can decompose five sub-objectives according to the needs of the driver, and by completing these five sub-objectives, we can complete the main objective.

For any car driving use case, there are different external environments that can influence the parameters of the car driving use case. These external environment parameters include static elements, dynamic elements, traffic participants, weather elements and lighting elements. The details of these elements are shown in Table

Main goal: Setting up automated driving
Sub-goal 1: Judge environmental situation
Subgoal 2: Activate automated driving
Sub-goal 3: Present the status of automated driving
Sub-goal 4: Perceive traffic scenario
Sub-goal 5: Remind/notify driver about his responsibility

Fig. 11.5 Analysis of use case/scenario related to car driving

Table 11.4 Elements of the external driving environment

	Driving task information	Perceptual recognition, path planning, human–computer interaction and networked communication driving task information
Static environmental elements	Obstacle	Positive barriers; Negative barrier
	Surrounding landscape	Flowers and trees; Architecture
	Transportation facilities	Road auxiliary facilities; Road traffic markings; Road traffic signs
	Road	Bridges and culverts; Ramp; Intersection; Road meter; Road section
Dynamic environmental elements	Dynamic indication facility	Traffic light; Variable traffic signs; Traffic police
	Communication environment information	Signal strength; Electromagnetic interference; Signal delay
Traffic participant elements	Other vehicles	Vehicle; Non-Motor Vehicle
	Pedestrian	Pedestrian; Runners; Disabled
	Animal	Cats, dogs, etc
Meteorological elements	Ambient temperature information	–
	Illumination condition information	Light intensity; Light angle
	Weather information	Rain; Thunder; Fog; Haze; Wind; Hail

Main goal: Holding a conference call while driving
Sub-goal 1: Confirm participants
Subgoal 2: Select platform for the meeting
Sub-goal 3: Choose participants from the platform
Sub-goal 4: Start the conference call
Sub-goal 5: Environment noisy; need to mute
Sub-goal 6: Add more participants

Fig. 11.6 Analysis of use case/scenarios not related to car driving

11.4. For each of the intermediate sub-targets, there may be further parameters to consider. In this way, for each sub-objective, different environmental parameters will affect it differently, and it is important to note that the driving use cases are not exhaustive.

For use cases not related to car driving, we can also take a use case here, such as in Fig. 11.6, where we take "holding a conference call while driving" as a goal. Again, we can decompose the six sub-goals from this main objective according to the needs of the driver, and by completing these six sub-goals, we can complete the main objective.

Similarly, use cases that are not related to car driving have their own external environmental factors, which can also be divided into static elements, dynamic elements, traffic participant elements, weather elements and light elements. The static element is not a road facility, but rather an internet-like signal; the dynamic element may be manual/automated driving, road complexity; the participant element may be interaction with other people, personal daily arrangements, etc.

11.7 Task Analysis

Generally speaking, Task Analysis involves identifying the task, collecting task data, analysing the data to gain a deeper understanding of the task, and then describing the different steps in the task. In the literature, at least 100 different approaches to task analysis can be found. Most task analysis is done to understand the sequence of actions that need to be performed to achieve the goal. Task analysis is the decomposition of a task according to the way the user needs to operate the system (physical actions) and complete the task. Depending on the purpose served by the task analysis, the analysis methods used and the analysis process can vary considerably. If the task analysis is done for writing a product manual, then the steps to manage the product are strictly reproduced. If it is at the early stages of product design, then the task analysis method can be used to find the best design solutions and the information framework behind them. In a sense, it is somewhat like the scenario analysis process in Sect. 11.6, where the main purpose is broken down into sub-purpose processes, except that the sub-purposes in scenario analysis are necessary to complete the main purpose, but they are relatively independent of each other, and also, it has only one level. Task analysis, on the other hand, is an action decomposition, which can be at multiple levels, while being related to each other. Table 11.5 summarises the advantages and disadvantages of different approaches to task analysis (Stanton et al. 2005).

Table 11.5 Analysis of the advantages and disadvantages of different task analysis methods

Methods	Advantages	Disadvantages
HTA—Hierarchical Task Analysis	(1) its output can be used as input to many human factors projects; (2) the method is widely used in various fields; (3) there is a very detailed analysis of the activities to complete the task	(1) mainly descriptive data are provided; (2) no data on the cognitive-psychological aspects of the task operation; (3) for complex tasks, the analysis process can be time-consuming and labour-intensive
GOMS-Goals, Operators, Methods and Selection Rules	(1) a detailed hierarchical analysis of the activities of the task, including the mental (cognitive) operations; (2) the main application is in the design of human–computer interaction (HCI)	(1) no value for applications other than HCI; (2) the analysis process can be time consuming and labour intensive
VPA—verbal protocol analysis	(1) rich source of data; (2) glimpse of mental (cognitive) processes from textual descriptions; (3) easy to use	(1) data analysis is complex; (2) it is not easy to describe cognitive processes in speech; (3) it is not natural to intersperse language in operations
Task decomposition	(1) flexible in approach, allowing for a variety of analysis methods; (2) can cover a wide range of aspects of interface design, such as operator error, usability, interaction time, etc	The operation is somewhat complicated and time consuming
The sub-goal template Method	The data generated are very useful	The validity of the technique needs to be further verified
Tabular Task Analysis	(1) Flexible approach, can be analysed on demand; (2) Wide range of applications	Time consuming, therefore, less used

Of these methods, we focus on the HTA method, the most common of all task analysis methods, which is used for a wide range of purposes, including interface design and evaluation, training, functional distribution, job description, labour organisation, instructional design, error prediction and labour load. It can be conducted applying the following steps.

Step 1: Define the task: The task to be analysed should be clearly defined and the purpose of the task analysis should also be clarified.

Step 2: The data collection process: The data are collected by means of observation, interviews, cognitive walks, etc. These data are used for the analysis of the task, including the technology required to complete the task, the interaction between people, machines and teams, the limitations of the task, etc.

Step 3: Determine the objective of the task: The overall objective of the task needs to be determined in the first instance. For example, to switch on the autopilot mode when driving on the motorway.

Step 4: Define the sub-goals of the task: through the operation of these sub-goals, the overall goal can be accomplished.

Step 5: Further decomposition of the sub-target: this decomposition process can be one step, or several steps, until a physically manipulable level is reached.

Step 6: Planning and analysis: Once we have completed all the sub-goals and decomposition work, we can start planning the relationships between the different steps. The relationships between the different steps can be of multiple types, as shown in Table 11.6 (Stanton et al. 2005).

A comparison of the use case/scenario analysis methods shows that steps 3 and 4 are very similar to the determination of primary and sub-purposes in scenario analysis. It is important to emphasise here that scenario analysis is as independent as possible of the technology available, whereas task analysis, on the other hand, takes into account the process of interaction and the interaction technology chosen.

Figure 11.7 shows an example of a Hierarchical Task Analysis. This case depicts an illustration of the task of how to play music in the car. The results of the task analysis are also often presented in tabular form, as in Table 11.7.

Table 11.6 Types of task analysis step planning (HTA plan)

Planning (plan)	Case studies
Linear relationships	Operate step 1, then 2, then 3
Non-linear relationships	Steps 1, 2, and 3, in any order
Simultaneous	Operate step 1, then 2 and 3 simultaneously
Branch relationships	Step 1, if X appears, you can proceed to 2 and then 3. If X does not appear, the operation is terminated
Circulation method	Do step 1, then 2, then 3, and so on until X appears
Select method	Operate step 1, then 2, or 3

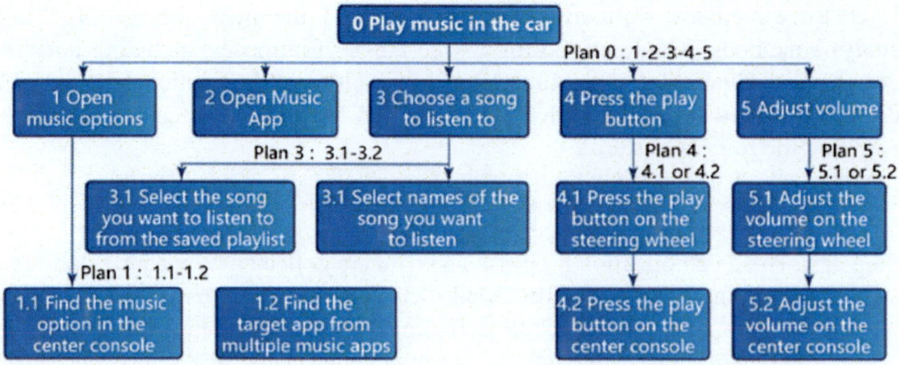

Fig. 11.7 Association diagram representation of task analysis

Table 11.7 HTA task analysis table representation

Mission main objective: play music in the car Plan 0: Play music in the car. Procedure: Do Plan 1, to Plan 2, to Plan 3, to Plan 4, to Plan 5 in order
Plan 1: Open the music option. Steps: Do 1.1 first, then 1.2 1.1 Find the music option in the centre console 1.2 Find the target app from multiple music apps
Plan 2: Open the music app
Plan 3: Select the song you want to listen to. Steps: Do 3.1 first, then 3.2 3.1 Select the song you want to listen to from the saved song list 3.2 Search for the name of the song you want to listen to
Plan 4: Press the play button. Procedure: Do 4.1, or 4.2 4.1 Pressing the play button at the steering wheel 4.2 Pressing the play button at the centre console
Plan 5: Adjust the volume level. Procedure: Do 5.1, or 5.2 5.1 Adjusting the volume on the steering wheel 5.2 Adjusting the volume on the centre console

It is important to note here that the results of the task analysis are not unique and that there can be multiple methods of analysis. If the analysis is for an existing product, then the operational steps need to strictly reflect the operational steps of the design. If the task analysis is done for a design, a variety of step-by-step decomposition diagrams can emerge. With this decomposition diagram, we can continually optimise and simplify the design and operational steps.

References

5GAA (2020) C-V2X use cases volume II: examples and service level requirements. White Paper. 5GAA Automotive Association, Munich

Carroll JM (2000) Five reasons for scenario-based design. Interact Comput 13(1):43–60
Cockburn A (1997) Structuring use cases with goals. J Object Orient Program (Sept-Oct 97)
Cooper A (1999) The inmates are running the asylum. Macmillan Publishing Co., Inc
Cooper A, Reimann R, Dubberly H (2003) About face 2.0: the essentials of interaction design. John Wiley & Sons, Inc
Corbin J, Strauss A (2008) Basics of qualitative research, 3rd edn. SAGE Publications, Inc
Dey I (1999) Grounding grounded theory: guidelines for qualitative inquiry. Academic Press
Fallman D, Waterworth J (2005) Dealing with user experience and affective evaluation in HCI design: a repertory grid approach. In: CHI 2005, pp 2–7. ACM, New York
Fontana A, Frey JH (1998) Interviewing-the Art of Science. In: Denzin NK, Lincoln S, Yvonna (eds) Collecting and interpreting qualitative materials. SAGE Publications, pp 47–78
Fransella F, Bell R, Bannister D (2004) A manual for repertory grid technique. Wiley
Gutman J (1982) A means-end chain model based on consumer categorization processes. J Market 60–72
Hassenzahl M, Platz A, Burmester M, Lehner K (2000) Hedonic and ergonomic quality aspects determine a software's appeal. In: CHI 2000: proceedings of the sigchi conference on human factors in computing systems (pp 201–208). ACM, New York
Hassenzahl M, Tractinsky N (2006) User experience-a research agenda. Behav & Inf Technol, 25(2):91–97
Hertzum M, Clemmensen T, Hornbæk K, Kumar J, Shi Q, Yammiyavar P (2007) Usability constructs: a cross-cultural study of how users and developers experience their use of information systems. In: Aykin N (ed) HCII 2007. LNCS, vol 4559. Springer, Heidelberg
Jacobson I, Christerson M, Jonsson P, Overgaard G (1992) Object-oriented software engineering-a use case driven approach. Harlow, UK
Kelly GA (1955) The psychology of personal constructs. Norton, New York
Kelly GA (1969) A mathematical approach to psychology. In: Maher B (ed) Clinical psychology and personality: the selected papers of george kelly.
Osgood C, Suci G, Tannenbaum P (1957) The measurement of meaning. University of Illinois Press
Patton MQ (2002) Qualitative research & evaluation methods. Sage, Thousand Oaks Ca
Stanton NA, Salmon PM, Walker GH, Baber C, Jenkins DP (2005) Human factors methods—a practical guide for engineering and design. Ashgate
Taylor-Powell E, Renner M (2003) Analyzing qualitative data. University of Wisconsin-Extension. G3658–12. http://www1.uwex.edu/ces/pubs

Chapter 12
Generative Methods

Abstract In this chapter, we go into different generative methods. We discuss the process of going from initial ideas to elaborated concepts and discuss different methods that designers may apply to generate and select concepts that are both innovative and feasible. Furthermore, standards and guidelines that may support the design process are listed. Finally, we go into the conditions that may foster the development of ground-breaking design visions.

Design is a creative process. It is a process of generating ideas and concepts. The outcome of the design process needs to satisfy certain constraints. There are technical constraints (not everything is possible), there are constraints deriving from the user requirements. And there may be company constraints (style guides, legacy and so forth). Before doing so, we briefly define the main terms. An ***idea*** is a thought or image formed in the mind. For the design context, it is more convenient to say that it is an external representation of an idea, that is, a formulated thought, or an image of an object formed in the mind. Likewise, a ***concept*** is an external representation of an idea after elaboration. It has internal structure, describing what it is (the main features) and how it works (main technical characteristics). Initially, the concept is rudimentary. As the design process evolves, more and more details are added and the representation becomes more complex.

Typically, given the project brief, one starts a process that consists of a number of iterations each consisting of diverging activities followed by converging activities, as shown in Fig. 12.1. In Iteration #1, the designers begin to generate a variety of innovative design ideas through a brainstorming approach. The aim is to gather as many ideas as possible. During this process, it is important to avoid premature selection and dismissal of ideas, as otherwise the design space will be narrowed down too early and important opportunities may be missed. The creative ideation process allows designers to explore their own limitations, boundaries and assumptions, as well as rethink and reorganise their understanding of design issues. Once enough ideas have been gathered, the designers then select, consolidate and condense them according to the nature of the problem to be solved, the context, the technical limitations, the time and staff constraints, and so on. Typically, a few proposals are selected as most promising proposals. In Iteration #2, the promising proposals

© China Machine Press 2023

F. Chen and J. Terken, *Automotive Interaction Design*, Springer Tracts in Mechanical Engineering, https://doi.org/10.1007/978-981-19-3448-3_12

Fig. 12.1 Generation of the solution to a design problem, a process that consists of a sequence of iterations each consisting of a diverging phase followed by a converging phase

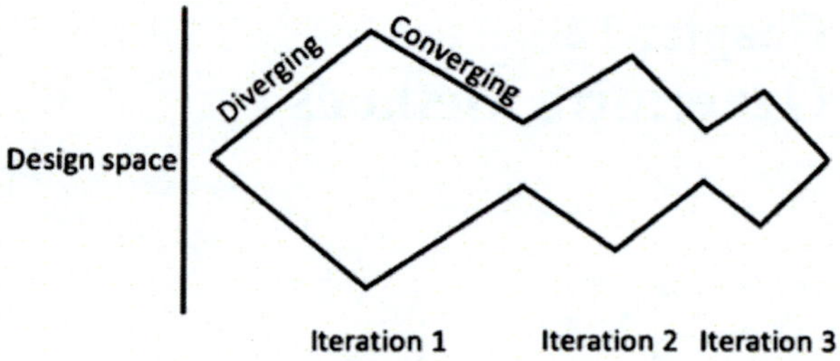

are elaborated by working out further details (divergence), issues are identified and the set of proposals is narrowed down further to one or a few (convergence). The result of the selection, consolidation and condensation process is one or more paper design prototypes or other kinds of low-fidelity prototypes, and a design rationale, i.e., a documentation of the motivations for the prototypes. The result of this stage is discussed with management for approval and with potential users to identify design strengths and weaknesses, and then different design solutions are tried out to refine the design. This is the process of conceptualisation. This process of diverging activities followed by converging activities is continued until a solution is obtained that meets the requirements. At the end of each iteration, reviews and/or evaluations are set up to reflect on the results of the iteration and/or collect user feedback, varying from initial quick-and-dirty evaluations to more formal user evaluations later on in the project (see Chap. 13).

There are many different methods used in the interaction design process. A useful source for creative techniques for the different stages (generation, selection) is Thinkertoys by Michalko (2000). The next sections will provide a brief introduction to a few generative methods.

As stated before, initially, a design concept can be expressed by one or a few sketches and/or a small textual description. The concept generally needs to show what the product to be designed is intended to do, how it should behave, and how it serves the user goal. In the course of the design process, the design concept is elaborated, adding more and more details about how the concept works and what the user experience will be like. This concept can also be explained to the user in order to collect feedback. The key guiding principles of conceptual design are (Sharp et al. 2019):

- Keep an open mind, but don't forget the users and their backgrounds.
- Discuss ideas with as many other people as possible.
- Use low-fidelity prototypes to get quick feedback.
- Iterate, iterate, iterate again.
- Considering different options and thinking iteratively from different perspectives can help to expand the solution space.

12.1 Design Methods

Different creative techniques, serving different purposes, are appropriate for the different stages in the design process. Therefore, before choosing a creative technique, we need to determine what goal we want to achieve. A distinction may be made between techniques that are intended for generation of ideas and concepts and techniques that are more suited for selecting ideas and concepts. The reason to include techniques for selecting ideas and concepts under creative techniques is because the selection process also involves creative activities, consisting of generating criteria by which to evaluate proposals.

12.1.1 Idea Generation Techniques

Many different techniques are available and suggestions for how to apply the techniques can be found on the web (see e.g., Mindtools.com/pages/main/newMN_CT.htm).

Classical brainstorm. In this generative technique, the design team starts from the design brief and tries to generate as many ideas as possible, in a relatively unconstrained way. Often, there are two stages: an individual stage and a team stage. After the individual stage, members exchange their ideas, and, taking inspiration from the ideas of the other members, add ideas through free association and reversal. The advice is to try to be extreme. There are many variations on this basic scheme. For instance, reverse brainstorming asks how things can be made worse instead of how they can be made better. Again, the goal is to stimulate creativity and encourage team members to be extreme.

Random inputs (De Bono). Given the problem definition, random words are used as a starting point for generating ideas.

Seven Essential Innovation Questions (SEIQ, O'Connor). The team starts from the problem with the pain points of the current product or system and asks seven questions: What could we look at in a new way? What could we use in a new way? What could we move, changing its position in time or space? What could we interconnect in a different way? What could we alter or change? What could we make that is truly new? What could we imagine to create a great experience?

SCAMPER comprises several creative techniques: substitute, combine, adapt, modify, put to another use, eliminate, reverse. There is a facilitator who decides what the right moment is to switch from one technique to another.

In the **Concept fan** technique (De Bono), the design team starts from the problem and writes down potential solutions to the problem. In turn, these proposed solutions are used as input for generating further ideas. If this does not give the desired solution, then the process is repeated by redefining the problem more broadly and generating solutions for the redefined problem. This process continues until a satisfactory proposal is obtained.

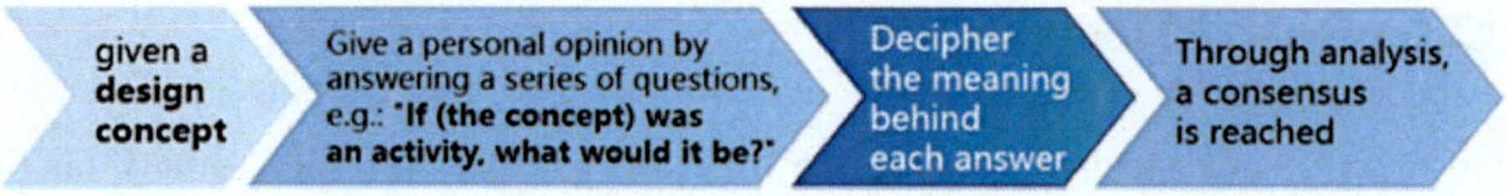

Fig. 12.2 Flowchart of the concept portrait method

In **Provocation** (De Bono), potential solutions are generated that are provocative and too extreme to be realistic, by applying techniques such as negation and exaggeration. In the next phase, the provocative ideas are translated into more realistic ideas by extracting principles and focusing on differences.

Concept portraits (CP). The "conceptual portrait" (Gkouskos 2016) method is suited when one or more concepts are available. The aim is for the design team to arrive at consensus. It is a way to analyse concepts for complex and deep problems that often have different meanings and generate different understandings. The concept portrait approach enables better sharing of these concepts for the design team at an early stage of the design process. Concept Portrait is based on a pastime game where players are asked to make word connections between different concepts. An example question is "If (the concept) were an activity, what would it be?" The diagram below (Fig. 12.2) illustrates the steps involved in the concept portrait approach.

In short, the design team starts with a concept and then makes connections by answering six questions about the given concept. These questions can include connections between the given concept and places (e.g., countries, public spaces), animals, famous people, objects, etc. The design team must prepare the questions before the start of the session and must answer them quickly and without much thought. Once everyone has completed their individual associations, members take turns encouraging the rest of the team to make their own choices. Finally, the card method (see below) can be used to qualitatively analyse the results of the CP through examples.

Future workshop. Future workshops are used in an integrated approach to conceiving designs for a future world. The approach works by shifting the designer's attention from current problems that may hinder creativity to envisaging solutions for future problems. The focus on the future also helps designers to overcome the limitations that may be caused by existing technologies and technical configurations. The approach involves several stages: first, the design team defines a design problem. The team is then placed in an imaginary future world in which the design space differs significantly from reality. Using some features of the future as inspiration, the designers generate design solutions to the problems they have identified and finally develop a plan to realise these solutions (Gkouskos 2016).

The Future Workshop method can be used by adding pre-made scenarios to the future world. The scenario details a series of activities that the user will perform in the future. For example, if the future is a utopian world where teleportation exists, the scenario would detail a Swedish user getting ready for work each morning and then teleporting to his company's office in Brazil. The scenario helps the designer to imagine how people will operate and experience the event in the future, while also demonstrating possible user needs and desires for the system (Gkouskos 2016; Carroll 2000; Nielsen and Svensson 2006).

12.1.2 Idea Selection Techniques

Once many ideas are available from the generative session(s), by applying the techniques summarised above, the process continues by narrowing down the set of ideas to select a small number of ideas for the further design process. To begin with, if there are many ideas, voting is an efficient method to discard ideas and narrow down the set of ideas. Alternatively, the quadrant method can be applied.

Quadrant method. All proposals are put in a two-dimensional space with relevant dimensions, for instance innovativeness/originality and feasibility. Only the ideas in the upper right ranking high for originality AND feasibility are selected. Once the set of ideas is narrowed down to a manageable number, one of the following methods can be used for further selection.

PMI analysis. For each idea, positive (Plus) and negative (Minus) points are identified. Properties that cannot be classified as positive or negative may be labelled as Interesting. The classification is used as a basis for selection.

Pugh matrix or **controlled convergence method**. The analysis of the ideas may proceed using the so-called Pugh matrix or controlled convergence method, and this may result in selecting one concept as the starting point for the further design process. One side of the matrix lists criteria, the other side of the matrix lists the ideas. Criteria may be taken from the user requirements (e.g., usefulness, ease of use, pleasurability), from company values (e.g., sustainability, branding/distinctiveness), the technical department (feasibility) etc. Each idea is evaluated for each criterion, and values of 1, 5 and 9 are assigned: 9 if the idea scores high for that criterion, 1 if it score low and 5 if it scores neither high nor low. Criteria may be given weights, for instance using a 1-3-5 scale. The totals for each idea are calculated and the best ideas win. As a next step, the ideas scoring best may be improved: if a good idea scores low for one or more of the criteria, the design team may look for inspiration from other ideas that score high for that criterion.

12.2 Concept Elaboration

During the conceptual design phase, three important questions need to be answered.

1. What the product needs to achieve: Understanding what the product needs to accomplish is the basis for generating a conceptual design. For example, a driver needs to answer a phone call while on the road. To achieve this task of receiving and making calls, the car's interactive interface must be able to provide several functions: the function of receiving calls, the function of dialling numbers, the

function of remembering phone numbers, and so on. It is also necessary to ask yourself which of these functions will be operated by humans and which will be operated automatically by the system.

2. How the functions are related: Some functions may be temporarily related, for example, one must be executed before another, or two may be executed in parallel. They may also be linked together by many possible classifications. The relationship between tasks may be limited by which devices are used in them.
3. What information must be provided in what context: what data are needed to perform the task? How does the system transform these data? For example, if the driver wants to listen to music, we need to provide him with a catalogue of music to choose from.

In the further elaboration of the concept, from an interaction design perspective there are three questions that need to be answered: Which interaction mode best supports the user's activities? Is there an appropriate interface metaphor to help the user understand the product? Which interaction paradigm will the product follow?

1. **Interaction modes**

   Which interaction mode is best suited to the product being designed depends on the activities that the user will engage in when using it. This information is determined through user requirements research. Interaction modes refer to the way in which the user operates when interacting with a device. As technology evolves, more and more interaction modes will be available and multimodal interaction will become increasingly popular, especially in cars, where traditional menus will be replaced by one-button access, and where voice and gesture interaction technologies will become more common.
2. **Interface metaphor**

The purpose of interface metaphors is to combine familiar knowledge and new interaction concepts in a way that helps the user to understand the system. Choosing the right metaphor and combining new and familiar concepts requires care, and it needs to be based on a sound understanding of the user and their environment. For example, consider a system that coaches and gives feedback to the driver about his/her driving habits. One could show messages and graphs on the mid console screen or send them to the user's mobile phone for later inspection. However, taking the coach function as a starting point, one could develop an embodied conversational agent that presents the feedback in a friendly, gentle, and non-patronizing way. The use of this metaphor would then induce certain expectations about the cognitive and emotional intelligence of the system, thus telling the user what to expect and how to interact with the system.

When choosing different interface metaphors, we often ask several questions.

1. Does this interface metaphor provide a framework for solving the problem?
2. Is the chosen interface metaphor relevant to the user in terms of the problem to be solved?
3. Is this metaphor easy to express?
4. Can your users understand this metaphor?

5. How scalable is this metaphor?

A critical reflection on these questions may guide the choice of an appropriate metaphor. For instance, in connection with question 4, one might consider that, as smartphones become more powerful and functional and users have become familiar with the interface metaphors used in mobile phones, borrowing such metaphors for in-car systems may reduce learning costs.

3. **Interaction paradigm**

The most common interaction paradigms are windows, icons, menus and pointers, similar to those used in personal computers. Of course, as technology evolves, other paradigms are coming into view, such as in-car wearable devices, tangible user interfaces and holograms, all of which will bring new experiences to human–computer interaction and present completely different design challenges.

12.3 Scenario-Based Design

Scenario-based design is a common approach to interaction design for cars, where a series of user scenarios are used to map out the user journey and storyboard. It is primarily used to further clarify and identify user requirements and to generate initial design concepts, and as such, it serves as the basis for the overall design, reflecting the technical implementation possibilities, and is the best tool to facilitate communication and consensus between members of the design team from different backgrounds.

The description of a scenario through storyboarding provides a flexible approach to system or device design, helping designers and design teams to propose, evaluate and modify design concepts. The scenario-based approach is convenient for developing and presenting the hypothetical future context of the new design system. It uses storyboards to describe the future operation of the device/system in question. This includes presenting a design concept and using the who, what, when and why. Once a scenario has been created, new design ideas and proposals can be added to the storyboard and the design is modified as a result. There are several steps involved.

***Step* 1:** Scene identification. The description of the scenario should be complete and include the purpose of the scenario, the objective environment and the behaviour, but preferably also the input system, the interaction interface and output information, the user characteristics, the context in which the scenario takes place, the various goals, behaviours and outputs of the people involved. The method of description can also be in the form of a table. Scenarios are often generated from user research.

***Step* 2:** Generating storyboards. Storyboarding is the process of taking the content of the scenarios analysed above and linking it together through storytelling. The process of storyboarding is also the process of conceptualising the design and generating low-fidelity prototypes. By breaking down a scenario into smaller steps, this helps the designer to consider the details of the design in more detail. Figure 12.3

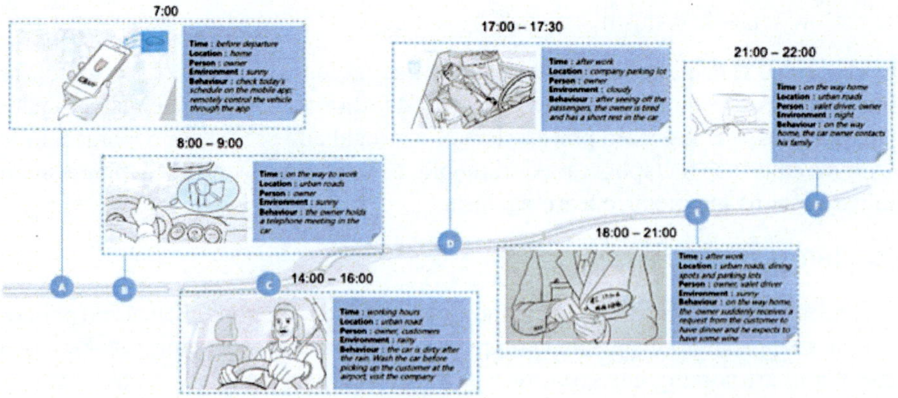

Fig. 12.3 Typical storyboard

gives a case study depicting the complete journey of a private company owner as he commutes to and from work, receives clients and entertains them at drinks during his daily working day.

12.4 Prototyping

The purpose of prototyping is to elaborate and test various design concepts and design ideas. Prototyping allows members of the design team to see the details of the design and how the design problem is solved. Prototyping is also important for obtaining feedback from users and for verifying the viability of the technology. There are low and high-fidelity prototypes, and different types of prototypes are used at different stages of the development process. Figure 12.4 illustrates the design process for in-vehicle interactive systems and the production and use of different levels of prototypes. In general, low-fidelity prototypes are used early in the design process, and high-fidelity prototypes (e.g., limited software implementations) are used in the middle to late stages of the design process. Paper prototyping is often used to test different design concepts. Paper prototyping, as the name suggests, is the hand-drawing of a design concept on paper, while also showing the interaction and results of the system by pasting various drawings of the keys, the steps, and the display showing how the system changes after each step on the paper. Paper prototypes have a number of advantages over other low-fidelity prototypes, starting with their ease and flexibility, the ability for people to test and discuss designs, and the low cost of continuous improvement. By getting feedback from users about paper prototypes, user requirements can be better captured and represented to ensure that no major mistakes are made in later designs. High-fidelity prototypes use materials that look more like the final product. As a result, various design tools may be employed.

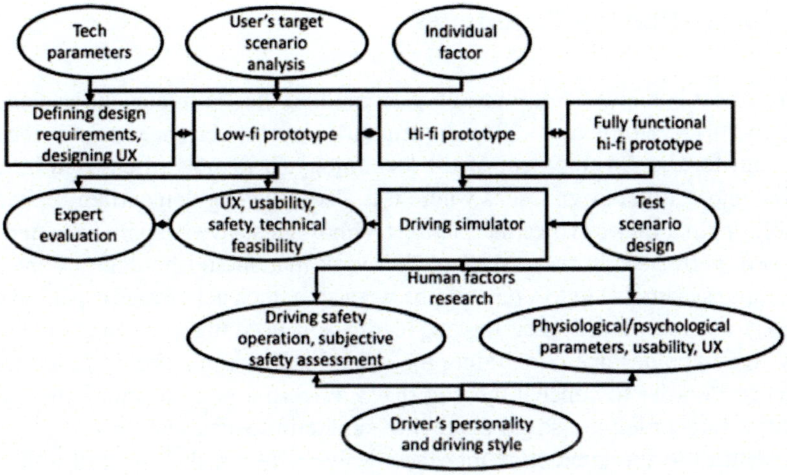

Fig. 12.4 Production and use of different levels of prototypes in the design process of in-vehicle interactive systems

Both types of prototypes have advantages and disadvantages, which are summarised in Table 12.1 (Rudd et al. 1996).

Rettig (1994) argues that more effort should be spent on low-fidelity prototypes because there are many problems with high-fidelity prototypes: they take too long to build, reviewers and testers tend to comment on the interface design part without much regard to the content, and developers are reluctant to change something they have spent a long time (weeks, even months) to carefully design. A software prototype may set too high expectations.

Table 12.1 Comparison of the advantages and disadvantages of low and high-fidelity prototypes

Type of prototype	Advantages	Disadvantages
Low-fidelity prototype	Low development costs and high speed Multiple design concept reviews Facilitate dialogue between design parties Focus on the core ideas and content Can test the needs of the market	Errors are not easily detected Lack of meaningful detail for coding Limitations in the evaluation of usability and user experience
High-fidelity prototype	Full functionality Interaction integrity Clear interaction path Very similar to the final product Usability and user experience testing	High development costs and long lead times Not easily used to research user needs

12.5 Role-Playing Prototypes

Wizard of Oz is a low-fidelity prototyping method that simulates the use of a technology before it is implemented, to investigate the user experience and provide further user input for the development of the technology. It is most often used in relation to voice interaction design issues where it is also called "green curtain" experiment. Possible applications of voice interaction technologies have been investigated before they were perfected. In these studies, the work that should be done by the speech recognition system is left to a person (the wizard) in the background ("behind a green curtain"), and the wizard's hearing and feedback is used to represent the feedback of the system to be designed, including the voice feedback in a pre-designed way (the protocol). In order to conceal the fact that a Wizard is used to mimic the system's actions, pre-recorded messages deploying synthetic speech may be used.

In contrast to the interactive interface design process, this method is also often used to test the usability, user experience and design requirements of a design before the formal coding of the program begins. A key point of this operation is that the user is unaware that it is not the system that is operating in the background, but a human being that is replacing it.

12.6 Standards and Guidelines to Guide Design

While a user-centred design approach emphasises the importance of involving users in the design process to gather feedback on design recommendations, not all design decisions require detailed evaluation by users. Other ways of making fundamental design decisions may deploy standards published by the International Standards Organisation (ISO), the Society of Automotive Engineers (SAE), the National Highway Traffic Safety Administration (NHTSA) and the Japanese Automotive Standards Organisation (JASO). These organisations and some of the world's leading companies such as Google and Microsoft have also published various design guidelines. These standards and guidelines provide guidance on the design of automotive systems and associated HMIs and the evaluation of these systems. They are based on extensive research and design practice and should therefore be consulted by experts at the outset of design to ensure that no low-level errors are made.

While it can be assumed that standards and guidelines relating to human information processing capabilities and limitations are universal in nature and therefore should not require separate guidelines for different countries and regions, there may still be cultural differences between people in different countries, particularly if the functions applied by the designer take advantage of underlying cultural practices. Therefore, it is not self-evident that standards can be simply translated from one culture to another. In other words, standards should be applied with care and critical reflection to identify elements that may vary from culture to culture. Table 12.2 contains a non-exhaustive list of standards from different organisations.

Table 12.2 Standards and recommendations for design of automotive systems and HMIs

Name	Title
ISO 26262-1	Road vehicles—Functional safety
ISO/PAS 21448	Road vehicles—Safety of the intended functionality
ISO 1503	Spatial orientation and direction of movement—Ergonomic requirements
ISO 3958	Passenger cars—Driver hand-control reach
ISO 4513	Road vehicles—Visibility—Method for establishment of eyellipses for driver's eye location
ISO 15005	Road vehicles—Ergonomic aspects of transportation and control systems—Dialogue management principles and compliance procedures
ISO 15006	Road vehicles—Ergonomic aspects of transport information and control systems—Specifications for in-vehicle auditory presentation
ISO 15007	Road vehicles—Measurement of driver visual behaviour with respect to transport information and control systems
ISO 15008	Road vehicles—Ergonomic aspects of transport information and control systems—Specifications and test procedures for in-vehicle visual presentation
ISO 16352	Road vehicles—Ergonomic aspects of in-vehicle presentation for transport information and control systems—Warning systems
ISO 16673	Road vehicles—Ergonomic aspects of transport information and control systems—Occlusion method to assess visual demand due to the use of in-vehicle systems
ISO 16951	Road vehicles—Ergonomic aspects of transport information and control systems (TICS)—Procedures for determining priority of on-board messages presented to drivers
ISO 17287	Road vehicles—Ergonomic aspects of transport information and control systems—Procedure for assessing suitability for use while driving
ISO 17361	Intelligent transport systems—Lane departure warning systems—Performance requirements and test procedures
ISO 17488	Road vehicles—Transport information and control systems—Detection-response task (DRT) for assessing attentional effects of cognitive load in driving
ISO 20545	Intelligent transport systems—Vehicle/roadway warning and control systems—Report on standardisation for vehicle automated driving systems (RoVAS)/Beyond driver assistance systems
ISO 21956	Road vehicles—Ergonomics aspects of transport information and control systems—Human machine interface specifications for keyless ignition systems
ISO 21959	Road vehicles—Human performance and state in the context of automated driving

(continued)

Table 12.2 (continued)

Name	Title
ISO 22902	Automotive Multimedia Interface
ISO 26022	Road vehicles—Ergonomic aspects of transport information and control systems—Simulated lane change test to assess in-vehicle secondary task demand
ISO/TR 23049:2018	Road Vehicles—Ergonomic aspects of external visual communication from automated vehicles to other road users
SAE J1050	Describing and Measuring the Driver's Field of View
SAE J2365	Calculation of the Time to Complete In-Vehicle Tasks
SAE J2395	Definitions and Measures of Driver Visual Behaviour
SAE J2399	Adaptive Cruise Control (ACC) User Interface
SAE J2400	Forward Collision Warning Systems User Interface
SAE J2808	Blind Spot Monitoring System (BSMS) User Interface
SAE J2830	Comprehension Testing of In-Vehicle Icons
SAE J2831	In-Vehicle Alphanumeric Messages
SAE J2944	Operational Definitions of Driving Performance Measures and Statistics
SAE J2972	Definition of Road Vehicle Hands-Free Operation
SAE J2988	Speech Input and Audible Output
SAE J3016	Surface Vehicle Recommended Practice, Taxonomy and Definitions for Terms Related to Driving Automation Systems for On-Road Motor Vehicles
SAE J3114	Human Factors Definitions for Automated Driving and Related Research Topics
NHTSA-FMVSS 571.101	Controls and Displays
JASO Z015-10	Road vehicles—Ergonomic aspects of transport information and control systems (TICS)—Procedures for determining priority of on-board messages presented to drivers
JASO Z013-86	Checking procedure of driver hand control reach for passenger cars
EU C(2008) 1742	Commission Recommendation of 26 May 2008 on safe and efficient in-vehicle information and communication systems: update of the European Statement of Principles on human–machine interface (notified under document number C(2008) 1742)

12.7 Envisioning

While incremental innovation (see Sect. 10.8) is primarily aimed at solving problems with current systems that users run into, and therefore may be called user-driven, in the case of radical innovation user involvement is more problematic, as users are not trained to think into the future and reflect upon radical innovations, let alone conceive them. In that case, design teams need to develop a vision about the value of the product or system, or product-service system. The design vision consists of a description of the convictions of the design team about how value may be created for users (the "why"), an outline of the envisaged product or product family, and a specification of how it connects to the strengths of the companies (i.e., why it is valuable for the company to initiate the project).

There are no clear methods to generate design visions, but two general statements may be made. In the first place, it requires sensitivity and reflection. The design team needs to be sensitive to what is going on in industry, the market and society and needs to engage in reflective activities to strengthen and consolidate the insights gained through interaction with industry, market and society. According to Verganti (2009), successful envisioning does not happen in a vacuum, but requires a continuous dialogue with a network of industrial, market and societal actors, as shown in Fig. 12.5. In the second place, this process of developing a vision and getting it landed in society may well take several years.

Finally, it is important to emphasise the following. While the above suggests that, in the case of radical innovation, users may play only a minor role in the design process, from Fig. 12.5 it is clear that the development of a design vision also presupposes a good understanding of people. In the second place, as was argued in Sect. 10.8, in the case of radical innovation users still may provide useful feedback

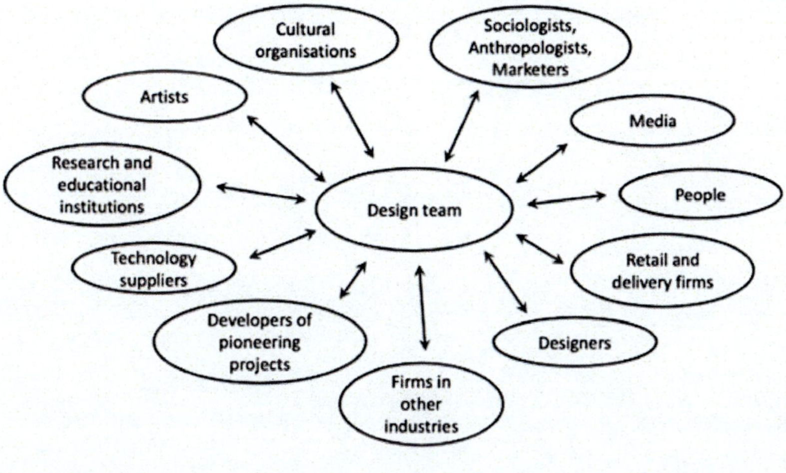

Fig. 12.5 Network of market and social parties with which design teams exchange information that lays the foundation for the envisioning activities of the design team. After Verganti (2009: p.12)

on design visions, if they are helped to think into the future. Finally, once concepts are elaborated and choices must be made about how users may interact with the system, design visions must be confronted with feedback from users gathered through evaluation sessions.

References

Carroll JM (2000) Five reasons for scenario-based design. Interact Comput 13(1):43–60

Gkouskos D (2016) User experience insight steering experience design through meaningful incorporation. Chalmers University of Technology, Göteborg, Sweden

Michalko M (2000) Thinkertoys. Gestión

Nielsen KA, Svensson L (2006) Action and interactive research: beyond practice and theory. Maastricht Shak. Publ

Rettig M (1994) Prototyping for tiny fingers. Commun ACM 37(4):21–27

Rudd J, Stern KR, Isensee S (1996) Low vs. high-fidelity prototyping debate. ACM Interact Mag 76–85

Sharp H, Rogers Y, Preece J (2019) interaction design—beyond human-computer interaction, 5th edn. Wily

Verganti R (2009) Design driven innovation: changing the rules of competition by radically innovating what things mean. Harvard Business Press

Chapter 13
Validation/Evaluation

Abstract Central to the human-centred design process is the reflection on intermittent design proposals and collecting feedback throughout the design process. In this chapter we consider methods for evaluating design proposals in different stages of the design process, both in the early and later stages. Methods for conducting evaluations are discussed, both for conducting expert reviews and user evaluations.

In order for users to effectively evaluate the design of an interactive product, designers must produce a number of prototypes for their ideas. In the early stages of development, these prototypes may be made of paper and cardboard or other means (mock-ups) to facilitate exploration and testing, and as the design progresses and the various ideas become more detailed, they may be polished and eventually formed into physical parts and interactive prototypes including functionality, metal or 3D printed plastic, resembling the final product.

Usually, a distinction is made between *formative* evaluation and *summative* evaluation. Formative evaluation is intended to inform the design team. In this case, the evaluation may be quick-and-dirty, in particular in the early phases of the design, and the goal is to obtain feedback on early ideas ("to see whether we are on the right track"). The data in the early stages will primarily consist of qualitative data. Summative evaluation is intended to inform the stakeholders, and to show that certain targets have been achieved. Usually, summative evaluations are more formal, and the data will include quantitative data, such as grades or scalar judgements. For instance, one may have agreed with the stakeholders that the new version of the system should have a significantly higher Mean Opinion Score (MOS) than the existing system. Or one might have agreed that the outcome of the design process has a certain positive score on the System Usability Scale (see Sect. 13.5). Although the feedback from the quick-and-dirty early evaluations is important to inform the design team, except for Sect. 13.1, the methods below are mostly used towards the end of the design process, when a prototype with functionality is available.

© China Machine Press 2023

F. Chen and J. Terken, *Automotive Interaction Design*, Springer Tracts in Mechanical Engineering, https://doi.org/10.1007/978-981-19-3448-3_13

13.1 Concept Evaluation

In the case of incremental innovation, the design process usually starts from complaints by users about the current system, observations of problems that users run into or opportunities for new features that are supposed to enhance the user experience. In this case, the normal user-centred design process applies. In the case of radical innovation, however, one or more concepts originate from a design vision. In this case, as was stated above, users may still provide useful feedback if they are helped to think into the future and reflect on radically new concepts.

One method aiming to help users reflect on radically new concepts is the Co-Constructing Stories method (Buskermolen and Terken 2012). The underlying idea is that developing a new concept does not only entail generating the concept itself, but also a story of why this concept would be valuable to users. While the initial story is generated by the design team, it is based on assumptions, and these assumptions need to be validated and enriched on the basis of feedback from users. This feedback is collected in interviews that consist of two phases. In the first phase, users are sensitised to the topic by asking them questions about relevant current activities, and they are encouraged to tell stories about their activities, through questions such as "Could you tell the most recent experience?"; "Could you tell the best (or worst) experience?" and so forth. In the second phase, the concept is introduced, and they are asked to reflect on the concept and tell imaginary stories about how it would be to use the concept. Questions helping them to do so are "How would the story about the recent/best/worst experience be different if you would have had the concept?". In this way, they are helped to think into the future in two ways. In the first place, the reflection links to their already existing experiences. In the second place, the reflection is helped by making things concrete, instead of eliciting feedback about abstract proposals that are devoid of connections to everyday life. A similar method, which also embodies the idea that helping potential users to think into the future requires making the future concrete, has been proposed by Pettersson (2017).

13.2 Interface Assessment

There are different methods of interface evaluation for different products at different design stages, each with a different purpose, including usability, user satisfaction, incorrect operation, interface layout, label usage, and controls and displays used. The output of the evaluation can be used to iterate on the design to improve problematic interfaces, increase usability, user satisfaction and reduce user errors, thereby improving the interaction experience of the design.

According to ISO 13407, interface user reviews play different roles at different stages of a product's lifecycle, from assessing the design concept during the design phase to assessing the impact on performance during the operational phase. The different approaches have their own advantages and disadvantages. Table 13.1

Table 13.1 Comparison of evaluation methods for interactive interfaces

Method name	Description	Advantages	Disadvantages
Checklists	Subjective reviews operated by interaction design experts	For the assessment of human operational and cognitive processes, direct measurement and easy to execute	The context of use is ignored. Data are more subjective
Heuristic evaluation	Expert subjective reviews operated by interaction design experts	Commonly used, easy to execute, results can be used directly	No statistical validity, unstructured, more subjective data
Interface Survey	Expert subjective reviews conducted by interaction design experts	Based on various standards and design guidelines, easy to follow and more comprehensive	Needs available, relatively full performance OS, more time consuming
Cognitive walk-through	Expert subjective reviews conducted by interaction design experts	Conventional method, easy to follow, results have direct relevance to design	Needs available, relatively complete performance of the system under test
SUS System Usability Scale	Review of user operation	A commonly used, generally accepted method for comparing different interfaces	Limited output
Self-assessment model	Review of user operation	Simple, intuitive user experience measurement model	No feedback on specific design issues
User Experience Curve	Review of user operation	The system is used by users over a period of time and they record their feelings and problems	Takes a while to complete

provides a summary of some subjective testing methods (Stanton et al. 2005). In the next sections, we provide a brief introduction to each method. The reader is advised to read the relevant articles and books to learn about the details of the methods.

To begin with, a distinction may be made between evaluation by experts and evaluation by potential users. In expert reviews (the first four in Table 13.1), sometimes members from the design team, but preferably user interface experts from outside the team go through the interface in a structured way, identifying problems and—where possible—suggesting improvements. It should be noted, however, that experts can only identify potential problems. Some problems they encounter may not actually occur when users interact with the system. Also, they may miss problems that users may run into during a user test. Therefore, expert evaluation should not be done to replace user evaluation. However, expert evaluation may be conducted before actually presenting the system to potential users, in order to identify and cure major bottlenecks that would annoy potential users in a user test.

13.3 Checklist

Checklists are simple, easy to complete and low cost. The key issue is the creation of the checklist. The content of the checklist depends on the characteristics of the interface being checked. The method is therefore also flexible. Table 13.2 shows a typical checklist for the interaction design of a visual interface. It is important to note that the contents of the table and the principles of evaluation are based on the user experience requirements and the various standards and design guidelines. Therefore, the contents of Table 13.2 should not be used as a basis for the development of such a checklist.

13.4 Heuristic Assessment

This method requires the evaluators (usability design experts or professional evaluators) to identify the tasks that can be completed through the interface. Once the tasks have been identified, the evaluators work on each task one by one on the interface, recording the process and the effect it produces, and evaluating the interface for the relevant content. Heuristics (from Greek "Eurèka", which means "find" or "discover") are rules, insights or principles that have been learned from experience, instead of being derived from theory. Heuristics can be self-determined, but usually they are taken from Nielsen's ten design heuristics (Nielsen 1994) or Shneiderman's eight golden rules (2018). These rules are described in detail in Sects. 2.6 and 8.7. They are also easily accessible from the web, so we will not repeat them here.

It is important to emphasise that the interaction design does not have to be evaluated using all heuristics, but only the more relevant ones. For example, the most common one is 'ease of learning'. Nowadays, people don't spend a lot of time learning how to operate systems, so 'ease of learning' is very important. Similarly, consistency and fault tolerance are requirements that most systems need to meet in their design.

During the evaluation process, the evaluators keep a detailed record of each task as it is performed. Once the assessor has completed a series of operational tasks, he or she goes back and makes a summary of each item based on the records.

The results of the assessment can be expressed in writing. The content of the presentation depends on the purpose of the test. For example, if a prototype car has been produced and its usability needs to be evaluated so that the designers can make final improvements, the report needs to contain sufficient detail. For example, a review of a car music playing system might look as follows.

Table 13.2 A typical checklist for the design of a visual interface interaction

Inspection items	A	M	S	N	Description
Does each page have a clear title or description? Does the user understand the main content of the page?				√	Some screens are untitled
Is important information clearly visible on the page? (Are different colours, or brightness, or other markings used?)			√		
When the user enters information on the screen, is it clear: How should the information be entered? Where is the information entered displayed? What format should I enter it in?				√	
Does the system prompt when the user enters an incorrect message on the screen? Can the user change the information?			√		
Is the information logically organised on the screen? (e.g., same type of app on one page, or related functions arranged in a logical way)			√		
Are the different types of information clearly separated on the screen? Does the position of the information displayed not jump around randomly?				√	The different messages are often divided into lists
Is there too much information on the page? Are users unable to find the information they need?			√		
Is the contrast between important information and the screen background colour clear?				√	
Does the use of colour make sense? Does it make the display clearer?			√		
Do the pages look neat and tidy?		√			
Are the icons and diagrams (e.g., graphs and charts) clearly drawn? Are there notes?				√	

A = Always; M = Most of the time; S = Some of the time; N = Never/No

The buttons on the steering wheel that control the volume work very well.

The song list is very well picked by the up and down rotating buttons on the steering wheel.

The need to choose between several different playback software is rather cumbersome.

The interface is inconsistent from one playback software to another, making it a poor experience.

We need to remember that the purpose of our reviews is to improve the design, so such records, as detailed as possible, should facilitate the designer's understanding and improve the design.

13.5 Interface Survey

The Interface Survey approach is based on the classification of interfaces according to their different physical properties. Of course, this will change as more and more screens and interaction concepts are introduced into automotive design.

- The controls can be further subdivided into buttons, paddles, steering wheel, accelerator, pedals, etc.
- The display, which can be further subdivided here into centre screen, dashboard, rear view mirror, HUD, etc.
- Driver monitoring system.
- Environment

Interface surveys are used to point out deficiencies in an interface or concept design. Each category is described in detail below.

Control and display measurement

Control and display surveys are used to evaluate a particular control and display interface. The analyst first records all the parameters of the interface controls and displays and then creates a list containing descriptions of the use of each control, its position, the type of control and any other relevant details such as movement (e.g., up/down, rotation, left to right, etc.).

Display

Each display should be investigated in the same way, e.g., display type, display content, location, etc. This list classification is hierarchical, which includes system, sub-system and sub-system parameters. This facilitates the execution of the relevant tasks. If required (depending on the scope of the analysis), relevant standards and appropriate design guidelines can also be used as criteria for checking.

Driver monitoring systems

Line of sight measurement is the analysis of the driver's line of sight, including distance, angle and the way the eyes move, where they are looking and what they represent.

Environmental surveys

Environmental surveys measure the state of the environment, such as levels of noise, lighting, temperature and humidity.

13.6 Cognitive Walk-Through

Cognitive walk-through is a method used by interaction design specialists to detect system problems. The first step is to identify a set of tasks that the system is capable of performing and then the analysts operate the tasks one by one, usually with at least three analysts working individually or together. The first step in this analysis is to identify the users of the system, their characteristics, capabilities and their purpose, the tasks to be performed, etc. The analysts then begin to operate the relevant tasks step by step, bringing the characteristics of the real-world environment into play as they do so. At each step of the process, they are asked to answer the following four questions.

A. *Can the user be expected to try to do this action?*
B. *Is the interface element needed to perform the correct action clearly displayed to the user?*
C. *Is there a clear relation between the interface element needed to perform the correct action and the element itself?*
D. *Is the feedback appropriate?*

When the walk-through analysis has been completed, the analyst should give the following answers.

A. Identify the problems that have been identified and answer the reasons why they are occurring.
B. Suggest changes and other issues that may arise.

An example of how to conduct a cognitive walkthrough can be accessed through https://www.youtube.com/watch?v=Edqjao4mmxM.

The various design testing methods mentioned here all serve a common purpose, which is to improve the usability and user experience of the product. Therefore, there are some commonalities in their preparation, as shown in Table 13.3.

13.7 System Usability Scale

There are many different questionnaires that require the user to evaluate a system through scalar judgements. Some scales enable one to elicit judgements about different aspects. For instance, the User Experience Questionnaire (UEQ, Laugwitz et al. 2008) measures different aspects of the user experience, such as attractiveness, novelty, stimulation, perspicuity. Usefulness, Satisfaction and Ease of Use (USE, Lund 2001) measures both user experience and usability. AttrakDiff (Hassenzahl et al. 2003) measures pragmatic quality (usability) and hedonic quality (user experience). It should be noted that the formulation of questions in such scales may have to be adjusted to make them fit to the domain.

A very quick and easy to use questionnaire designed to assess the usability of a particular device or product is the System Usability Scale (SUS, Brooke 1996).

Table 13.3 The three principles of preparing a usability review

Defining the task-user-interaction system	The **task** refers to the task that the user needs to perform in the system under test. The **user** refers to the characteristics of the possible user group of the system, different influencing characteristics need to be taken into account, such as age, gender, educational background, operating characteristics, etc. Their operating mental models need to be taken into account. **Interaction system** refers to the system being tested
Defining the environment and context of use	The usability of a system cannot exist in isolation from the context in which it is used, and the context determines the criteria for measuring its usability. For example, in-vehicle systems are tested in such a way that the driver's eyes cannot be taken off the road for long periods of time while driving. And there is also the difference between highway driving and driving on mixed roads in urban areas
Defining system usability metrics	Before the system is evaluated, the content of the evaluation and the usability criteria that the system needs to meet need to be determined

It contains ten questions concerning the usability of the system that users have to answer by choosing a response from a five-point scale ranging from Strongly agree to Strongly disagree. The results can be used to analyse the usability of a product or device and can even be used to evaluate the design of different pages. It does not suggest specific improvements, but can often be used to evaluate whether a design is sufficiently usable. The method is based on the participation of real users. The steps are as follows.

Step 1: Create a list of tasks for the device being analysed.
Firstly, the analyst should develop a list of tasks for the product or device to be evaluated. Due to the time constraints of the analysis, the task list should not be exhaustive but be representative of the full functionality of the device. This can be achieved through the HTA task analysis method.
Step 2: User action.
Real users follow the list of tasks to complete all of them
Step 3: Complete the SUS questionnaire
Once the user has completed the appropriate task list, they should be given the SUS questionnaire and instructions for completing it, and the equipment should be analysed in the light of their comments. The order of the 10 questions in the SUS cannot be changed.
Step 4: Calculate the SUS score for the analysis equipment.
Once completed, the SUS questionnaire score is calculated in order to produce a usability score. Scoring the SUS questionnaire is a very simple process. Each item in the scale is scored on a scale between 1 and 5. Item scores are calculated as

follows: The score for the odd-numbered items is the questionnaire score minus 1, and the score for the even-numbered items is 5 minus the corresponding score. The sum of all 10 scores, processed as above, is then multiplied by 2.5. The final figure represents the usability score of the device in the analysis and ranges from 0 to 100.

If we pre-determine that the usability of the system is acceptable if it has a usability score of 60, then if the user scores it below this, the design is not sufficiently usable. It is also possible to compare the usability of different products across the board.

13.8 Self-assessment Model

The Self-assessment manikin SAM (SAM) is a relatively simple method for evaluating user experience. It is a graphical emotion assessment method developed by Lang and Bradley (Lang 1980; Bradley and Lang 1994) to assess emotional changes and feelings during manipulation, and is commonly used to assess the user experience of voice interaction systems, but can also be used to evaluate other types of interfaces. The SAM measures the emotional response to various stimuli using facial expressions and physical changes in graphically depicted human emotions. It consists of three rows of scales, each assessing a different affective attribute: self-esteem, arousal and dominance, as shown in Fig. 13.1. The first row is to evaluate the sense of self-esteem, which denotes whether a person has positive or negative emotions. The second row measures arousal, i.e., whether you feel increasingly excited or bored and unmotivated during the operation. The third row tests the dominance of the system, which measures whether the operation process feels controlled, or cared for.

The assessment requires the user to select the picture in each row expressing best how she feels. Once the user has completed the series and answered the various assessment forms, the SAM can be used to measure how the users feel about the overall experience. SAM has been used extensively in many studies (Grimm and Kroschel 2005; Larsson 2010).

13.9 User Experience Curve

The UX curve (Kujala et al. 2011) is a user evaluation method that aims to measure changes in user experience over time. The approach is to ask the user to record the problematic points that make for good and bad experiences on a daily basis over a period of time when they use the product in their own daily situation. A curve is then plotted to indicate the positive and negative experiences the user has had with the product. The method is very simple, simply providing the user with a pen and paper and drawing a time line as shown in Fig. 13.2. Positive experiences will be labelled as points above the line and negative experiences will be labelled as points below the

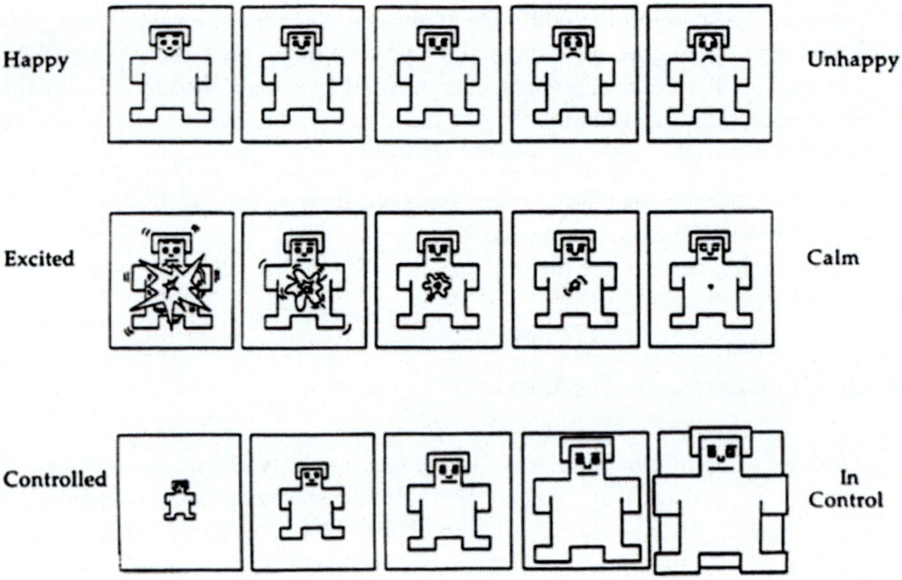

Fig. 13.1 Self-assessment manikin scale (from top to bottom) Self-esteem, arousal and dominance. *Source* Lang (1980), Bradley and Lang (1984). SELF ASSESSMENT MANIKIN © Peter J. Lang 1994. Reprinted with permission

line. The user can assess the importance of an experience and mark it on the paper accordingly. A line is then used to connect all the points, resulting in a UX curve. The curve provides information about the user's experience over time, without the need to constantly track the user as they experience the interaction with the product.

The evaluation methods mentioned here are all static methods that are suited for evaluation of user interfaces when used in isolation, i.e., not in combination with other tasks, and most of them do not take into account the driving scenario and the main task that the driver needs to perform in order to interact with the vehicle cabin. The dynamic evaluation methods for in-vehicle systems are described in detail in Chap. 14.

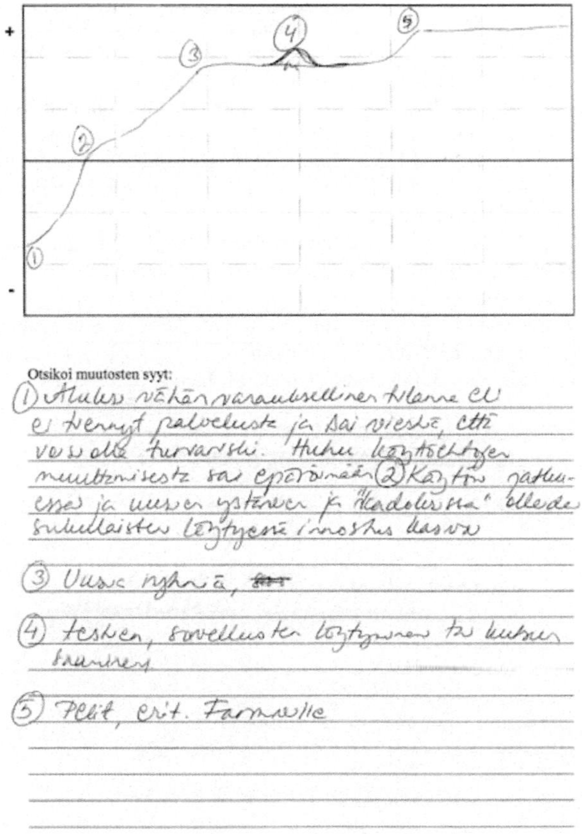

Fig. 13.2 User Experience Curve. Translation: "(1) First, I was little reserved or I did not know about the service and I got information that it can be a safety risk. A rumour about changing conditions of use made me hesitate. (2) I found new friends and "lost" relatives and my enthusiasm increased. (3) New groups. (4) Finding test, applications or getting invitations. (5) Games, especially Farmville." Courtesy S. Kujala. Reprinted with permission

References

Bradley MM, Lang PJ (1994) Measuring emotion: the self-assessment manikin and the semantic differential. J Behav Ther Exp Psychiatry 25(1):49–59

Brooke J (1996) SUS: A "quick and dirty" usability scale. In: Jordan PW, Thomas B, Weerdmeester BA, McClelland IL (eds) Usability evaluation in industry, pp 189–194. Taylor & Francis, London, UK

Buskermolen DO, Terken J (2012) Co-constructing stories: a participatory design technique to elicit in-depth user feedback and suggestions about design concepts. In: Proceedings of the 12th participatory design conference: exploratory papers, workshop descriptions, industry cases-volume 2, pp 33–36

Grimm M, Kroschel K (2005) Evaluation of natural emotions using self assessment manikins. In: Automatic speech recognition and understanding, IEEE workshop, pp 381–385

Hassenzahl M, Burmester M, Koller F (2003) AttrakDiff: Ein Fragebogen zur Messung wahrgenommener hedonischer und pragmatischer Qualität. In: Mensch & computer, pp 187–196. Vieweg+ Teubner Verlag

Kujala S, Roto V, Väänänen-Vainio-Mattila K, Karapanos E, Sinnelä A (2011) UX Curve: a method for evaluating long-term user experience. Interact Comput 23(5):473–483

Lang PJ (1980) Behavioral treatment and bio-behavioral assessment: computer applications. In: Sidowski JB, Johnson JH, Williams TA (eds) Technology in mental health care delivery systems. Ablex Publishing, Norwood, NJ, pp 119–137

Larsson P (2010) Tools for designing emotional auditory driver-vehicle interfaces. Auditory Display 5954:1–11

Laugwitz B, Held T, Schrepp M (2008, November) Construction and evaluation of a user experience questionnaire. In: Symposium of the Austrian HCI and usability engineering group, pp 63–76. Springer, Berlin, Heidelberg

Lund AM (2001) Measuring usability with the USE questionnaire. Usability Interface 8(2):3–6

Nielsen J (1994) Heuristic evaluation. In: Nielsen J, Mack RL (eds) Usability inspection. Wiley, New York

Pettersson I (2017) Travelling from fascination to new meanings: understanding user expectations through a case study of autonomous cars. Int J Des 11(2):1

Shneiderman B, Plaisant C, Cohen M, Jacobs S, Elmqvist N, Diakopoulos N (2018) Designing the user interface-strategies for effective human-computer interaction. Pearson Education Limited

Stanton NA, Salmon PM, Walker GH, Baber C, Jenkins DP (2005) Human factors methods—a practical guide for engineering and design. Ashgate

Chapter 14
Driving Simulator Applications

Abstract In the context of driving, driving simulators are an indispensable tool for conducting concept evaluations. In this chapter, different types of driving simulators are discussed, and advantages and limitations of the use of driving simulators are considered. Furthermore, we go into the practicalities of setting up driving simulator tests, and different types of measurements that may be elicited in driving simulator tests are summarised.

Driving simulators are an indispensable tool in the study of automotive interaction design and come in a variety of forms, from simple computers with a steering wheel for games to giant high-fidelity, multi-dimensional and dynamic driving simulators. Different car companies, research institutes and schools are constantly creating and installing new driving simulators, depending on their needs and on the development of automotive technology. The variety of driving simulators and the problems that need to be studied using them are endless. The use of driving simulators presents a variety of challenges. In the small space of this chapter we can only touch on some of the basic issues, but the technical aspects of driving simulators are beyond the scope of this book. The treatment here draws heavily on the Handbook of Driving Simulation for Engineering, Medicine, and Psychology (Fisher et al. 2011a, b).

Driving simulators are used for a wide range of purposes, for the training of new drivers, for the psychological rehabilitation of road accident victims, for skill reconstruction, for treatment methods for a number of diseases such as Parkinson's syndrome and chronic insomnia and, of course, for the design of cars and research into related issues. Here we focus on the application of driving simulators for the design and study of vehicle interaction.

The advantage of using a driving simulator over driving on real roads is that it can simulate all kinds of road scenarios, all kinds of traffic conflicts, all kinds of environmental changes, all kinds of different automated/assisted technologies and all kinds of interaction designs. The main purpose of the simulator is to observe the driver's actions in these environments in order to complete specific tasks, and to make detailed observations and quantitative records of the driver's driving behaviour and task actions, which can be repeated several times under the same conditions.

© China Machine Press 2023

F. Chen and J. Terken, *Automotive Interaction Design*, Springer Tracts in Mechanical Engineering, https://doi.org/10.1007/978-981-19-3448-3_14

14.1 Uses, Advantages and Disadvantages of Driving Simulators

As mentioned earlier, driving simulators have a variety of uses and here we give a rough summary.

- ***Various types of training*** and evaluation of the results of training and education, that is, as part of the training and testing for those who need to obtain a driving licence and those who need to regain it. This is perhaps the area where driving simulators are most widely used and where the largest number of people use them. Simulators for training are not only used in the automotive industry, but also in the three main areas of water, land and air, and in many industries that require training in the operation of complex systems, and there is a great deal of potential to be explored in this area.
- ***Traffic safety***, traffic accidents, traffic control systems research and evaluation. Before the construction of roads, in the development and design phase, you can use the simulator to study and test, so that you can avoid problems found only at a later stage of construction, therewith reducing investment costs. For traffic safety education work for the general public, driving simulators can let people immerse themselves in the real experience of the accident process, so as to improve the educational effect.
- ***Evaluation of vehicle design***. Evaluation of new vehicle technology and equipment; systematic research into driving ability disorders: this is currently a major application for driving simulators, but it requires a high level of driving simulator technology.
- Understanding the ***limitations of human driving control***: the applications of this will be the main focus of this chapter, so we will not go into more detail here.
- ***Games***: There are countless games related to car driving, some of which, in turn, have been used in human factors research in driving.
- ***New car technology***: Nowadays there is an increasing amount of in-car technology, and even car design professionals are not always sure how these new features work, so it can be very effective to use a driving simulator to create a showroom to recommend new designs and technologies to the public.
- Study of the effects of various ***drugs*** on human control, information processing, concentration and fatigue.

The use of a driving simulator, as opposed to doing tests with a real car, has the following advantages.

- It is possible to put the driver in various ***extreme scenarios*** without fearing that the driver will be harmed. For example, traffic accident scenarios, drug effects, fatigue, various types of distractions, extreme weather and the use of new technologies, are dangerous scenarios in which the driver's reactions are observed and it is investigated how the new design system can help the driver. It is difficult to simulate these scenarios with a real car. Not to mention that placing drivers in extreme scenarios is ethically unacceptable.

- ***Replicability***: In the laboratory, special events and scenarios, as well as different influencing factors can be strictly controlled and repeatedly experimented with, in a flexible and cost-efficient manner, while accurate quantitative records of the driver's behaviour can be made. Because many participants can be studied under the same conditions, estimates of the effects of the variables under study are less affected by individual differences and become more reliable.
- Many of the ***confounding factors*** that influence driving can be progressively cleared out through experimental design, such as traffic conditions, lighting, pedestrian influences, accidental occurrences, etc.
- Basic research in ***cognitive psychology***, such as reaction speed, perception of complex systems, multimodal interaction, etc. Even simple driving simulators can also be used for many research tasks. The emotional reactions of drivers in simulators are comparable to real driving and can therefore also be used for emotional research.

However, the use of a driving simulator has many shortcomings relative to driving a real car on real roads, and therefore it is not a complete replacement for real car driving tests. These shortcomings are.

- There is a natural difference between the psychological effect on drivers of traffic accident scenarios simulated on a driving simulator and being involved in a traffic accident while driving on a real road.
- Driving in a driving simulator is different from driving in a real car, so that participants in experiments need to familarise with driving in the simulator. It is an open question how long the familiarisation should take.
- The chaotic factors encountered in driving cannot be replicated by the simulator in a completely realistic way, so the results of the research in the driving simulator need to be verified in real road driving.
- The driving simulator does not simulate all the scenarios that may take place on real roads.
- Different simulations are available for different studies and it is difficult to give a complete answer here.
- A person can drive for two hours on real roads without feeling fatigued, but on a driving simulator it is generally recommended to drive for no more than half an hour continuously. Generally when arranging scenarios, the density of events that occur on a driving simulator will be much higher than driving on real roads.
- There will be differences between the subject's driving behaviour in the driving simulator and his driving behaviour in his own car. The driver's behaviour will be influenced by the laboratory environment.
- The use of driving simulators to train new drivers in driving skills is not a complete replacement for real car driving training.
- Driving simulators often induce motion sickness, which is an unresolved problem at present. The more immersive the simulator, the more intense the motion sickness. This motion sickness can affect the results to a certain extent, and it results in drop-out of participants.

These limitations affect the external validity of the results (the generalisability of the results to real road situations). Therefore, awareness of these limitations is important in interpreting the results.

14.2 Experimental Data Acquisition for the Driving Simulator

What data are collected from driving simulator experiments depends entirely on the purpose of the study and is also related to the level of technology of the driving simulator. The skill of collecting and analysing data is a test of the researcher's knowledge and experience, so here we can only give a general overview of the various types of data possible. The specific definitions of each parameter in the driving simulator are to be found in the SAE standard document (J2944-201506: Recommended Practice, Operational Definitions of Driving Performance Measures and Statistics). This standard provides definitions and guidelines for performance measures and statistics relating to the functioning of vehicles driven on the road. It sets standards for the measurement, calculation and statistics of relevant parameters in SAE and ISO reports, journal articles, papers, technical reports and presentations related to vehicles and driving, in order to make the results of various experiments comparable. It is important to emphasise here that companies and laboratories that do research related to automotive driving build their own driving simulators. Driving simulators may come from different suppliers and some laboratories develop their own hard/software systems. Regardless of the source, it is recommended that the definitions of the indicators are strictly based on this standard. In the reports that are provided, and especially in published scientific articles, it must be stated that the parameters are defined with reference to this standard. The standard provides lateral and longitudinal positioning of road vehicles and includes mainly data measurements and statistics related to driver/vehicle responses. For metric and statistical information related to eye movements, readers are referred to ISO 15007-1 (ISO 2002) and SAE J2396 (SAE 2007). It is important to note here that the data output from each simulator should ideally be defined strictly according to this standard for their measurements; otherwise, conclusions drawn may not be comparable to similar studies.

Different types of data can be measured in an experiment with a driving simulator. Table 14.1 gives a broad framework, as opposed to specific experiments, and the data measured will vary depending on the research question. The methods of data analysis also vary. In general, the amount of data that can be output from each driving simulator is large and varied. The general practice is to collect as many raw data as possible if it does not compromise the speed of data collection, and to further process the data during the data analysis phase, about which we will talk in Chap. 15.

It is important to note that not all of the parameters in Table 14.1 need to be measured for every experimental study; the choice of parameters depends on the

Table 14.1 List of data commonly used in driving simulators

Classification	Data type	Variables
Driving simulator data	Longitudinal control	Speed, change of speed, acceleration, distance from the vehicle in front
	Reaction time	Probe Detection-Response Time PDT, Brake Response Time BRT, Time to Contact
	Collision (crash)	Number of collisions, collision status, TTC (time to collision)
	Lateral control	Lateral Position LP, Standard Deviation of LP, lane exceedance LANEX, Time to Lane crossing TLC, number of steering wheel changes per unit time, angle of change, reversal rate RR, distance travelled per unit time
Eye movement data	Eye movement measurement	Glance, eye-off-road-time, fixation, percent Dwell time in a fixed area
Psychological data	Subjective scales	Subjective Workload Measure NASA-TLX, Situation Awareness Global Assessment Technique (SAGAT), Driving Activity Load Index (DALI), Self-Assessment Scales, Big Five Personality Scales and many other types of data
Physiological data	Physiological measurements	Heart Rate HR, HR Variability, respiration, ECG, skin conductivity, facial expression from video recordings
In-vehicle task performance data	Task performance indicators	Error rate, safety, line of operation, task completion rate Video recordings and screen recordings

problem to be studied, but the parameters given in this table are the most common. For driving simulator experiments, multiple parameters are generally required to be evaluated because different drivers adopt different approaches and strategies to driving in response to different driving tasks. For example, when driving and simultaneously using the phone, an operation that is known to have an impact on driving safety, different people drive differently when on the phone, some slow down, some keep the car as close as possible to the right of the lane line, some drive in a zigzag pattern, some press the left lane line. Even the same person, at different points in time, may be driving differently.

14.3 Measurement of Physiological Parameters

Physiological indicators are natural indicators of mental activity and workload, because any work or activity, both mental and physical, has a basis in physiological activity, which is reflected in different physiological indicators. The measurement of physiological parameters in driving simulator experiments has become a routine measure (Brookhuis and de Waard 2011).

There is a large body of research showing that both physical and mental work causes changes in heart rate (De Waard and Brookhuis 1991; Mulder 1986, 1992); stress causes changes in skin electricity (Boucsein 1992), blood pressure (Rau 2001) and respiration (Mulder 1992; Wientjes 1992). Heart rate variability (HRV) (Mulder et al. 2004) is more sensitive to psychological load than heart rate itself. Brain waves can reflect event-related brain activity (Kramer 1991; Kramer and Belopolsky 2004), and certain facial muscles map onto facial expressions in response to external influences (Jessurun 1997). The measurement and judgement of different physiological indicators are discussed below.

1. ***Electrical skin reaction***

   The most commonly measured parameter on the skin is electro-dermal activity (EDA). EDA includes the well known galvanic skin response (GSR), galvanic skin potential, peripheral autonomic surface potential, etc. Electrical skin recording is actually a very old physiological method dating back to the nineteenth centuries (Boucsein 1992, 2004). GSR is caused by sweat produced by glandular activity influenced by the autonomic nervous system. Most researchers consider GSR to be the result of a combination of arousal, stress–strain and emotion, as an expression of autonomic nervous system activity. We measure GSR to understand the degree of orienting response and adaptive habituation, and in research it can be used to estimate the information processing capacity of the human brain during a task and to establish arousal or to understand stress levels, especially for negative emotions (Boucsein 1992). A progressive increase in GSR reflects an increased level of arousal and represents the body's readiness to begin action (Boucsein 2004). GSR has been used to monitor the workload and (emotional) mental strain of a task. GSR is easy to measure but not always easy to interpret. The main disadvantage of its use in driving simulators is the need to fix electrodes on the palms of the hands (and sometimes the feet), as these locations give the best results. This can have an impact on driving. Other possible locations are the shoulders, but this may be experienced by participants as intrusive.
2. ***Electromyography***

   Electromyography (EMG) records the electrical signals emitted during muscle contraction and can be used to detect muscle activity. The muscle evokes potential changes during contraction, but it requires advanced semi-conductor technology to extract important information from it. The electrode signal has other noises that need to be cleaned up (Goebel 2004). In the last few decades, advanced electronic instrumentation and powerful analytical methods have allowed EMG to be applied in research. Muscle forces are made up of the activity of many motor units, each with a different discharge rate (5–50/s). Surface electrodes are used to collect the sum of the different unit potentials under the skin. The average EMG signal amplitude increases with muscle contraction. However, when applying this parameter, it is necessary to be able to precisely locate the area and activity of the muscle to be measured. The shoulder and neck muscles contract when a person is tense, so EMG is often recorded in these areas.

3. ***Electrocardiogram***

The activity of the heart is regulated by an autonomic rhythm emanating from the sinus arteries of the heart, which is jointly regulated by the sympathetic and vagal nervous systems. An electrocardiogram (ECG) allows the activity of the heart to be measured relatively easily by means of three electrodes attached to the body's chest. Alternatively, pulse oximeters can be attached to the index finger to measure the heartbeat through Photoplethysmography (PPG), and are therefore less intrusive. However, PPG gives less reliable measurements than sensors attached to the chest, and PPG measurements are very vulnerable to motion artefacts (Lu and Yang 2009).

The activity of the heart is associated with physical work, physiological demands, and mental effort. In the driving simulator experiment, the duration between each heartbeat is used as an indicator of the intensity of the mental effort. Heart rate (HR) is the number of heartbeats in a fixed period of time (usually one minute), while the average heart rate or inter-beat interval (IBI) is the value over a specified period of time. The duration of the heartbeat is variable, with different oscillation patterns, and the time difference between two wave crests in the recordable ECG is also known as heart rate variability (HRV) (Kramer 1991), which reflects the time variation between each heartbeat and is also known as the R-R interval or heartbeat interval, as shown in Fig. 14.1. A decrease in HRV is demonstrated during the task when the subject has to exert

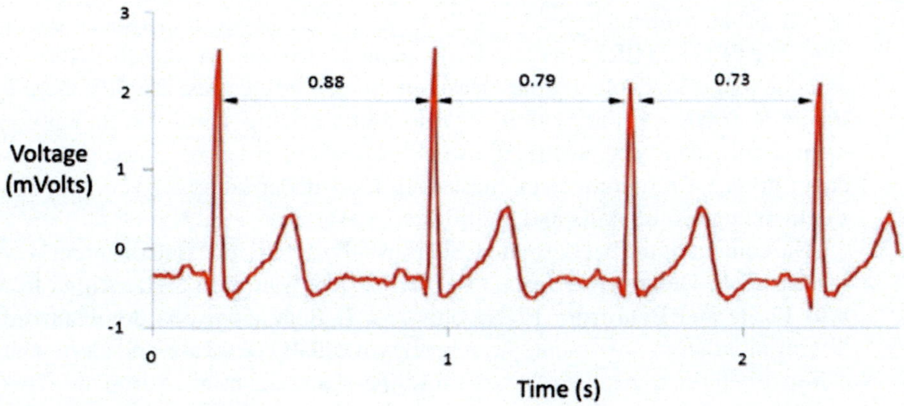

Fig. 14.1 Heart rate variability measurements

mental effort, especially at the 0.10 Hz frequency, which is usually clearer than in the resting situation, the effect depending on the amount and type of mental load (Mulder 1992; Mulder et al. 2004).

4. ***Blood pressure***

   Ambulatory blood pressure monitoring (ABPM) was originally developed for clinical purposes as a dynamic assessment indicator of strain in workload related tasks (Rau 2001). ABPM allows automatic, repeatable, non-invasive recording of arterial blood pressure with the aid of a portable recorder; including systolic and diastolic blood pressure, mean arterial blood pressure, etc. Therefore, the use of ABPM in simulators is feasible, especially in driving task studies.
5. ***Breathing***

   Mental load results in an increase in respiration rate and a decrease in respiration variability. There are two parameters for measuring respiration: respiratory depth and frequency (Wientjes and Grossman 2004). Depth of breath is usually expressed in terms of tidal volume (i.e., the volume of air expelled in a single breath); respiratory rate is the number of breaths per minute. The volume of air expelled per minute is the product of tidal volume and respiratory rate, a value that is usually associated with the body's metabolic activity. Other parameters that evolve from this include the duration of each phase of the respiratory cycle (inspiratory and expiratory time), total cycle time, average inspiratory rate (tidal), and duty cycle time. Measuring gas exchange involves calculating the volume or measuring the oxygen consumption per unit time (VO_2), as well as the amount of carbon dioxide (VCO_2) produced. In this way energy consumption can be calculated (Wientjes and Grossman 2004). However, not all parameters are suitable for use in driving simulator studies.
6. ***EEG and event-related potentials***

   Analysis of the 'raw' EEG or background EEG (i.e., in the range of approximately 1 and 30 Hz) can reflect the level of activation in the brain. Event-related potentials (ERPs) are sequences of transient brain voltage oscillations that can be distinguished from background EEGs as the brain's response to a specific stimulus. The EEG and ERP are recorded from the scalp using (AgAgCl) electrodes, and due to the low voltage, the measurement procedure must amplify them by orders of 1000×. If the EEG is measured in a driving simulator, its environment should ideally be electrically shielded to avoid excessive noise, otherwise data filtering is required. Even measuring EEG under laboratory conditions is a demanding test of skills and facilities.

   The content of the background EEG is usually subdivided into different wave regions, with Delta waves from 1 to 5 Hz, Theta from 5 to 8 Hz, Alpha from 8 to 12 Hz and Beta from 12 Hz onwards. If Beta activity is dominant in a driving simulator experiment, the subject will usually be awake and alert, while a drop in activity to Alpha indicates drowsiness, and a continued drop into Theta indicates drowsiness or sleeping. Therefore, the EEG is considered to be the best indicator of an operator's alertness and vigilance in driving situations (Åkerstedt 2004).

ERP is usually defined in terms of polarity (P or N) and waiting time (in milliseconds) relative to a particular stimulus. It reflects many different perceptual, cognitive and motor processes in the brain and thus provides a useful signal for how the human brain breaks down and processes information in complex task situations (Fabiani et al. 2000). ERPs have been used to study brain activity such as operator performance in vigilance conditions, cognitive processes, mental load, fatigue, etc. for over 40 years.

7. ***Eyelid movement***

Eyelid activity, especially the degree of slow eyelid closure, can be used to detect sleepiness (Åkerstedt 2004). The percentage of eyelid closure or PERCLOS (Wierwille and Ellsworth 1994; Wierwille et al. 1994), is inversely associated with visual vigilance (Mallis and Dinges 2004). PERCLOS is achieved by video scoring of slow eyelid closure.

8. ***Eye movement***

The study of eye movements has become increasingly important in understanding information acquisition in vision (Fisher et al. 2011b). The position in which the eye rests (i.e., where the eye is looking) is directly related to the information being acquired and processed. This is because when the brain is processing a piece of information, the time spent gazing at an object is closely related to the difficulty of processing information about that object. The duration of gaze and the location of gaze are important to analyse. In driving scenarios, eye movement data are already routinely collected.

Standardised terminology for eye movements in driving research is defined by ISO 15007-1 (2002) and SAE J2396 (SAE 2007). We need to adhere to these definitions. There are basically three states of eye movement: fixation, sweeping (saccades) and smooth pursuit movements. Consider first the static situation, such as reading a road sign while stopped at a red light, where the eye is only fixating and sweeping; the eye spends most of its time fixating, remaining essentially stationary, and these fixations are interspersed with rapid eye movements called sweeping movements or saccades. During sweeping, the vision captures information from the outside world at random (sometimes called saccadic suppression), when there are blurred motion images on the retina that are obscured by the visual information of the subsequent gaze. Thus, in the case of fixation gazes interleaved with saccadic movements, the only meaningful information is that captured during gaze. We are usually not interested in the details of the sweep. The duration of the sweep is essentially dependent on the distance of the sweep (longer sweep distances take more time).

In the presence of moving objects, motion tracking is an important part of eye movement and the observer will even focus on the position of moving objects in the environment by moving his or her body. Smooth tracking movements occur when the driver tries to keep both eyes focused on a moving object. These movements are much slower than the saccadic movements and, more importantly, the incoming visual information is not suppressed during such movements. For the driver, moving objects on the road appear to move in a near-linear direction outside a certain distance range. When the human eye is tracking a moving object in front of it, the eye does not move because the object is always directly in front of it, a situation that is often mistaken for gaze. However, as the vehicle approaches it, the moving object moves more and more away from the centre of vision.

14.4 How Do I Choose the Right Driving Simulator?

There are many differences between driving simulators, but in essence, it is a question of degree of fidelity (Table 14.2). The degree of fidelity reflects the difference with real driving, so the question that many researchers ask is: what degree of fidelity do I need for my research? Is the higher the level of fidelity, the better? As we all know, the higher the level of fidelity, the more expensive the simulator will be, and the cost per use will be high.

Of course, the above classification is not absolute, the current driving simulators are very malleable and one may connect a high definition monitor with a very simple simulator for games and 3D virtual driving simulators are increasingly favoured.

So before choosing a simulator, first consider what your research question is and what factors will affect the results. When choosing a simulator, it is best to focus on one that can simulate these factors to meet the functions you need to test and record the measured parameters, and ignore the other factors for now. For example, do you need to know if the infotainment system you are designing will produce enough driving distractions to threaten driving safety? Such questions are verified by driving simulator experiments. In a driving simulator, a driver is given a certain driving scenario and is asked to use the infotainment system to perform specific tasks in different road conditions, while his driving behaviour, eye movement data, and task actions are measured. In such a test, whether the driving simulator is dynamic or not does not have an essential effect on the test results. Therefore, it is possible

Table 14.2 Driving simulators can be classified according to their level of fidelity

	Low fidelity	Medium fidelity	High fidelity
Base	Fixed	Fixed or with small movements	Dynamic plinths
Screen width (degrees)	20	150	360
Signal legibility	Not good	Medium	Good
Night visibility	Does not have the capacity	Not good	Yes

to choose a relatively simple static driving simulator. But it is important to note that the more advanced the driving simulator, the better the data will be.

Of course, if the problem to be studied involves special lighting conditions (e.g., night, dawn or fog), high definition displays, driver response to vehicle dynamics, perception of visual afterglow, etc., then there are special requirements for driving simulators. Different levels of autonomous driving technology are being developed and the human behaviour in relation to different interaction designs needs to be investigated, but in many cases the driving simulators available may not be able to meet the required testing requirements due to technical limitations.

14.5 Comparison of Simulator Driving and Real Road Driving Results

The benefits of using a driving simulator in terms of experimental control and collecting measurement data as opposed to driving a real car on real roads are obvious, so the most frequently asked question is: to what extent do the results obtained from driving simulator experiments reflect real road driving? This question is also known as simulation validation.

We can consider this validation on two levels, one is relative validity, i.e., the measured trend of change is the same as the trend of change in real road driving. For example, when there is an obstacle in front of the vehicle, or a car overtakes at close range, the driver's reaction in the simulator, either by slowing down or accelerating to change lanes, is the same as that observed on the real road. The other is absolute validity, which, as the name suggests, means that the data measured on the driving simulator are the same as those measured on the real road. These two validities are different for different driving simulators and the manufacturers of the simulators are supposed to give detailed explanations, but the reality is much more complex and it is often up to the researcher to verify this. Because of this difficulty, most researchers consider relative validity to be more significant than absolute validity. Therefore, the majority of studies using driving simulators are concerned with the relative validity of the data. The relationship between relative and absolute validity is illustrated in Fig. 14.2.

A number of comparisons have been made between driving simulator data and real-world data, but the results can vary due to the simulator technology itself. A brief review of the results of such comparisons in Fisher et al. (2011a, b) shows that people drive faster in the simulator than on the real road. The lateral control on the simulator is not as good as on the real road. On the simulator, subjects do not pay as much attention to dangerous driving as they do on the real road and therefore have accidents much more frequently than they do on the real road.

Scientists have different attitudes towards the significance of driving simulator data and real driving data, some agreeing and some disagreeing. There are many influencing factors here, and comparative studies are not easy to develop. Although

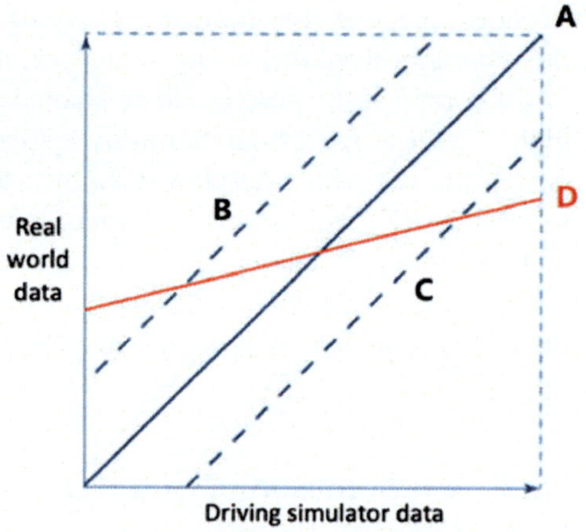

Fig. 14.2 Relationship between absolute and relative validity, where line A represents absolute validity, while lines B and C represent relative validity, and line D indicates invalid results

it is possible to partially reconstruct real road scenarios on driving simulators, they are generally simplified road scenarios, whereas on real roads there are many uncertainties that cannot be easily controlled and some driving data cannot be measured, making comparative studies difficult. From the published articles, it appears that some experiments produce very similar results and some do not. When reading such articles, it may be necessary to look at the detailed experimental comparison design to determine the significance of the results in question.

In the EC project HASTE, Swedish scientists (Engstrom et al. 2005) compared two different driving simulators, one high-fidelity dynamic and one low-fidelity static, using a visual search and memory load task. The two simulators were identical except for the dynamics, and the software architecture was the same for both simulators. They found that most of the results obtained with both simulators were similar, the most significant difference being in the lateral control of the driver, which seemed to be worse with the static simulator.

How much universal significance do the results of driving simulators have? This is the same question as other questions about the extent to which the results of laboratory studies are indicative of the real world. If the tasks that drivers have to perform are comparable, the traffic environment is comparable to the real world, and the population of subjects chosen is representative, then the results should also be comparable. Again, this needs to be viewed in two different ways. One phenomenon that is not easy to explain is that in many experiments on mobile phone use and driving and talking on the phone there are often crashes, whereas in the real world almost everyone drives and talks on the phone at the same time, but crashes due to mobile phone use are uncommon. Possibly, when driving in real world circumstances, drivers tend to regulate their mobile phone use according to the traffic situation, avoiding phone use in dense traffic situations, whereas in experiments scenarios are set up to investigate the effects of mobile phone precisely in critical situations.

So, can we say that if the experiment was conducted in a real vehicle, on a real road, it is going to be of more value than the data from a driving simulator? No, because there are many uncertainties on the road in a real car, and one or more of these uncertainties may directly affect the data, but they are not the factors of interest to the researcher. Therefore, the conclusions drawn from them are not necessarily more meaningful than those from a driving simulator. So, it is the causal analysis behind the data that is most important. The factors that pose a threat to internal validity and external validity in the driving simulator experiment are summarised in Table 14.3 (Fisher et al. 2011a, b).

As can be seen from Table 14.3, many of these factors that pose a threat to the results relate to issues of research methodology and experimental design. This aspect will be discussed in Chap. 15.

14.6 Scenario Making

In the context of driving simulator studies, the term 'scenario' refers to the environment that is offered to the participant, including the type of road, the weather conditions, the traffic conditions, other road users, traffic signs, the events, the environment. Scenario production is like writing the script for a play: it determines what the subjects will do on the driving simulator, how they will do it, what factors will influence it and so on, and we need to use it to achieve the desired test results. The production of scenarios is particularly important when using a driving simulator for scientific research. It is important to emphasise that using a driving simulator is a study of the behaviour of subjects under controlled conditions and it is very different from driving a real car, so it is not true that the closer the scenario of a driving simulator is to a real road scenario, the better. The design of the scenario is highly relevant to the question that the researcher needs to study. The key point of scenario creation is to highlight the key elements that need to be studied and ensure that they are manageable and repeatable. At the same time, factors that apply on real roads but are not relevant to the topic under study can be ignored in the scenario production. Some factors, which may have an unwanted effect on driving data, such as curves, or roads that go up and down hills, need to be considered carefully.

Two factors make the design of the scenarios in the study difficult. One is that we do not fully understand driving behaviour: human driving behaviour is very complex and it is difficult to fully replicate road driving on a driving simulator. In the driving scenarios, in addition to the car, we also include other vehicles and other road users, such as bicycles and pedestrians. These virtual road users are programmed to behave in a way that is up to the programmer and is designed to create a driving environment and set driving conditions that are relevant to the research question. For example, pedestrians are designed to suddenly cross the road to test the driver's reaction time, or vehicles in front are brought to a sudden stop to see how the driver reacts. The degree of complexity of simulating such dynamic road scenarios will vary from one driving simulator to another, and is a particularly complex issue when considering the capabilities of the various driving aids on board, or the ability to drive automatically. The second factor is the diversity of driver behaviour. Drivers' driving behaviour

Table 14.3 Analysis of factors that threaten internal validity and external validity in the driving simulator experiment

Threat factors	Description	Solutions
Subject selection	Anomalous data are generated due to personal health factors in some subjects, masking the true effect of the data (this is a very common problem)	During the course of the experiment, it is important to observe the data in detail and, if this is found, to remove the data as invalid. Covariate analysis can also be used to remove data. In general, it is advisable to screen subjects to prevent such events from occurring
General significance of the results	The task chosen for the experiment, the background of the users and/or the test environment is different from the one you are interested in	This issue needs to be considered very carefully in the design of the experiment and needs to be as consistent as possible, if not quantitatively, then at least qualitatively
Dropping out of the experiment due to dizziness in the driving simulator	Dizziness caused by the simulator itself, or by the task being measured, is a common problem with driving simulators	Choose your subjects carefully and try to avoid turning movements, which can cause dizziness
No randomisation of subjects, experimental conditions, or test events during the experiment	The order in which the independent variables appear in the test must be random for different subjects, otherwise systematic bias, and predictability problems, will arise	During the experiment, the experiment must be arranged strictly according to the principle of randomness, otherwise the validity of the data is flawed
Learning effects due to carry-over effects	A subject has a learning effect if he repeats the experiment several times; the order of the experimental conditions can have an asymmetric effect on the results, or systematic bias	If a learning effect may be expected, subjects should not be allowed to participate in repeated experiments, and ways to counteract this effect and eliminate sequential effects should be considered in the experimental design
Too few subjects in each test unit	Too few participants can affect the stability of the data	Increase sample size and change experimental design

is influenced by a variety of factors, such as their mood, their nervousness about the driving task, or the way they avoid boredom, which makes them take different approaches when dealing with road conditions. They will constantly adjust their speed, their position at the lane line, the distance to the car in front of them, and even the level of risk taking. Not only do they vary greatly from one individual to another, but even from one drive to another by the same driver. The technical performance of the driving simulators themselves, the scene animation effects they produce, can also have an impact on driving behaviour. This makes each driving simulator experiment unique in its own way.

Any driving scenario is divided into two parts: dynamic and static. The static part refers to the road itself: the main city road, the motorway, the mixed road, the straight or the curved road, the trees, the buildings, the junctions, etc. The dynamic part refers to the various virtual vehicles, animals, pedestrians, etc. that travel on the road and how they move. As a general rule, a driving simulator experiment should not require the subject to drive for too long—30 min is appropriate. In this way, we usually arrange for multiple events to occur during the half hour, or for similar events to be repeated in seemingly different circumstances. Sometimes even events are simulated that may easily cause traffic accidents, so that they would cause far more accidents on the driving simulator than in normal driving. One of the most critical aspects of dynamic scenario design is that the experimenter needs to understand what kind of roads you expect the scenario you are studying to take place on? What are the surrounding traffic conditions like? What would you expect the driver's actions to be? Minimise any distractions other than the reactions you need to measure.

Driving simulators have been used for scientific research for more than 50 years. There was a desire to standardise the scenarios so that comparisons could be made between different studies. However, this attempt was not successful. The reason for this is that there are many different problems that need to be studied with driving simulators, and different scenarios and experimental designs are needed for different problems. So here, too, we do not try to do such a useless exercise. Rather, we are trying to give you more information about the principles and knowledge. At the same time, the practical experience of each individual researcher is essential.

14.7 Psychological Factors in Simulator Driving

There is one important factor in real driving that is difficult to imitate with a driving simulator, and that is the purpose of people's trips. In normal life, people don't drive for the sake of driving, every time they travel, they have a purpose, and these purposes affect the driver's state of mind, which in turn affects driving behaviour. For example, when a father receives a call that his son has been hospitalised due to illness and he must rush to the hospital immediately, his driving mentality at this time will be very different from the state he drives to and from work every day, he will ignore some dangerous factors on the road because he is thinking about the state of his child, and even produce phenomena such as running red lights and speeding. The behaviour in such a state of mind is very difficult to simulate in a driving simulator.

It is because we have a purpose for each trip, which cannot be simulated on a driving simulator, that the instruction, the wording of the guidance before the experiment is very important when doing guidance to the subjects. Each subject comes to the laboratory and they think, what is the purpose of this test? What is expected of them by the experimenter? If we say to the subject, "You will drive in the simulator as you have always driven", we are not taking into account the variability of the driver's behaviour due to the purpose and circumstances. If you say "you will drive on the simulator just as you have driven in the past after a party", then you are beginning to take into account the influence of the purpose of the trip on behaviour. In driving simulator experiments, it is important to give the subject specific details about the purpose of the trip, otherwise it may not be effective.

We often have subjects being required to complete a number of subtasks in the driving simulator operation. We then need to make it clear to the drivers that safe driving is the priority, otherwise they will focus mainly on completing the subtasks. It follows that careful design is needed of the pre-driving instructions to the subjects. Researchers must provide clear guidance and assistance, otherwise obvious errors may result due to different driving data.

Sometimes, in order to increase participant engagement, we may use monetary or material incentives to influence them, but it is important to be very careful about how this incentive is framed. For example, in the above experiment, if we say that the better the subtask is completed, the greater the reward will be, then subjects will neglect driving safety and prioritise the completion of the subtask. Penalty regulations for traffic violations, such as speeding and other offences, can be delivered in a way that simulates the state of affairs when driving on real roads.

14.8 Data Processing Issues in the Driving Simulator

Typical driving simulators generate dozens of variables at a rate of 30–240 Hz, so the data derived from driving simulator experiments must be simplified into meaningful information that provides a basis for evaluating driver behaviour. Data simplification is the process of transforming raw data into meaningful, trustworthy metrics that can then be analysed. Because the research questions are designed individually for each experiment, with different dependent variables, scenarios, etc., the process of data processing and simplification needs to be tailored for each study. Due to the large volume of data, the conventional way of collating tables can be too time-consuming and labour-intensive. Therefore, it is often a useful investment to design data processing programs to assist in the data collation process. However, this step has to be taken very carefully, as it can easily lead to errors. Sometimes, human driving behaviour does not follow the pre-designed procedure and during data processing, even with a pre-designed procedure, the processing of each set of data needs to be carefully checked.

The question of how to process the driving simulator data needs to be considered at the beginning of planning the experiment, rather than waiting until the data collection is complete. This is because during the design of the experiment, various annotations that are needed for data processing are set aside. Here a problem arises: we may not know what will happen during the experiment. What will the data look like? How will the data be processed? What answers do we want to find from the data? This is where the importance of pre-experimentation (piloting) comes into play. The better the process is, the fewer problems will arise later in the experiment and the smoother the data processing and analysis will be.

The initial data processing and simplification process should be based on the principles that guide the research questions and the theoretical underpinnings of the experiment. Specifically, the variables and their definitions need to be clearly described and the data processing and simplification process should be able to capture the specific dependent variables. The development of the data simplification process should go hand in hand with the experimental design and scenario development. Pre-experiments are a very important part of the process before scenario development is completed and formal experiments begin. The pre-experiments are not only to verify that the experimental steps are reasonable and what precautions to take during operation, but also to check that the data simplification procedure is designed correctly and that it facilitates the data analysis process later on. A good pre-experiment can save a lot of time in the later analysis of the data and provide relevant information for the possible results of the study. This is the more important, as experiments providing invalid data or data that cannot be appropriately analysed are not only a waste of researchers' time but also of participants' time.

References

Åkerstedt T (2004) Ambulatory EEG methods and sleepiness. In: Stanton N, Hedge A, Hendrick HW, Brookhuis KA, Salas E (eds) Handbook of ergonomics and human factors methods. Taylor & Francis, London

Boucsein W (1992) Electrodermal activity. Plenum Press, New York

Boucsein W (2004) Electrodermal measurement. In: Stanton A, Hedge HW, Hendrick K, Brookhuis A, Salas E (eds) Handbook of ergonomics and human factors methods. Taylor & Francis, London

Brookhuis KA, de Waard D (2011) Measuring physiology in simulators. In: Fisher DL, Rizzo M, Caird JK, Lee JD (eds) Handbook of driving simulation for engineering, medicine, and psychology, pp 17-1–17-10. CRC Press

De Waard D, Brookhuis KA (1991) Assessing driver status: a demonstration experiment on the road. Accid Anal Prev 23:297–307

Engström J, Johansson E, Östlund J (2005) Effects of visual and cognitive load in real and simulated motorway driving. Transp Res Part F: Traffic Psychol Behav 8(2 SPEC. ISS.):97–120. https://doi.org/10.1016/j.trf.2005.04.012

Fabiani M, Gratton G, Coles MGH (2000) Event-related brain potentials. In: Cacioppo J, Tassinary L, Bertson G (eds) Handbook of psychophysiology. Cambridge University Press, New York, pp 53–84

Fisher DL, Pollatsek A, Horrey WJ (2011b) Eye behaviors: how driving aimulators can expand their role in science and engineering. In: Fisher DL, Rizzo M, Caird JK, Lee JD (eds) Handbook of driving simulation for engineering, medicine, and psychology, pp 18-1–18-22. CRC Press

Fisher DL, Rizzo M, Caird JK, Lee JD (eds) (2011b) Handbook of driving simulation for engineering, medicine, and psychology. CRC Press

Goebel M (2004) Electromyography (EMG). In: Stanton N, Hedge A, Hendrick HW, Brookhuis KA, Salas E (eds) Handbook of ergonomics and human factors methods. Taylor & Francis, London

ISO 15007-1 (2002) Road vehicles— measurement of driver visual behaviour with respect to transport information and control system—part 1: definitions and parameters

Jessurun M (1997) Driving through a road environment. University of Groningen, Groningen, the Netherlands

Kramer AF (1991) Physiological metrics of mental workload: a review of recent progress. In: Damos DL (ed) Multiple-task performance. Taylor & Francis, London, pp 279–328

Kramer AF, Belopolsky A (2004) Assessing brain function and mental chronometry with event-related potentials (ERP). In: Stanton N, Hedge A, Hendrick HW, Brookhuis KA, Salas E (eds) Handbook of ergonomics and human factors methods. Taylor & Francis, London

Lu G, Yang F (2009) Limitations of oximetry to measure heart rate variability measures. Cardiovasc Eng 9(3):119–125

Mallis MM, Dinges DF (2004) Monitoring alertness by eyelid closure. In: Stanton N, Hedge A, Hendrick HW, Brookhuis KA, Salas E (eds) Handbook of ergonomics and human factors methods. Taylor & Francis, London

Mulder G (1986) The concept and measurement of mental effort. In: Hockey GRJ, Gaillard AWK, Coles MGH (eds) Energetics and human information processing. Martinus Nijhoff Publishers, Dordrecht, the Netherlands, pp 175–198

Mulder LJM (1992) Measurement and analysis methods of heart rate and respiration for use in applied environments. Biol Psychol 34:205–236

Mulder LJM, de Waard D, Brookhuis KA (2004) Estimating mental effort using heart rate and heart rate variability. In: Stanton N, Hedge A, Hendrick HW, Brookhuis KA, Salas E (eds) Handbook of ergonomics and human factors methods. Taylor & Francis, London

Rau R (2001) Objective characteristics of jobs affect blood pressure at work, after work and at night. In: Fahrenberg J, Myrtek M (eds) Progress in ambulatory assessment, pp 361–386. Hogrefe & Huber, Seattle, WA

SAE (2007) Definitions and experimental measures related to the specification of driver visual behavior using video based techniques J2396_200007

Wientjes CJE (1992) Respiration in psychophysiology: measurement issues and applications. Biol Psychol 34:179–203

Wientjes CJE, Grossman P (2004) Measurement of respiration in applied human factors and ergonomics research. In: Stanton N, Hedge A, Hendrick HW, Brookhuis KA, Salas E (eds) Handbook of ergonomics and human factors methods. Taylor & Francis, London

Wierwille WW, Ellsworth LA (1994) Evaluation of driver drowsiness by trained raters. Accid Anal Prev 26:571–581

Wierwille WW, Ellsworth LA, Wreggit SS, Fairbanks RJ, Kirn CL (1994) Research on vehicle-based driver status/performance monitoring: development, validation, and refinement of algorithms for detection of driver drowsiness. Final Report. US Department of Transportation. NHTSA. DOT HS 808 247

Chapter 15
Behavioural Research Methodology

Abstract In the context of automotive innovation, design processes are usually subject to different types of constraints. Validations therefore involve thorough research efforts. In this chapter, we provide a very succinct introduction into behaviour research, describing the process of setting up an experiment, generating hypotheses, designing the experiment and collecting and analysing experimental data.

Evaluating the effects of new vehicle technologies or new interaction designs on driver behaviour usually requires that we conduct experiments, and these studies cannot be conducted without a driving simulator or a car on a real road. Behavioural research methodology is a complex subject that cannot be summarised in a single chapter. However, there are quite a few books to read on the subject, so only a brief discussion of conventional research methodology and experimental design principles will be given here, to avoid making relatively low-level mistakes in research. The experimental methodology discussed here is not only applicable to experiments with driving simulators, but also to any experiments in a general sense. For more in-depth study, please refer to specialist reference books such as Research Methods in Human–Computer-Interaction (Lazar et al. 2010), or a more specialised book, Design and Analysis—A Researcher's Handbook (Keppel and Wickens 2004).

15.1 Steps in Conducting an Experimental Study

The process of conducting experimental research generally takes the following steps.

1. Identify a research question.
2. Translate the research questions into hypotheses to be tested and parameters that can be used to test the hypotheses.
3. Identify quantifiable results and indicators to argue for or against the validity of the hypothesis.

© China Machine Press 2023

F. Chen and J. Terken, *Automotive Interaction Design*, Springer Tracts in Mechanical Engineering, https://doi.org/10.1007/978-981-19-3448-3_15

4. Design the research protocol and develop the research steps; identify the independent variables and the range of variation in the study that can be controlled by the experimenter.
5. Determine the parameters of the dependent variable to be collected, and the appropriate statistical analysis methods to be performed.
6. Calculate the sample size based on the degree of individual variation as known from previous experiments or pilot experiments.
7. Acquire resources, set up research environments (e.g., driving simulator scenarios, etc.) and conduct pilot experiments. Adjust the set up when needed and repeat the pilot experiments.
8. Conduct the formal experiment; create a secure database and collect data.
9. Conduct descriptive and formal statistical analyses to test the validity of hypotheses.
10. Interpret and report the results and suggest relevant future research questions.

These are the conventional steps, written in chronological order. This order is not absolute, as particular steps depend on the problem solved and results obtained in the previous step, and members of the research team will need to revisit one or more of the previous steps before deciding on the next one. Furthermore, there are strong links between these steps, the details of which we discuss next.

In general, the attributes studied can be divided into three categories (Lazar et al. 2010), as shown in Table 15.1. A common type of the descriptive approach to driver behaviour is the observation of the driver controlling the vehicle. The researcher sits next to the driver and observes their driving activities while asking questions that provide insight into the driver's psychological activities during some of the behaviours. Other methods are to install several cameras inside and outside the vehicle to record road conditions, driver behaviour, eye positions, facial expressions and so on. One of the best-known studies of this type is the 100-car naturalistic driving observation (Neale et al. 2005), which found that the driver's driving behaviour changed if the driver was doing other activities in the car, such as eating, talking on the phone, or listening to the radio. However, this observation does not

Table 15.1 Classification of study attributes

Research category	Content of interest	Typical description	Methodology used
Descriptive	Describe a situation or series of events	X is happening	Observations, field trips, focus groups, interviews
Correlational	Identify the relationship between different variables	X is associated with Y	Observation, fieldwork, questionnaire research
Experimental	Identify the causal effects of variables on behaviour	X is the cause of Y	Experiments under controlled conditions

explain why this is the case and the inherent relationship between the different activities and driving behaviour. Correlation studies are designed to establish a relationship between the independent and dependent variables, for example, when a certain amount of change in factor X occurs, a certain amount of change in Y occurs. In the example above, extensive data analysis has shown that when a driver takes his or her eyes off the road for more than two seconds, the driving risk is high. And this interval of two seconds is irrespective of what exactly is being done. Also, the longer the eyes are taken off the road ahead, the greater this driving risk will be. But this approach does not provide a good explanation of the causal relationship, especially between multiple factors. For example, the intrinsic causality between different road conditions, different traffic situations, human fatigue factors, different in-vehicle tasks and so on. They only find statistical correlations.

Experiments in the laboratory are designed to establish causal relations. If we are not able to find a causal relationship and to explain that relationship scientifically, we will not be able to make predictions for other similar scenarios. In this chapter, therefore, our focus is on how to do experiments.

The aim of any experimental research is to find causal relationships between different factors and thus to find scientific laws. A proper research method can help one to find the real causal laws. Research methods can be divided into three main categories, real experiments, quasi-experiments and non-experiments. If a study can use random assignment of participants to experimental conditions (more on this later), this is a real experiment, which is usually done in a laboratory. If a study contains several different conditions and different populations, and participants cannot be randomly assigned to conditions, the study is a quasi-experiment. This type of experiment is generally used for studies with various natural groups, for example to study the effect of a new management system on the productivity of a company's personnel. In this study, we compare the new management system with the old one and identify two different departments within the company, each using one system, to examine productivity. Finally, if it were just an observation of a particular group's behaviour, it would not be an experiment. In this section, we will only discuss real experiments.

A real experiment will generally conform to the following characteristics.

- It has at least one hypothesis, and the purpose of the experiment is to test that hypothesis.
- There should normally be at least two experimental groups (test condition group and control group)
- The dependent variable is continuous data that can be measured
- The results can be analysed statistically
- The results are reproducible, i.e., if the same experiment is done using a different population of subjects, in a different place and time, the same results can be obtained.

15.2 Development of the Research Hypothesis

Before any experiment can begin, we first need to understand what the question is that we need to investigate. This question is not posed randomly. It is important to first understand the context and motivation for asking the question, and what implications there are for having answered it. Then we look through the literature to find out if the question has been studied before. Do the conclusions they draw meet your needs? If so, there is no need to repeat the same research, unless you just want to verify other people's results. Replicating other people's findings gives the results a stronger foundation. If your question has not been studied, then you also need to do background literature reading and checking to see if the relevant knowledge is already available.

Once the research question has been established, we first need to establish one or more hypotheses. A hypothesis is a precise description of the expected outcome, which can be directly verified by experimentation. It is derived from the theoretical background. The hypothesis describes that the independent variable (the influencing variable) has such and such effect on the dependent variable. The theoretical background describes why it is believed that this is the case. It is important that the hypothesis is built with precision and can be evaluated by statistical analysis of the experimental data.

A hypothesis (H_1) is usually evaluated against the so-called null hypothesis (H_0), which states that the factor being tested has no effect. The purpose of an experiment is to refute or invalidate the null hypothesis through statistical analysis of the data, thereby supporting the alternative hypothesis H_1.

To further illustrate what the null and alternative hypotheses are, let's take a case in point. It is often argued that the mobile phone has become an indispensable item in people's lives and that people want to have access the phone everywhere and all the time. What is available on a mobile phone should also be available in a car. So, the question arises: is it safe to put such features in a car? We need to experiment on a driving simulator to evaluate this. If we use A to represent this function, for example the ability to send and receive text messages via a text messaging service (TMS) while driving, then the research question becomes: Is it safe to drive while using TMS? The messages can be presented either visually (A1) or by Text-to-Speech (voice reading) (A2). Driving safety is generally measured using speed control, lateral control and visual tracking (see Table 14.1). Now the hypotheses are formulated as follows.

H0: A1 and A2 have no impact on driving safety, i.e., no impact on any of the indicators for speed control, lateral control, visual tracking, etc, compared to driving without using TMS.
H1: A1 and A2 impede driving safety, i.e., on one of the indicators of speed control, lateral control, visual tracking, etc, compared to driving without using TMS.

The purpose of the experiment is to confirm whether H0 is correct or H1 is correct. In each experiment, there should be at least one such pair of hypotheses, but there can be more than one such pair. In the above case, for example, it could also be assumed that

> *$H0_2$: A2 does not induce more mental load than normal driving.*
> *H2: A1 induces more mental load than normal driving.*

A good hypothesis generally requires that the following conditions are met.

1. The description is very accurate and clear.
2. The focus is confirmable by this experiment.
3. It clearly describes the control group and experimental conditions.

Preferably, the hypotheses are stated such that they express measurable outcomes. For instance, for H2 and $H0_2$ the hypotheses should state how mental load will be measured. This is known as the process of operationalization.

15.3 Variables in the Experiment

In experimental design, independent variables are the variables that are manipulated by the experimenter. If we are controlling more than one independent variable in an experiment, for convenience we will refer to each independent variable as a Factor, and a multi-factor experimental design will be called a Factorial design. It is important to emphasise here that each factor is independent of the other and is not related to, influenced by or constrained by the other.

For each independent variable, we need to find or identify the critical feature of the independent variable. Generally speaking, this critical feature can be measured in terms of some parameter, and a change in this value will presumably lead to a change in user behaviour. Using the example above, we need to experimentally answer the following question.

Case 1.

Does sending and receiving text messages via *TMS while driving pose a safety hazard?*

This question can be decomposed into several questions: (1) Is there a relation between driving safety and the way the information is taken and read? (2) Is there a relation between driving safety and the length of the information? (3) Is there a relation between driving safety and the state of road traffic? This question has several independent variables.

Variables related to TMS.

> *In the case of **independent variable A**, messages are sent either as text or speech.*
> *In the case of **independent variable B**, the length of the message is manipulated in terms of word count or seconds. Three text messages and three speech messages of different word count or duration can be set.*

Scenario variables (driving operations to be performed by the driver) regard traffic roads.

Independent variable C*, two different roads, one highway and one mixed urban road.*

The experimental design is such that the driver is asked to follow a car in front of him on different roads, while completing a series of subtasks. The driver has to fetch and read 6 TMS messages on the same road. Taking into account the stability of the data, even for the same message length, the same fetching and reading pattern may need to be repeated 2 or 3 times with different texts (to avoid learning effects).

There are three types of independent variables: quantitative (e.g., number of words in the above example), qualitative (e.g., auditory and visual in the above example), and categorical, which are not controlled by the researcher (e.g., gender, age, personality, etc.).

Dependent variables are the variables that will change as a result of changes in the independent variables, i.e., the outcome variables that are of interest to the researcher. In the above problem, the researcher's concern is driving safety, and the way to measure driving safety is through driving behaviour. So, in this experiment, we would need to record a range of driving activities, such as speed maintenance, change in distance between cars, lane line maintenance, etc. In Table 14.1, those commonly measured parameters are listed.

In such experiments, the most notable aspect is the confounding effect. Confounding occurs when the observed association between the dependent and independent variables can be attributed, in whole or in part, to a third variable. The notion of confounding factor can again be illustrated by the example above. The way the message is presented, the length of the message and the state of the road traffic are not the only factors that can affect driving safety. Many other factors may also have an impact, such as the driver's age, eye sight, familiarity with short messages, message font and font size, lighting, proficiency with driving simulators, etc. However, none of these factors are of interest to the researcher, and therefore they constitute potentially confounding factors. In any experiment, there are such confounding factors that, if not properly controlled, can have a direct impact on the data.

There are four ways to control for confounding variables: (1) Make the confounding variable a constant value. For example, by setting the experiment at the same time, in the same environment, with the same experimenter and using the same procedures; by keeping message font and font size constant; and so forth (2) By setting counterbalanced factors to counteract the effects of the confounding variable. For instance, as stated above, if participants have to go through different experimental conditions, often a learning effect occurs. By having different groups of participants go through different orders of conditions, the learning effects are counterbalanced. (3) By making the confounding variable an independent variable in the experiment. If it is believed that font size or voice (male vs. female) have an effect on performance, it may be made an independent variable. However, including more independent variables makes the experiment more complex, and very complex designs usually give poor and unclear results. In that case, it is preferable to evaluate the effects of such confounding variables in separate experiments. (4) By making it a random variable. Randomness encompasses two important aspects, one being that subjects should be obtained randomly and without bias from the population under study (random sampling). That is, there is no deliberate, or intentional, directed selection of subjects, but rather theoretically everyone in the population we are studying has an equal chance of being a subject. In the experimental arrangement itself, the assignment of participants to experimental conditions is also randomised, and the order of conditions is also different for each person in the group. In other words, for a completely randomly arranged experiment, no one, including the experimental operator, can predict what the next experimental condition will be for that subject.

15.4 Experimental Design

When doing experimental design, we need to understand two issues.

1. How many independent variables do we have in this experiment?
2. How many levels are there for each independent variable?

Figure 15.1 shows the experimental design for different variables.

If we change the question in Case 1 to

Case 2.

Does visual text information affect the driver's driving during driving on a motorway in conditions where there are no other road users?

Here, we have only one independent variable, and that is the visual information. Visual text messages can be long or short, so in designing the experiment we might choose 3 different lengths of text (for example 5 words, 10 words, and 15 words), which is the number of experimental conditions.

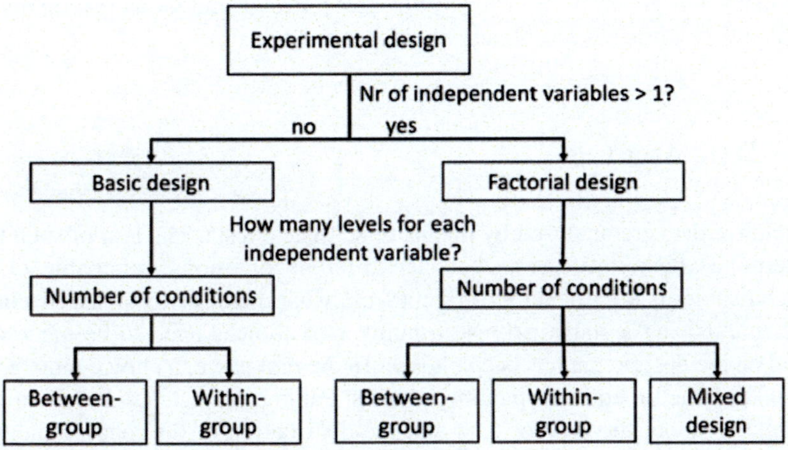

Fig. 15.1 Experimental design for different variables

Table 15.2 Comparison of the advantages and disadvantages of the between-group and within-group designs

	Between-groups design	Within-groups design
Advantages	Clarity Avoiding learning effects Good control of some confounding factors, such as fatigue from excessively long experiments	Efficiency, because the same group is used multiple times Inter-individual variation is well controlled Easy to obtain significant differences
Disadvantages	Many subjects required, because the same group is used only once Individual differences may have a significant impact on outcomes Statistically significant differences are not easily obtained	Difficult to control learning effects Prone to fatigue, which can have an impact on experimental results

In Case 1, there are three independent variables: independent variable A (presentation mode), independent variable B (message length) and independent variable C (road type). The number of experimental conditions for independent variable A is 2, the number of experimental conditions for independent variable B is 3 and the number of experimental conditions for independent variable C is 2.

There are two types of experimental designs, (1) Between-group designs, where each group of subjects participates in only one condition of the experiment. For example, in Case 2, one group of subjects may obtain only 5 words of textual information in the driving simulator experiment, while another group may obtain only 10 words of textual information, and so on. (2) Within-group designs, where the same group of subjects takes part in all conditions, e.g., in Case 2, each subject receives three different lengths of messages. A mixed-design is a mixture of these two experimental arrangements. There are advantages and disadvantages to each of the different experimental designs. Table 15.2 provides a comparison of these advantages and disadvantages.

As can be seen from Table 15.2, unless the learning effect is relatively large, an experiment will generally use the within-group design model as it requires fewer participants and it is easy to obtain significant results.

15.5 Data Analysis

Quantitative data are analysed by means of statistical tests. The aim of such tests is to determine whether differences between different conditions are reliable, or, to use the technical term, significantly different from what might be expected by chance.

When choosing a statistical test, a number of choices need to be made, which depend on the design that has been chosen. In the first place, a choice must be made between parametric and non-parametric tests. Parametric tests are based on certain assumptions about the type of data and the distribution of the measurements. For parametric tests, the data need to be at interval or ratio level (such as temperature and reaction times). This is not the case with data from scalar judgements, which are ordinal data. Also, parametric tests assume that measurements on a group of participants are normally distributed around the mean, so that measures such as standard deviations can be used. If this is the case, then the test is valid. If this not the case, the outcome of the test may not be valid, and non-parametric tests should be used, which do not make these assumptions about the distribution of the data. Statistical packages such as SPSS test whether the assumptions are satisfied as part of the standard testing routine. In the past, it was believed that parametric tests were preferable over non-parametric tests because of their greater sensitivity, but in fact with smaller sample sizes non-parametric tests are as sensitive as parametric tests, or even more sensitive, in case of gross violations of the assumptions for parametric tests.

If a non-parametric test is chosen, further choices need to be made. For between-group designs ("independent samples") different tests apply than for within-group designs ("related samples"). If the independent variable has two levels, that is, if there were two conditions, for instance, the test condition and the control condition), then a Mann–Whitney U test can be applied if an independent samples design was used, and a Wilcoxon matched pairs-signed rank test can be used if a related samples test was used. Yet other tests apply if the independent variable had more than two levels.

A disadvantage of non-parametric tests is that they are poorly equipped to deal with more complex factorial designs involving two independent variables. Although there are some non-parametric tests that may be applied in such cases, in order to give reliable results, they usually bring requirements about the number of participants that are not satisfied in most cases.

So far, we have acted as if there is always one dependent variable, but this is not always the case. If there are more dependent variables, in principle separate analyses may be conducted for each dependent variable, but this does not take into consideration that are may be correlations between dependent variables. Therefore, if the assumptions for parametric tests are satisfied, it is preferable to conduct one multivariate analysis of variance, in which such correlations between dependent variables are taken into consideration.

An important concept is effect size. As the term says, it refers to the size of the effect of an independent variable on the dependent variable. In general, the smaller the effect size, the larger the number of participants needs to be in order to get a statistically significant result. While for larger effect sizes a sample size of 20–30 may already be sufficient, with small effect sizes a sample size of over 100 participants may be needed to get a statistically significant result (if a between-subjects design is used, this means one has to recruit over 200 participants!). While from a scientific perspective small effect sizes may be interesting and relevant, from a design perspective only larger effect sizes are interesting. This is because small effect sizes mean that the difference between conditions, for instance between the control group and the condition with the concept or interface that is put to test is small and subtle. In that case, one might say that for daily practice it doesn't really matter which concept or interface users are presented with.

Testing whether differences between conditions are significant is important, but from a theoretical perspective it is often equally valuable to construct models about the relations between variables, for instance, how variables A and B influence variable C. There are several modelling techniques available, such as Structural Equation Modelling, and Machine Learning Techniques have also become popular for modelling. We refer the reader to the literature.

15.6 Interaction Effects

When we have multiple independent variables, we need to consider potential Interaction effects between independent variables. Let us illustrate this with a simplified version of Case 1.

Case 3 .

> *Does visual reading of text messages on roads at 80 km/h have different effects on drivers of different driving ages?*
> *Independent variable.*
> A: Number of visual characters, a1: 5 characters, a2: 15 characters
> B: Driver's driving experience, b1: less than 1 year, b2: more than 5 years
> *Dependent variable.*
> Driving speed maintenance

Here, we can have two hypotheses.

> *H0a: Driving speed is maintained independent of the number of visual characters*
> *H0b: There is no relationship between driving speed maintenance and driving experience*

We use a mixed design for this experiment because it is impossible for the same subject to be a novice and an experienced driver. Therefore, we have two groups of subjects, one with less than one year of driving experience and one with more than five years of driving experience. These two groups drive on the driving simulator and complete the text-reading task, trying to maintain a speed of 80 km/h. Two possible outcomes are shown in Fig. 15.2.

As can be seen in Fig. 15.2, if the outcomes are as in the left panel, with the two sets of data parallel, there is no interaction effect, i.e., driving age does not influence the effect of message length on driving speed. If the outcomes are as in the right panel, then there is an interaction effect, i.e., driving age influences the effect of message length on driving speed. As a side remark, it should be noted that there does not have to be a cross-over effect for an interaction between the two variables to exist: if the lines are not parallel, there is an interaction effect, whether the lines cross each other or not. If the lines do not cross, the interaction is called ordinal. If

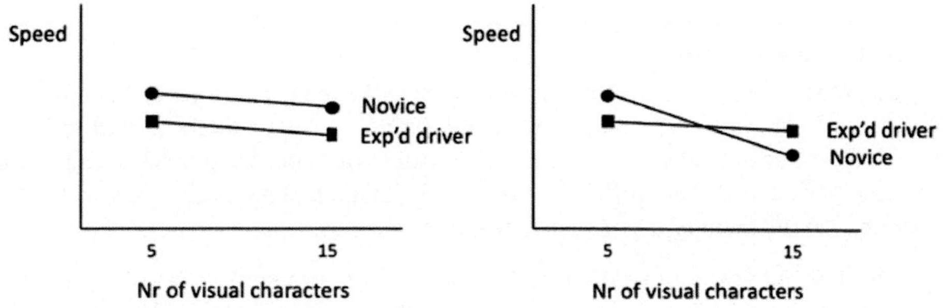

Fig. 15.2 Effect of nr of characters in visually presented text messages on driving speed for novice and experienced drivers. Left: the outcomes for novice and experienced drivers are parallel. Right: the outcomes for novice and experienced driver are not parallel; this is the interaction effect

they cross, the interaction is called dis-ordinal; in the latter case, variable A has one effect in condition 1 of variable B and the opposite effect in condition 2 of variable B. In any experiment where there is an interaction effect, each condition must be considered separately and no statement can be made about the general effect of an individual independent variable. In the third place, there may be interaction effects where variable A has an effect in condition 1 of variable B and no effect in condition 2 of variable B. Therefore, in an experiment with multiple independent variables, the first consideration in the data analysis is the interaction effect. And the larger the number of independent variables, the more complex the interaction effects may be and the harder it is to interpret them.

Suppose that we have 3 independent variables, A, B and C, each with 2 parameter settings (levels). The complete experimental design is then

$$2\,x\,2\,x\,2 = 8 \text{ conditions}$$

If it is a within-group design, then each subject would need to go through eight conditions. In total, at this point, there is the potential for interaction effects as follows (x means that the factors interact with each other).

$$A\,x\,B\,x\,C;\ A\,x\,B;\ A\,x\,C;\ B\,x\,C.$$

The creation of interaction effects can make the interpretation of the outcomes relatively complex, so it is generally not advisable to have more than three independent variables.

15.7 Bias

For each dependent variable, or the measured value obtained in the experiment, two components are involved.

Measured or observed value = true value due to the effect of the independent variable + random error (random error).

This random error is unavoidable and uncontrollable in experiments, but its effect on the real data is also random, so it can be removed statistically to find the real values. Some errors, however, are caused by problems in the design and arrangement of the experiment and are difficult to remove by statistical methods. We will briefly describe the different types of systematic bias as follows.

1. Systematic deviations caused by the measuring instruments. Due to instruments that do not measure with sufficient precision, are inaccurate, have not been calibrated, etc.
2. Bias caused by improper arrangement of the experimental sequence. If complete randomisation is not done, bias can arise. For example, sometimes when doing driving simulator experiments, for convenience, everyone starts with the control group, or everyone follows the same order of conditions, and the sequence of events can cause systematic bias in the data because we don't know if the former scenario will have an effect on the behaviour of the driver in the latter scenario.

   In particular, when conducting driving simulator experiments, one should be aware that driving in a simulator is different from driving in a real car. Therefore, usually the experimental session starts with a familiarisation trial, giving the participants the opportunity to get familiar with the simulator and with driving in the simulator. However, how long should the familiarisation last? It would be better to have objective performance criteria, so that the familiarisation terminates only when the performance criteria are met. If this is not the case and the familiarisation is too short, the participants will continue learning during the actual experiment, and the results for later conditions will in part reflect the improved performance due to learning. If all the conditions are administered in the same order to all participants, no conclusions can be drawn about the effect of the independent variable(s) because of this confounding.
3. Bias brought about by the subjects. This is also a serious problem. The subjects we choose, for example, may not be representative of the population of interest. When we do experiments in universities, for convenience we often use students as participants, and in companies, for various reasons, we use colleagues from within the company as participants. Such subjects produce results that are not representative of the broad population and do not satisfy the conditions for random sampling.
4. Bias caused by the experimenter. The experimental staff can influence the results of an experiment in a number of ways, such as his attitude towards the subjects being different, or the instructions to the participants being different (in particular if they are spoken instructions), the equipment being installed and commissioned differently, etc. Therefore, in experiments, we emphasise that the same experimental staff should do the same work, that the instructions for the participants are written down, and that consistency should be maintained when explaining them.

5. Bias caused by environmental factors. This environmental factor includes both physical and social factors. The physical environmental factors are temperature, light, noise, vibration, etc., while the social environmental factors include who is around the experiment and the relationship between this person and the subject.

All five of these deviations need to be strictly controlled during the design and operation of the experiment.

References

Keppel G, Wickens TD (2004) Design and analysis: a researcher's handbook, 4th edn. Pearson Prentice-Hall, Upper Saddle River, NJ

Lazar J, Feng JH, Hochheiser H (2010) Research methods. Wiley

Neale VL, Dingus TA, Klauer SG, Goodman M (2005) An overview of the 100-car naturalistic study and findings. Traffic Saf 1–10. Retrieved from http://citeseerx.ist.psu.edu/viewdoc/download?doi=10.1.1.172.2366&rep=rep1&type=pdf

Part V
Human Factors of Automated Driving

Chapter 16
Human Factors of Automated Driving

Abstract In this chapter, we discuss the human factors of driving automation. First, we go into the objectives of driving automation, and conclude that, from a safety perspective, attempts to automate driving certainly make sense. However, full automation will not appear in the market overnight, and consumers will be exposed to intermediate levels of automation. We discuss issues with different levels of automation, and what they mean for design. Furthermore, we discuss human factors issues that apply to all levels of automation: how to make sure that people understand and trust the automated system; what needs for shared control may be expected; what may be expected with regards to acquisition and loss of driving skill; what ethical issues may arise and how should these be dealt with; and, finally, do interests of individual customers and society in automated driving converge; if not, how can design contribute towards a solution. Finally, we go into design consequences and opportunities of the fact that automated vehicles will interact with other road users.

In order to discuss the human factors of automated driving, we first decompose the driving task into three levels, based on the components of its activity, using Michon's (1985) model introduced in Chap. 5. The highest level is the strategic level, which involves decisions about the destination of the trip, the route (whether to take a fast route, a scenic route, etc.) and the required arrival time. The tactical level involves the manoeuvres required to execute the overall plan, such as the maintenance of distance to the leading vehicle, overtaking, left and right turns, etc. The operational level concerns the planning and execution of the actual operations required to execute the overall plan and the associated manoeuvres. These operations are carried out through the available controls such as steering wheel, accelerator and pedals. This process also includes constant input from environmental factors and constant feedback on the effects of manoeuvres: the driver needs to observe the progress of events based on the vehicle's position in relation to other vehicles and the environment in order to be able to make subtle adjustments during manoeuvres and to anticipate upcoming actions.

For automated driving, the Society of Automotive Engineers SAE has proposed a taxonomy of automation levels (SAE 2021), as shown in Fig. 16.1 (see also Fig. 1.1). Here, to further illustrate the significance of the three levels described above, we

© China Machine Press 2023

F. Chen and J. Terken, *Automotive Interaction Design*, Springer Tracts in Mechanical Engineering, https://doi.org/10.1007/978-981-19-3448-3_16

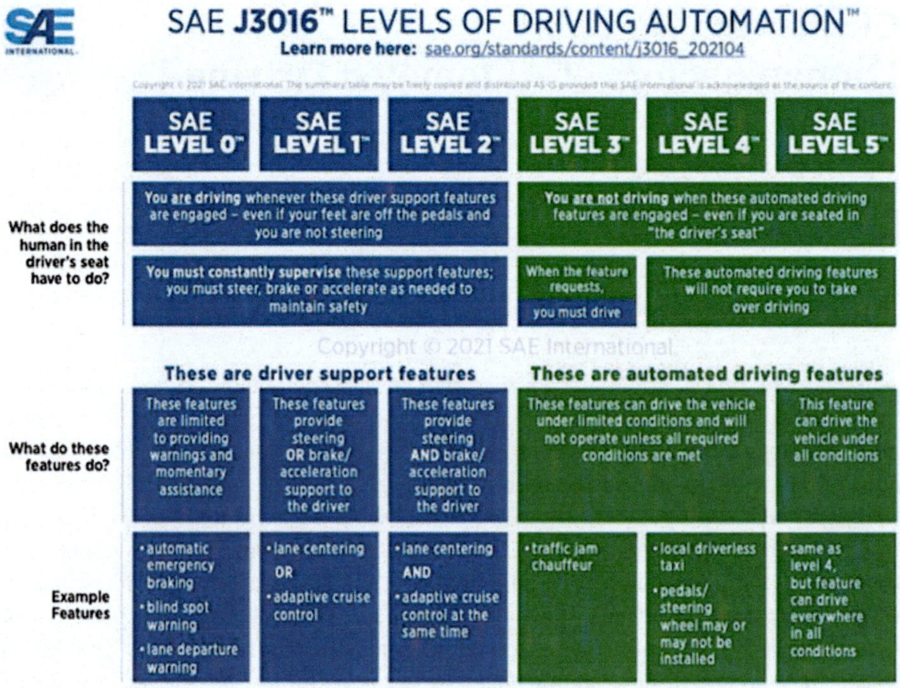

Fig. 16.1 Levels of driving automation. © SAE, 2021. https://www.sae.org/binaries/content/assets/cm/content/blog/sae-j3016-visual-chart_5.3.21.pdf

interpret the levels of automated driving from a human factors perspective. Level L0 is included for taxonomic completeness; it indicates that no task is performed automatically and the driver is required to perform all tasks involved in driving. At Level L1 automated driving, only one task is automated. This is typically speed control at the operational control level, either through Cruise Control (CC), where a fixed driving speed is set by the driver, or through Adaptive Cruise Control (ACC), where the driver sets and drives at a fixed speed, but the system can adjust its speed and distance to a leading vehicle. At L2 automated driving, multiple functions can be performed automatically at the operational level. Level L2 automation is typically represented by a combination of longitudinal control through ACC and lateral control through automatic lane keeping. In addition, some systems allow the driver to initiate an automatic lane change, where the driver makes a decision and gives the system a command to initiate the automatic lane change, which is then performed automatically. L1 and L2 are also labelled as assisted driving, as the driver remains fully responsible for driving, while the functional automation at the operational level only assists with the driving task. This also means that the driver needs to constantly monitor the performance of the system, so that s/he can intervene if the performance of the system is not as required.

Level L3 and higher is where true automated driving begins, that is, where the system can take charge of driving tasks on its own. The difference between L3,

L4 and L5 levels lies primarily in the extent to which the system can handle road conditions. At L3, the automated driving system can operate well in relatively simple road conditions (e.g., normal driving on a motorway), but does not perform well when dealing with more complex road conditions (e.g., urban traffic) and abnormal situations (e.g., motorway road repairs, or accident scenes). If such complex and unusual road conditions occur, the system will issue a take-over request (TOR), asking the driver to regain control of the car's driving. Importantly, at L3 and beyond, automated systems are able to perform tactical control. In other words, the system is able to decide when to do a lane change operation and initiate the lane change itself. In addition, the system is able to monitor the driving situation of the vehicle in relation to other vehicles and the environment, thus freeing the driver from monitoring tasks. At L4, the system is able to handle more complex road conditions compared to L3, and can even cope with many unexpected road conditions. Another feature of L4 is that if the driver does not respond to the TOR (e.g., because he/she is fatigued and drowsy or feels unwell), the system is able to drive the vehicle automatically to a safe location, such as the emergency lane, and bring the vehicle to a halt and initiate a series of measures to maintain safety. At L5 the functions of the automated system are further extended, so that the system is able to handle all the complex road conditions that can occur. The system can steer the vehicle even without the driver. For this reason, Level L5 is also known as fully automated or driverless driving.

16.1 The Purpose of Automated Driving

There is a story of a minister who said to his king: "I have a way to make people travel dozens of times faster. From Amsterdam to Paris is 500 kms. In a carriage it is 100 km a day at most, so it takes 5 days, but if there is a means of transport that can travel 100 km an hour, it will only take 5 h to get there". The king said: "That's great! Let's invest in such a system!" But the minister went on to say: "Yes, but there is a cost to this system. More than a million people will die every year all over the world and tens of millions more are injured"! Do you think the king will invest in this system? Sweden was the first country to announce to the world that it wants to build a system with zero traffic fatalities. And Volvo has stated that it is unacceptable for a person to leave home in the morning to go to work and die under the wheels of a car on the road. As medical technology develops, more and more fatal diseases are being tackled and life expectancy is being increased, and fewer people die from all kinds of diseases. Nowadays, traffic accidents have become a major cause of death. According to the World Health Organisation, injuries caused by road traffic accidents rank eighth among the causes of death in industrially developed countries. Therefore, the elimination of the harm caused to human beings by traffic roads has become a common goal for all mankind.

In 2016, 1.4 million people died in road traffic accidents worldwide. Furthermore, analysis of accidents shows that between 70 and 90% of road accidents are caused by human error (violations, distraction, fatigue, alcohol, etc.). It is believed that automated driving technology can free people from the task of driving and therefore significantly reduce the rate of traffic accidents and fatalities. The implementation of these technologies is the original driver for manufacturers to invent new systems and offer new features that will improve driving safety, enhance the customer driving experience and gain a competitive advantage. Passive and active safety has become an important part of vehicle development and manufacturing, and government agencies around the world have been actively issuing regulations on vehicle safety requirements. Innovations in sensor and computer technology have made automation possible, and this has been seen as the next step in efforts to achieve safer mobility. As full automation (L5 level) does not appear overnight, lower levels of automation are being introduced to the market first so that partial automation can have a beneficial impact on traffic safety and provide a platform for testing to allow further development of automation technology.

But safety is not the only selling point of automation. Other selling points are sustainability, efficiency, convenience, comfort, productivity and mobility for all.

- ***Sustainability***: Because autonomous driving systems are superior to manual driving in terms of operational control and can avoid the speed changes and violent braking that are common in motorway traffic, they can do a better job of speed control. Also, automated driving systems allow for shorter spacing between vehicles in motion and therefore potentially result in less energy consumption.
- ***Efficiency*** : For the same reasons as mentioned under "Sustainability", automated driving systems have the potential to reduce traffic congestion by enabling vehicles to travel at shorter vehicle distances and on narrower roads, resulting in increased efficiency per unit area and improved road capacity. It should be noted that, in order to realise these benefits, vehicles should not only be automated, but also fully networked (V2X) to communicate with each other and with the infrastructure. In urban traffic, automated and connected vehicles can coordinate their movements, thus increasing the efficiency at intersections and traffic signals.
- ***Convenience***: Once the automated driving system is able to perform the driving task, it allows people to engage in other activities while the vehicle is in motion. Common activities mentioned in surveys are making phone calls, sending text messages and using the smartphone for other purposes such as playing games, listening to music, reading and sleeping. The concept of turning the car into a mobile office may also be realised.
- ***Comfort***: The automated driving system can be better adapted to the needs and preferences of the occupants in terms of operational control and interaction with other vehicles, thus increasing physical and mental comfort. Likewise, knowing that the system still works well under rigorous testing conditions and has been designed for accident prevention may improve the psychological comfort of the driver and occupants.
- ***Productivity*** : As driving tasks are performed by the automated driving system, users are able to use this time for activities related to work or everyday life. For example, people can start preparing for an upcoming meeting or deal with emails on their way to work.

- ***Mobility for all*** **:** Driverless cars make it possible for people who currently rely on others to get around to become independent. This concerns in particular people who are not allowed to drive for whatever reason (age—young and old, people with physical or perceptual impairments or those with temporary impairments (e.g., drunkenness)) or who are dependent on public transport. Such travel can be more customised.

However, things may not be so simple and ideal. One may question whether automated driving technology will be accepted wholesale. Will the technology be adopted in the way the manufacturer expects? If not, the above claims may not be realised. Therefore, to see whether automated driving and the gradual introduction of automation will indeed provide the claimed benefits, the human element needs to be taken into account. The discussion in the following sections arising from this question may provide suggestions and guidelines for technology development.

16.2 Human Factors Issues in L1-L2

The study of human factors in automated driving is best differentiated by the different levels of automated driving. At levels L1 and 2, the driver's driving task may be assisted by one or several systems, but the driver remains responsible for the correct execution of the overall driving task. These systems are known as assisted driving. At levels L5, the system is responsible for completing all subtasks in the driving task and the system is also responsible for the correct execution of the overall driving task. This level will be referred to as autonomous driving. L3 and L4 are a mixture of assisted and autonomous driving: for some parts of the driving, the driver must drive fully manually (L0) or assisted (L1 or L2). For other parts the system can drive fully autonomously and the driver may start to engage in other activities. Switching between levels depends on the situation. A typical scenario would be for the driver to leave home, drive in urban and/or rural areas using L0, L1 or L2, switch on the automated driving system (level L3) when entering a motorway and then set the mode to manual or assisted driving again when leaving the motorway to drive through the city or suburbs to the office.

In order to more fully understand the human factors aspects of this, the human factors issues in assisted and automated driving are considered separately for the different levels in Sects. 16.2 and 16.3. In all cases, relevant elements such as usability and user experience are also taken into account. Usability is related to human cognitive processes such as perception, understanding and attention. Here, there are several typical questions: Does the driver understand the system's function? How can the system be made to do specific things for them? Is the information provided by the system communicated at the right time? User experience relates to human emotions. Here, questions arise such whether people will appreciate the functionality of the system and how will they interact with it. User experience is also about the perceived usefulness of the system, e.g., do people feel that the system adds value for them? In

Sect. 16.4 we discuss the human factors issues related to driving automation across levels.

16.3 Human Factors Issues in L1

At level L1, the driver is responsible for the driving task, but there will be some system to assist in operating a sub-task. In principle, this could be any task, such as lane keeping or parking. But in practice, the automated task is usually speed control, which is most commonly performed through cruise control or the more advanced adaptive cruise control. The cruise control system allows the driver to set the speed to a fixed value, and the driver can gradually increase or decrease the speed (in steps of typically 2 km/h). The cruise control can be switched off by pressing the "Cancel" button or by pressing the brake. Once switched off, the cruise control can be resumed by pressing the "Resume" button. The driver can temporarily increase the speed by pressing the accelerator, which will return to the pre-set speed when the accelerator is released. Adaptive Cruise Control has a further control that enables the driver to set the distance to the leading vehicle (usually at three levels: long, medium and short). Some systems also allow the driver to set an upper speed limit, so that even if the driver increases the speed through the throttle, the speed limit will not be exceeded. CC and ACC are controlled through buttons on the steering wheel or a handle behind the steering wheel (depending on the brand/model). Icons on the instrument panel communicate information to the driver about the status of the system (Fig. 16.2).

The usability issue with this design is whether the driver understands the system. The driver needs to develop a mental model of how the system works and how it allows the driver to control the (fixed speed/adaptive) cruise control system. In addition, this understanding needs to lead to a skilled understanding of the function so that the driver is comfortable setting the cruising speed and the distance to the vehicle in front without excessive mental effort. During operation, his/her eyes should only have to glance at the system to guide the interaction with it. The buttons on the steering wheel appear to be easier to "glance" at than the handle behind the wheel. Standardisation helps to develop the driver's habit of operating the system blindly and also avoids the difficulty of having to get accustomed to another implementation of the function, which may occur when another implementation is encountered (for

Fig. 16.2 **a** icon for fixed speed cruise control; **b** icon for adaptive cruise control. In general, green indicates that the function is on, while white indicates that the function is available

example in a rental car). Standardisation allows the transfer of mental models from one vehicle (or make) to another.

With regard to Adaptive Cruise Control, the user also needs to know whether the system is full range ACC or only works at travel speeds above 30 km/h. Failure to know this technical limitation can lead to unpleasant surprises when ACC is used during traffic jams. Some users have reported that using ACC in traffic jams can ease traffic anxiety. In addition, users need to understand how the system adjusts speed according to the presence or absence of the car in front of them. Especially when approaching slow traffic upfront, the driver may want to change lanes and take over driving. If the left lane is occupied at this point, the system will slow down to accommodate to the speed of the leading vehicle. Once there is a gap in the left lane, the driver starts to change lanes, in which case the car will automatically accelerate because there is no leading vehicle in front of it, and the driver should know the details of this acceleration behaviour of the system. If not, a bad user experience may result and there may be hindrance for upcoming fast traffic in the left lane.

From an empirical point of view, users who have acquired the correct mental model of the system usually find fixed speed cruise control or adaptive cruise control very comfortable. Fixed speed/adaptive cruise control reduces the physical effort of holding the throttle in a fixed position for long periods of time and reduces the mental effort of having to check the speedometer frequently to avoid violating the speed limit. It does not reduce concentration, however, as they still have to take control of the steering wheel themselves and remain engaged in the driving. Furthermore, the automatic operation of the system can be overruled by the driver at any time, so in effect control is delegated to the system by the driver, rather than taken over by the system, and the driver always feels in control.

The cruise control may lead to behavioural adaptation, which usually refers to undesirable behaviour caused by a certain technique. If a driver approaches a slow vehicle in front and wants to overtake, but there is no gap in the left lane, s/he can maintain the set speed before overtaking and may tolerate a narrow distance from the vehicle in front. In addition, when overtaking with fixed speed CC, the driver may maintain the set speed rather than accelerating, and consequently take longer to overtake. These actions may make passengers, other drivers and/or other road users feel uncomfortable or even annoyed.

16.3.1 Human Factors Issues in L2

At L2 level, the automatic system is able to handle at least two subtasks. Typically, adaptive cruise control is used for longitudinal control and automated lane keeping for lateral control, so that in normal cruise mode, operational control can be performed automatically by the system. This situation is often referred to as "eyes on the road—hands and feet off the control system". As long as the driver is satisfied with keeping the vehicle in the same lane, he/she does not have to perform any vehicle driving tasks himself/herself, but s/he cannot disengage from the supervisory task of monitoring the performance of the system. Intervention is required if the system's autopilot operation does not perform as expected or if the driver wishes to perform an operation such as a lane change. The intervention can take place at the operational control level, for example when the vehicle is not driving in the correct lane (perhaps because the lane lines are not clearly marked) and the driver corrects this by steering; or at the tactical control level, for example if the driver wants to perform a lane change, he can either operate the turn signals himself and perform the lane change, or the system can perform the lane change. With regard to the latter, whether the execution of a lane change requires tactical (only the lane change command is given) or operational control depends on whether the lane change manoeuvre can also be automated technically.

With L2 autonomous driving, the driver remains fully responsible for the driving operation and is always monitoring the performance of the system, watching what is happening on the road and intervening when necessary. The value of L2 for the driver is that the system can take over operational control under normal operation, reducing the psychological load on the driver. However, as the driver's driving task becomes one of monitoring, the driver is required to monitor the operation of the system performance at all times without actively engaging in the driving task, which can cause serious cognitive problems. Figure 16.3 shows curves of human signal detection rate and reaction time in a signal detection task as a function of operating time. The left vertical axis shows the detection rate of the signal. The solid curve shows that the signal detection rate drops from 100% to less than 40% over a period of 30 min. The right-hand vertical axis shows the human response time to a detected event, which, as can be seen from the dashed curve, rises significantly over a 30-min period. Figure 16.3 suggests that, as the driver's role changes from a manipulative driving task to a surveillance alert task, their detection rate of anomalies in the vehicle

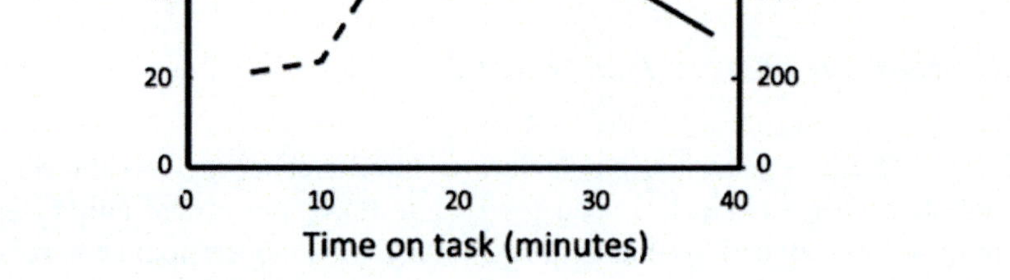

Fig. 16.3 Signal detection rate (solid line) and response latency (dashed line) as a function of versus task supervision time

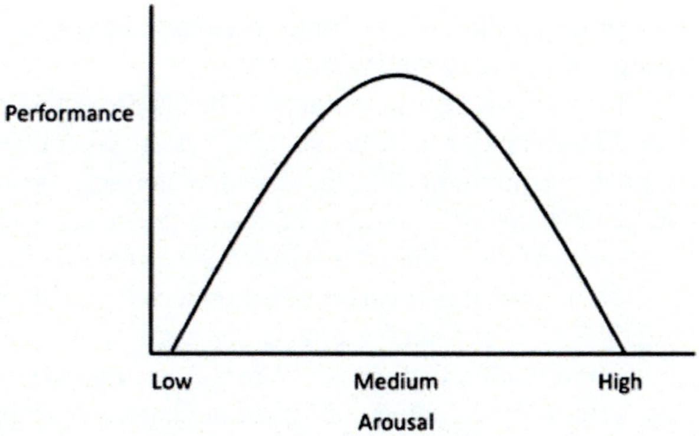

Fig. 16.4 The Yerkes-Dodson law: the relationship between arousal and performance

and on the road may decrease and their response to anomalous road conditions may become slower as the duration of the ride increases.

This phenomenon is explained by the Yerkes-Dodson law, which describes the relationship between arousal levels (or motivation) and performance (see Fig. 16.4). In the range from 'low' to 'medium', the performance is positively related to the driver's arousal, with arousal typically depending on how actively the driver is involved in the driving task. When the driver is not actively involved in the driving task and does not have to adjust the vehicle's position all the time, but simply has to monitor the vehicle's movement, this reduces the arousal and therefore makes it difficult for the driver to remain alert. The result is the degradation of the driver's ability to monitor and his/her ability to take over control when s/he needs to do so within the required reaction time frame. This leads human factors experts to conclude that automation at the L2 level is not the recommended form of autonomous driving.

In terms of user experience, users of L2 Autopilot systems (e.g., Tesla's Autopilot) often intuitively understand this level of autopilot as true automated driving. Initially, they will monitor the performance of the Autopilot system for a period of time (e.g., 10 min). During this time, they make two observations. Firstly, they find the monitoring task boring. Secondly, they notice that the system is working properly. This induces a high level of trust in the system ('over-trust') and a tendency to see the system as an L3 automated driving system that does not require supervision. As a result, they feel they can engage in non-driving related activities, such as texting, watching videos or playing games on their mobile phones (Lin et al. 2018, 2019). This state is very dangerous because, once the control needs to be taken over by the driver, the response time available to the driver is generally short and the driver may be unable to take over the driving properly within the given timeframe. To avoid this situation, manufacturers have developed systems that force the driver to keep his or her hands on the steering wheel, such as sensors on the steering wheel that detect whether the driver is holding the wheel and prompt the driver to hold the wheel when no hands are detected on the steering wheel. However, drivers have also developed countermeasures to this prompting system, by taping a heavy object to the steering wheel to trick the system into believing that his or her hands are indeed on the wheel, so that he or she can continue to engage in non-driving related activities without being prompted by the system.

Other researchers have taken other approaches. Seeing that the problems with Level 2 automation arise because the system takes over the operational control during normal operation, so that the driver is not actively engaged, they propose that the operational control should remain with the driver, so that the best that can be done is to assist the driver. This has resulted in warning and assistance systems such as Lane Deviation Warning (instead of automated Lane Keeping), Collision Warning, and Haptic Pedal for Speed guidance (Mulder et al. 2010). Yet another approach links to the observation that users of Level 2 automated driving systems tend to over-trust the system. By inducing controlled deviations from proper performance, developers attempt to counter-act the over-trust of the users. From a customer perspective this is not an attractive direction.

16.4 Human Factors Issues in L3-L5

16.4.1 Human Factors Issues in L3

In theory, L3 automated driving (also known as "eyes and hands free, but not brain free" automated driving) is a system that can perform all the sub-tasks that make up the driving task, so the driver no longer needs to monitor the system. However, it can only do so under certain road conditions. Usually, such road conditions involve less complex road conditions, such as normal traffic on a motorway. If these regular conditions are not met, for example if road repairs or an accident cause the road on a motorway to narrow or divert, or in urban or rural road conditions with pedestrians and two-wheelers, the system will ask the driver to take over driving again before entering its boundaries. As L3 level automated driving is the first form of automated driving likely to enter the market, the take-over issue has attracted considerable attention in research. The questions that need to be answered are: How much time is needed for a human take-over action? How should take-over requests be designed to ensure that the driver performs an accurate and timely take-over? The human take-over response is a series of processes, the sequence of which is shown in Fig. 16.5.

At T1, the system identifies the conditions ahead and determines that it is not suitable for automated driving and that the driver needs to take over control. At T2, the system issues a take-over request (TOR). At T3, the driver's first reaction is usually something like a hand on the steering wheel and a foot on the pedal. Often

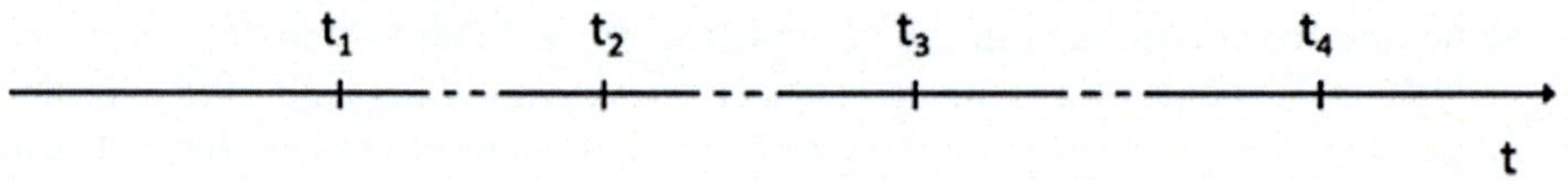

Fig. 16.5 Take-over sequence

the driver's first reaction is not necessarily the correct action, but because of this action, the automation is automatically switched off. At T4, the driver has stabilised the control of the vehicle and the car has the correct position and speed in the lane. The dotted line indicates that the interval between T_i and T_{i+1} has a variable duration. There are now several questions that the designer needs to answer: on what basis is it decided when the TOR should be released? What are the typical durations of intervals T2-T3 and T3-T4? What happens during these intervals? What are the factors that determine the duration of these intervals? How can design help ensure that these processes operate effectively? In answering these questions, a distinction needs to be made between two situations, one where the need to issue a takeover request is known in advance and only the timing of the request needs to be decided, and the other where this is not the case. For example, when the navigation system indicates that the vehicle approaches the exit where it will have to leave the highway, or when the system knows in advance, through V2X, that it is about to drive into a construction site where the road is narrowed by a road repair site, the system knows in advance that it will need to hand over control to the driver. These are expected take-over situations. The other situation is where something unexpected happens that the system doesn't know how to handle and therefore asks the driver to take over control. For example, if an accident occurs very close to the vehicle or if a sensor suddenly fails, this is an unexpected takeover.

Expected takeover

First, the system needs to determine when to issue a take-over request (TOR). The TOR needs to be issued early enough to allow the driver sufficient time to interrupt the non-driving related activity in progress and begin attending to the traffic situation. The time required for the driver to transition from non-driving activities to driving depends on the nature and state of the non-driving related task in which the driver is immersed. When drivers are engrossed in playing games or even sleeping, switching between non-driving and driving activities may take more time than when they are immersed in activities such as listening to music or texting. Furthermore, the perceived sensitivity to TOR signals depends on the driver's immersion in non-driving related activities, so the salience of TOR signals should be designed to match the immersive non-driving related activities, i.e., to allow the driver to perceive TOR signals in a timely manner, while not appearing too noisy when there are less immersive activities. When it is not possible to correctly determine the driver's state, the TOR signal might be timed well ahead of the time the driver needs to regain control, to compensate for the immersion in non-driving activities. Similarly, the TOR salience may be set to successfully engage the driver's attention even during immersive non-driving related activities.

From a design point of view, the main purpose of a takeover request is to enable the driver to divert his/her attention from non-driving related activities to the driving task. The TOR should therefore be multimodal, with at least visual warning signals combined with auditory signals. In addition, a haptic signal might be added, by vibration motors in the seat. If the system senses information about the activity the driver is engaged in, for example knowing that s/he is interacting with a smartphone, visual and haptic signals can be sent via the smartphone. Furthermore, the TOR may be built up stepwise with increasing salience, so that drivers who are not deeply immersed in the non-driving-related activity may already notice the non-obtrusive signal, while drivers who are more immersed may notice the TOR only when it becomes very obtrusive. This way, both situations can be handled gracefully. In such TOR situations, drivers should have the opportunity to confirm that the TOR has been noticed, for example by pressing a button or placing their hand on the steering wheel so that the signal can be stopped before it becomes annoying. It is important that this acknowledgement should not be regarded as an actual act of taking over, as the driver is not ready to take over driving at this point. In order to clearly distinguish the action of holding the steering wheel from the actual regaining of control, it may be necessary to establish the take-over action by an explicit deliberate action (e.g., by pressing two buttons on the steering wheel).

After issuing the TOR, the driver starts preparing to take over driving at T3. The process in the T2-T3 interval involves interrupting non-driving related activities, turning attention to the traffic situation, recognising what is happening, gathering information about the surrounding traffic situation (situational awareness building) and preparing to take action when control is in hand. Results from experiments show that the duration of the T2-T3 interval varies considerably. Some experiments showed durations as short as 1–1.5 s, but most showed durations between 4 and 8 s (Gold et al. 2013; Kuehn et al. 2017; Zeeb et al. 2015; Zhang et al. 2019). This is consistent with the general observation that simple reaction times are typically around 700–1000 ms when a person is engaged in another activity and may increase considerably for complex tasks (for detailed rationale, see Chap. 6, Sect. 6.4). The response time may be further increased if the operation of the response is complex and involves reconstructing the perception of the situation. It is clear from this that providing the driver with good situational awareness of road conditions becomes particularly important even in automated driving (Young et al. 2007).

When the driver is ready to take over control, the existing system requires explicit confirmation from the driver, for example, by pressing two buttons on the steering wheel. From that moment on, the movement of the steering wheel controlled by the driver can lead to a change in lateral position and speed. Assuming that the driver has developed situational awareness, s/he should be able to discern what the correct driving action is, for example, to stay in the same lane and slow down, or to prepare for a lane change. Usually, for such reactions, there is a trade-off between reaction speed and accuracy: faster reactions are usually more inaccurate, while more accurate ones are slower. In the case of take-over control, less accurate responses usually mean that the longitudinal and lateral control is less appropriate for the scenario at hand (Wickelgren 1977). Experiments have shown that it can take up to 30 s to fully stabilise the behaviour of a vehicle in terms of stable longitudinal and lateral behaviour (Merat et al. 2014). In the domain of process control, professional process operators are usually highly trained to avoid rushing to react in emergency situations. If the average driver knows they have enough time to regain control, through training they may also consciously suppress the tendency to react quickly, thus increasing the accuracy of their response.

Unexpected takeover

The situation is very different when the system is unable to inform the driver well in advance of the need to take over control, but rather requires the driver to take over immediately when a certain situation is detected. This can happen, for example, if a sensor malfunction occurs, because of sudden fogging or machine failure, or other accidents occurring a hundred metres in front of the vehicle (a vehicle travels 165 m in five seconds at a speed of 120 km/h). This is a paradoxical situation, as driving automation was motivated among other things because systems can react faster than human beings, who are not well equipped to deal with unexpected situations. In such circumstances automated driving makes sense because the reaction time of an automated system is shorter than that of a human, and it is precisely because human ability is not sufficient to cope with unexpected situations on the road in traffic that automated driving needs to be introduced. In an emergency situation, if people react quickly, as mentioned above, the accuracy of the reaction will be poor. Systems such as ABS (Anti-lock Braking System) and ESC (Electronic Stability Control) can support the driver in making quick reactions and improve the handling of the vehicle during sudden evasive manoeuvres, provided that the driver has enough time to check whether there is a vehicle in the flanking lane to prevent an evasive manoeuvre being performed.

In some cases, however, the driver may not have time to switch from a non driving related activity to a driving operation. Two solutions are currently proposed for this situation. The first is a technical solution. AEBS (Automatic Emergency Braking System) will automatically apply emergency braking in order to avoid hitting an obstacle. If the time budget is not sufficient to perform an evasive manoeuvre, AEBS may be the best method. The second solution uses an interaction design approach, based on the principle that people will react faster and more accurately if they are mentally prepared in advance to act. When designing the interface, even in automated driving, the system may continuously show the driver its own estimate of its ability to handle the situation on a scale from high lo low. If the display indicates that the ability of the system to deal with the situation is high, the driver may concentrate on non-driving related activities; when the ability goes down, the driver may switch to monitoring the performance of the system, so that s/he has good situation awareness and is prepared to take over driving at any time (Helldin et al. 2013). A key requirement for such a system is that the automated driving system has good self-awareness of its capabilities to handle the situation. Another solution is the use of ambient displays, which do not display information on displays that the perceiver has to focus upon, but in the environment (Löcken et al. 2015). The notion of ambient displays is based on the fact that people are able to process simple information in the periphery of their attention. Thus, an ambient display may provide information about the vehicles around while the driver is still engaged in non-driving-related activities, so that, if the driver needs to take over control, s/he already has basic situation awareness. Obviously, such displays offer no solution for the true emergency situations, due to the relatively slow reactions of human drivers in case of unexpected events, so that in these cases AEBS remains a last-resort solution. But anything that can be done to reduce the number of unexpected take-over situations, such as made possible by vehicle-to vehicle and vehicle-to-infrastructure communication and by smart interface solutions, deserves serious attention.

16.4.2 Human Factors Issues in L4 and 5

L4 is similar to L3, so the problems that exist in L3 also exist in L4. There are two differences, one being that, if the driver does not respond to a take-over request due to a sudden physical condition (e.g., heart attack), the L4 system should be able to maintain safety. In this case, the system should be able to direct the vehicle to a safe location, for example, by stopping at an emergency exit and calling emergency personnel. Another difference is that the number of road conditions that the system can handle is further increased, so that there are fewer situations where the driver needs to take over control.

The human factors issues arising from these changes are twofold. Firstly, users need to trust that the system can cope with the vast majority of road conditions. Secondly, the number of situations that require the driver to take over control has been significantly reduced, and this affects user expectations, particularly with regard to unexpected takeovers. If, for example, at L3, the system does not perform as expected once per trip, and, at L4, it does not perform as expected once in every three trips, the expectancy of such situations will further decrease, and this reduction will affect the driver's readiness to take over control. Of course, the driver's readiness to take over control from the system and hand over driving to the system also depends on the severity of the system's deviation from optimal performance. Small deviations may be ignored or forgotten, while large deviations may reduce the driver's willingness to use automated driving.

At level L5, the technical capabilities of the automated driving system have increased so much that it (in principle) is expected to be able to perform automated driving in all road conditions, to the point where vehicles without conventional controls such as steering wheels and pedals are created. This raises issues concerning trust, comfort, control and ethics. These will be discussed in the next section.

16.5 Designing Automated Driving Systems

While Sects. 16.2 and 16.3 discussed the issues associated with the individual levels of automation, in this section some lessons will be drawn across levels of automation.

16.5.1 Trust

In automated human–computer interaction, there is no issue more important than 'trust'. Research on this subject began as early as 1980 (Wiener and Curry 1980). Two concepts were introduced: one called Complacency or over-trust, and the other called the Cry wolf effect or under-trust (Parasuraman et al. 1993). A vast amount of research has been reported on trust in automation, notably Lee and See (2004) and Madhavan and Wiegmann (2007), who compared human-to-human trust and human–machine trust and found that there are a number of commonalities and many different issues. Hancock et al. (2011) provides a good overview of the trust problem between humans and robots. There are two issues that need to be distinguished clearly here, one is trust in automation and the other is dependence on automation. The former is a psychological issue and the latter is an operational issue. The two are related, but not in a one-to-one relationship. We may be dependent on automation for various subjective and objective reasons, but we may remain wary and not fully trust it. The user of the current L2 automated driving system, because of its imperfection, may have this type of relationship with the system. Alternatively, we may trust a system, but still prefer to operate it ourselves because of the enjoyment that comes from doing so. The highly automated driving systems of the future may elicit this kind of relationship.

There are several different factors that can affect the issue of trust and distrust. If the algorithms of automated processes are complex, the user does not understand the logic and process by which they make decisions, and the information is not transparent, giving the impression that the automated system is a black box. This can make it easy for the user to become distrustful. Another factor is trustworthiness. As no system can be guaranteed to be 100% correct at all times, any errors, or results that are different from what the user expects, will have an impact on trust. Measuring the relationship between system capabilities and the actual level of trust of the driver has become one of the central issues in the field of human–machine trust. The relationship between trust and mistrust is shown in Fig. 16.6. Researchers generally describe the relationship between trust and system capability by means of a two-dimensional coordinate system with system capability (Capability) as the

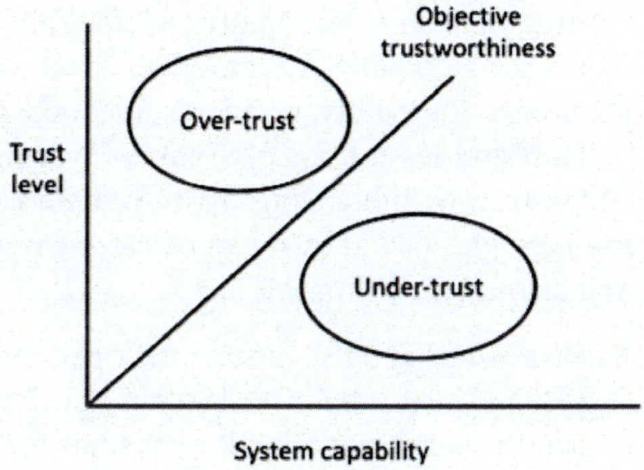

Fig. 16.6 Relationship between subjective user trust and system performance trustworthiness

horizontal axis and trust level (Trust) as the vertical axis. From a trust perspective, the diagonal line reflects the objective level of trustworthiness based on the system's capability in a given context, while the level of trust reflects the subjective level of trust that a person has in the system during actual human–computer interaction.

We can assess the appropriateness of the current state of trust by measuring the relationship between subjective trust and objective trustworthiness (Lee and See 2004). The relationship between the two is characterised by three states: appropriate trust, insufficient trust and excessive trust. Appropriate trust, also known as calibrated trust, refers to the driver's subjective level of trust being consistent with the objective trustworthiness of the system, as shown in the middle line of Fig. 16.6. Under-trust refers to the driver's subjective trust level being lower than the system's trustworthiness (lower right area of Fig. 16.6). Under-trust occurs when the driver underestimates the capability of the automated driving system and thus ignores the valid advice provided by the system or does not use the automated system functions (Disuse). Over-trust refers to the subjective trust level of the operator being higher than the objective trustworthiness of the system (Fig. 16.6, top left area). Over-trust occurs when the driver overestimates the capability of the automated driving system and thus does not monitor the current vehicle and road conditions in a timely manner, resulting in driver misuse of the automated functions (Misuse).

When a person observes a system for a limited period of time and is satisfied with the operation of the system, the trust may be in the upper left side, also known as over-trust. When a first failure occurs, the user's trust in the system may fall to the bottom right. Of course, this depends on the severity of the failure. With high automation trustworthiness, small failures may not even be detected. In case of a severe failure, the reaction to this 'first failure' is often inadequate (Rovira et al. 2007), and there is a temporary decrease in trust. After experiencing that the automation works well again for some time, however, trust is re-established (Lee et al. 2021). Therefore, it is important to build up the right level of trust in the system. Excessive trust creates complacent behaviour, i.e., when the automated system starts working, the operator stops paying attention to the operational status of the system, so that s/he loses the relevant situational awareness. And when an error occurs in the system and s/he starts to take over the operation, a wrong action may occur. Another problem caused by over-trust is that of automation bias. In a state of over-trust, the user assumes that the automatic system is doing the right thing, so that when it gives instructions, the user accepts them unconditionally and does not make his own judgements. There is a risk that if the automated system gives the wrong instructions and the user still follows them unconditionally, accidents may occur.

Hoff and Bashir's (2015) three-layer model is the most representative in terms of the factors influencing trust. Summarising the research on automated trust, the model suggests that trust can be categorised into dispositional trust, situational trust and learned trust (Fig. 16.7).

- ***Dispositional trust*** reflects the operator's pre-set tendency to trust an automated system and is influenced by operator characteristics, including factors such as age, gender and personality. It represents the fact that some people may have a benign

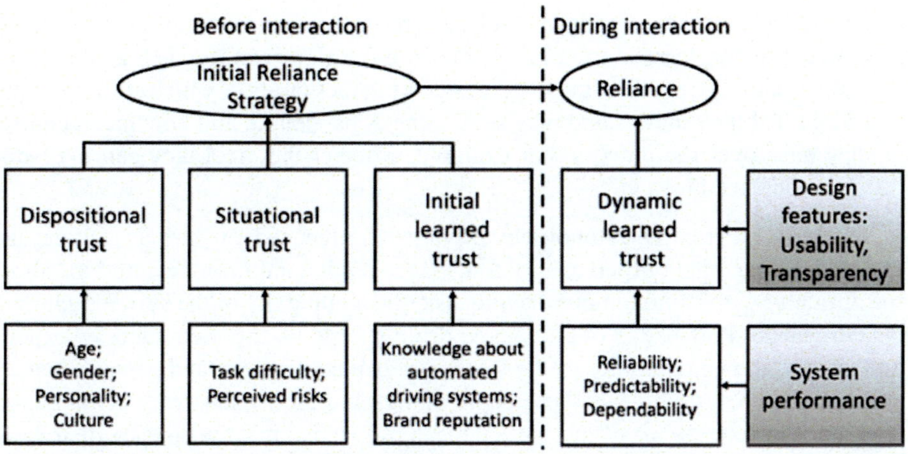

Fig. 16.7 Hoff and Bashir's (2015) trust model

belief in their environment, while others may be sceptical and have a fundamental distrust of their environment.

- ***Situational trust*** reflects the trust that an operator has resulting from the assessment of the particular situation. Situational trust is influenced by two sets of factors, one external and the other internal. The external factors mainly relate to the complexity of the system and the difficulty of the task. People will be less likely to trust an automated system to handle a task if the task is perceived to be difficult. Similarly, perceived risks and benefits play a role in determining situational trust. In the context of driving, one might argue that driving is a difficult task, especially when a variety of special situations may occur, and therefore requires a complex system to judge the state of the road and the driver. Internal factors mainly concern expertise in automated driving technology. As this expertise is often possessed by only a few people, the lack of expertise in autonomous driving can have two effects, depending on people's predisposed trust. Either they have a naive belief in the power of technology and therefore think that automated driving is great and can solve all sorts of road problems that they cannot, or they are sceptical about the technology and therefore think that the driving task is so complex that it cannot be successfully automated.
- ***Learned trust*** reflects the trust that an operator has based on the system's performance during use of the system and is influenced by the characteristics of the automated system, including system reliability and system errors. A person may already possess knowledge about the system or similar systems (or about automated driving systems in general), and this knowledge influences his/her expectation that the automated driving system can be successful (initial learned trust). Likewise, brand reputation contributes to this factor ("Brand X will never bring flawed technology to market!"). The second influence is dynamic learned trust, which comes from practical experience with the automated driving system. Dynamic learned trust is based on the performance of the system itself (in terms of

how reliably, predictably, dependably, etc. the system performs its tasks). Furthermore, dynamic learned trust is influenced by design features such as the transparency and ease of use of the system and its interaction style, which affect whether people feel they understand why something is happening and how much control they have over the process. For example, whether they feel they can intervene when they need to.

In summary, trust in autonomous driving is influenced by various factors: the individual's personality (personality traits, age, gender, culture), assumptions about the difficulty of the autonomous driving task, the complexity of the system required, his/her general knowledge of automated driving systems and their success; experience in actual use: reliability of the system, dependability, etc. and personal understanding of the system—whether people understand the reasons why certain automated driving operations occur. Survey results show that most people are distrustful of the possibility of self-driving cars and are sceptical that a fail-safe system is possible. They ignore or are unaware of the (partial) success that engineers have already had in developing technology, and they believe that driving tasks are so complex that they do not think driving can be successfully automated. To the extent that they may reject automated driving, one might say that these people have no trust in it. On the other hand, observations from users of automated driving systems suggest that most of these people believe that the system is able to circumvent failures. As mentioned above, users of L2 automated driving systems actually perceive them as L3 systems, so they neglect to monitor system performance and start engaging in task activities that are not related to driving. These people have developed over- trust in the automated system.

From a design perspective, the question is then what can be done to provide people with the right level of trust. Hoff and Bashir's model suggests that the factors that can enhance people's trust in an automated driving system are likely to be primarily the ease of use and transparency of the system and the way it interacts. Therefore, in order to design for trust, it is necessary to consider how the ease of use and transparency of the system can be influenced by design and which interactions should be used. Many people do not like to read manuals and instead want to explore how to use the system through constant trial and error on their own. So, for designers manuals can only be used as a last resort. For automotive systems, this will be no different. A recent study conducted by in the Netherlands showed that many lease car drivers were not aware of the presence of assistance systems such as lane departure alerts, adaptive cruise control (as opposed to cruise control) and distance alerts, where a brief look at the manual might have shown that these systems are available. This poses a challenge for designers to build user interfaces for self-driving cars, which need to be easy to use and transparent without having to consult a manual.

Two related developments have been proposed to influence people's trust in automated driving systems, which relate to transparency and the way in which the car communicates with the user. Starting with the observation that people have more trust in a medium that bears some resemblance to them, researchers have explored the use of anthropomorphism. Anthropomorphism means that an agent is designed such that it becomes more human-like, either by appearance or by behaviour, and preferably both. This makes it easier for people to attribute certain emotional values (e.g., 'friendly', 'intelligent', 'intelligent', 'intelligent') to the system, increasing trust in the system. A number of projects are currently experimenting with this, such as designing the front of a car to look more like a person smiling, or the lights to look more like eyes. Such designs are increasingly appearing on concept cars at various motor shows. Inside the car, the use of synthetic faces or programmed facial elements to explore anthropomorphic ways of displaying information is being adopted by many car companies.

Another related development includes so-called "explanation interfaces" (Ruijten et al. 2018), which explain to the driver why the system is performing certain actions. This design allows for intelligent and driver-friendly dialogue that not only involves the exchange of information, but also adapts to the needs of the listener and provides additional information when needed, with the 'explanation interfaces' clarifying why certain actions are performed in order to provide the driver with an understanding of what is happening and why (see Table 16.1). In this way, the operation of the system becomes more transparent. In addition, explanations can be provided through speech or voice, thus conforming to the principle of anthropomorphism.

An important question concerns how effective one might expect such design interventions to be. According to Hoff & Bashir's trust model, there are differences between people with respect to their disposition to trust automated systems. That is, there will always be believers and sceptics. One might say that it makes little sense to design for sceptics, as they might be hard to convince anyway, so that design efforts might better be spent to design primarily for people who believe that the driving task can be automated, maybe not for 100% of possible situations, but still safe enough for everyday life. In addition, it might be assumed that not all people will buy or start using an autonomous vehicle at the same time. Initially, automated vehicles will be used mainly by people who are characterised as innovators and early adopters in Rogers' Innovation Diffusion Theory (Rogers 1962). These early users will "pave

Table 16.1 Dialogue messages of an explanation interface

Situation	Voice messages
Automated driving system (ADS) makes way for bicycles	I'm giving right of way to a bike
ADS is waiting to pass at a traffic junction	I am waiting to pass
ADS slows down on cobbled roads	We are on a cobbled road with pedestrians, I had to slow down
ADS stops at yellow light	Traffic light is turning red
ADS is changing lanes to allow others to overtake	I let those who are faster than me overtake me
ADS waiting by the ramp	I need to wait for another car to pass before I can enter the motorway

Source Ruijten et al. (2018)

the way" for others, who will then hear about the experiences from the early users through (social) media and join later on. Furthermore, certification schemes might be set up that have certification marks, informing customers that the system meets safety requirements.

A final question about trust is whether over-trust is bad. If so, design efforts should be directed at designing against over-trust. In principle, one might say that properly calibrated trust is preferable to over-trust, so that people are aware of when the system will run into its boundaries. However, as automation progresses, the number of situations that the system cannot handle will further decrease, so that the likelihood of the system to deviate from satisfactory performance, measured as number of deviations from satisfactory performance per X km, will decrease. As a result, at more advanced levels of automation, systems may deviate from optimal performance only once every 2000, 3000 or 4000 km, or even less. In addition, hopefully most of the deviations from satisfactory performance will be minor deviations, and these may not even be noticed when the users are engaged in non-driving-related tasks, thus further reducing the probability of perceived failure. This low probability of perceived failure will enhance over-trust in the system, and it is hard to design against this tendency. The alternative, of providing users with precise knowledge of when and how often the system may make (minor) mistakes, is unattractive from a design perspective and unlikely to be successful, given the fact that people hate to read manuals and have trouble remembering the information from manuals and instructions. One possibility is that, similarly to the aviation domain and the process control domain, people will be exposed to and trained about relevant situations by means of driving simulators, so that they learn the boundaries of the system, therewith calibrating their trust. Such interactive instruction is usually more successful than information through manuals.

In practice, the situation mentioned above may still take a while. Recent reports (2022) from users and experts about a Beta Full Self-Driving (FSD) system available in the market indicate that experts and users alike think it may still take a decade or so for the FSD system to mature enough to be ready for the road. Currently, still many interventions are needed, and when the system prompts the driver to take over control (which makes it a Level 3 or Level 4 system rather than L5, which is fully automated driving), in fact the driver may find himself struggling to get the system hand over control.

16.5.2 Transparency

As already said, at lower levels of automation, drivers are engaged in the driving task all the time, and this creates a situation where they learn to understand the automated systems through observation and exploration. This way they build functional mental models of the system, so that they are able to tell what the system does, and how they need to instruct the system to behave in certain ways. At higher levels of automation, as already alluded to in the previous paragraph, it appears unlikely that automated driving systems can be made transparent to the users, so that they precisely know what the system can do and when it will run into its boundaries. At the same time, it is unlikely that people want to know. Once the system has a low probability of failure, it may be assumed that the average user will develop an attitude of "I have watched the system and it does what it should do", and that this is all such users want to know.[1] Furthermore, it appears that people's own experiences of mistakes have relatively little impact on their trust in the system.

A related topic that spans all levels of automation is *mode awareness*. Given that external conditions may result in setting different levels of automation, it appears important to design the system such that the user can easily determine which mode the system is in, or, at least, to design the system such that the user is prevented from committing mode errors and acts or fails to act in such a way that safety is jeopardized. Some efforts have been given to design Human Machine Interfaces to support mode awareness, but so far this remains a challenge (e.g., Feldhutter et al. 2017). Design efforts here require detailed understanding of the mode errors that people commit and their potential consequences, so that systems can be built that avoid serious consequences due to mode errors.

Finally, it has been suggested that the success of automation in the market is strongly dependent on its association with car sharing. The costs of software and hardware needed for full automation may make it unattractive for individual customers to buy a vehicle that has full automation. Instead, by subscribing to a car sharing program, one can share the costs with other subscribers as well. This situation may result in people using different brands and models all the time, so that it becomes the more important to build systems and associated user interfaces that people can understand easily and quickly. Standardisation might be one way, and therefore it's good that manufacturers participate in standardisation bodies such as the International Standards Organisation (ISO). At the same time, companies may desire to create innovations to differentiate from the competition and gain competitive advantage. Thus, it is not just a matter of complying to agreed-upon standards, but every manufacturer should take its responsibility and aim for systems and associated interfaces that are easy to understand and use.

16.5.3 Comfort, Driving Style and Motion Sickness

Different people have different driving preferences. Some people prefer a more calm and cautious driving style, while others prefer a sportier and more adventurous driving style. This raises the question of how self-driving cars should drive. A more sporty and risky driving style manifests itself by braking aggressively or accelerating hard at traffic lights, being more impetuous at intersections, violating speed limits, going through curves at relatively high speeds, etc. As safety is a top priority for technology

[1] Exceptions may be people who have a strong interest in technology and want to know all the details. They will be motivated to invest time and effort to find out how and why the system performs as it does.

developers, one might expect that self-driving cars will be equipped with a calm and defensive driving style. This raises the question of whether such a driving style is acceptable to people who favour a sporty driving style.

Experiments conducted to determine people's preferences for the driving style of automated driving systems have shown that both sporty and calm drivers prefer defensive driving styles over assertive driving styles (Yusof et al. 2016). As one of the participants in one of the experiments put it, he would not like to be "a passenger in a vehicle driven by someone who drives like me". This finding raises significant questions for those who aim to make a self-driving car drive more human-like through self-learning by artificial intelligence systems. Clearly, a calmer driving style reduces the amount of physical force exerted on the body, thereby enhancing physical comfort and reducing the incidence of motion sickness. In addition, a calm driving style is convenient if the occupant wants to engage in non-driving related activities. In addition, calm driving gives a sense of security.

This brings us to another related topic: physical comfort. Manufacturers have developed active suspension systems such as Magic Body Control (Mercedes) to reduce vertical accelerations that can cause discomfort. In addition, Magic Body Control includes a curve tilt function to reduce lateral forces. However, even in such systems, longitudinal and some lateral body displacement is still present, which may cause discomfort and induce car sickness, especially with sportier driving styles. Motion sickness has been found to be a risk in self-driving cars as occupants may want to engage in non-driving related activities such as interacting with their smartphones, watching videos or reading, all of which involve taking their eyes off the road, thus reducing situation awareness and increasing the likelihood of motion sickness.

16.5.4 Shared Control

As mentioned above, at level L5, the automated driving system is responsible for all aspects of the driving task, i.e., the automated driving system is responsible for all tactical decisions (manoeuvres to be performed) and operational tasks (vertical and lateral control). The strategic decisions (choice of destination, type of route and desired arrival time determining the cruising speed) remain the responsibility of the driver. The steering wheel or pedals may no longer be included here. As mentioned earlier, most traffic accidents are caused by human error and by taking the driver out of the driving task, the engineers believe that this will prevent these accidents.

From the user's point of view, it may be questioned whether this is a development that users applaud. Assuming that self-driving vehicles will adopt a defensive driving style and will not violate traffic rules, there are several situations where the behaviour of self-driving vehicles could lead to inefficient behaviour, deviating from what users expect and from what they consider desirable (Terken and Pfleging 2020). Given that defensive behaviour is characterised by playing safe rather than taking risk, self-driving cars will maintain longer gaps than manually driven cars. This will affect their behaviour at intersections and when entering a motorway from an on-ramp or when taking over. At intersections, they may have to wait a long time to pass. On the motorway, when they overtake, they may have to wait a long time to change lanes because they perceive the gap size in the left lane as unsafe, so they have to slow down, which makes merging with traffic in the left lane more difficult. Similarly, it has been reported that self-driving cars wait excessively long at busy pedestrian crossings, as they wait for large gap sizes. In all these cases, it might be desirable for the autonomous vehicle to be more assertive.

In addition, always obeying the rules of the road may lead to blunt, impolite, as well as inefficient behaviour. Figures 16.8 and 16.9 show two situations in which manual drivers may exhibit polite behaviour to accommodate other road users. In Fig. 16.8, car B in the R lane is waiting for the traffic signal to turn green. Once the traffic signal turns green, traffic starts moving again. Now, in the opposite lane, vehicle A wants to turn left into the car park and has to wait until the road is clear before he can do so. In automated driving mode, once vehicle C has moved forward, vehicle B will follow. However, in manual driving, the driver of vehicle B may respond to the situation by politely yielding to vehicle A and allow him to enter the car park and then move forward. Similar courteous behaviour by human drivers may

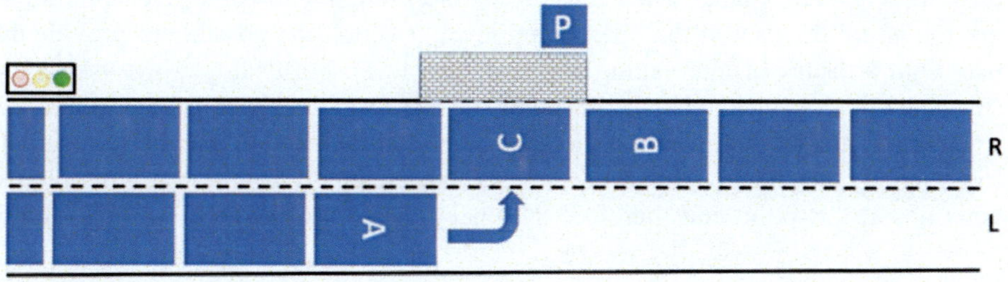

Fig. 16.8 Granting right of way to vehicles from the opposite direction

Fig. 16.9 Granting right of way to cyclists

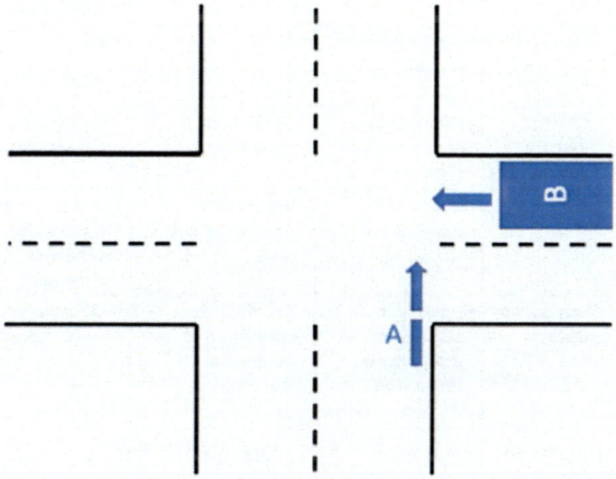

be observed if vehicle A wants to leave the parking lot. If the cars on the road are in automated mode, it may be hard for vehicle A to leave the car park.

In Fig. 16.9, cyclist A and vehicle B are approaching the intersection and vehicle B has right of way. However, the driver of vehicle B may decide to give way to the cyclist so that the cyclist does not have to brake and wait for vehicle B to pass. A similar situation occurs when a driver gives way to a pedestrian waiting on the pavement to cross first.

In both of these cases, if the self-driving vehicle obeys the rules of the road, it will not give way to a vehicle trying to enter a car park or to a cyclist or pedestrian. More generally, these situations require judgement based on a deep understanding of the traffic situation (discretionary judgement), which is likely beyond the capability of the automated driving system, which, in terms of the skill acquisition model of Dreyfus and Dreyfus (1980) may be able to operate at a competent or proficient level, but not at an expert level (see Fig. 16.10). In order to operate at an expert knowledge level, human judgement needs to be included in the decision cycle.

This may induce the desire in users of self-driving cars to be able to intervene and participate in decision-making. One possibility is to intervene at the operational control level so that the driver can control the pedals, use the steering wheel, apply the brakes or change lanes when entering an intersection. However, as it is preferable not to disturb the user at the operational level, it should be possible to provide the user with a means of intervening at the tactical level. For example, the user could select an option from a menu displayed on the centre console to tell the system when to enter an intersection, when to move or when to brake and to use the turn signals to initiate a lane change. However, the user may be involved in non-driving related activities at this point and therefore may need to be informed that an opportunity

	Novice	Advanced Beginner	Competent	Proficient	Expert
Perception	• Sees actions in isolation	• Limited situational perception	• Sees actions in terms of overall goals	• Sees overall picture	• Sees overall picture
Decision	• Rigid adherence to rules	• Analytic	• Analytic	• Decision making more confident	• Intuitive • Discretionary judgement
Coping with complexity	• No conception of dealing with complexity	• Only partial resolution	• Coping with crowdedness	• Holistic view of situation	• Holistic grasp of complex situations
Autonomy	• Needs close supervision	• Supervision needed for overall task	• Able to achieve most tasks using own judgement	• Able to take full responsibility	• Able to create own interpretations
Standard of work	• Unlikely to be satisfactory unless closely supervised	• Simple tasks completed to acceptable standard	• Fit for purpose	• Satisfactory standard achieved routinely	• Excellence achieved with relative ease

Fig. 16.10 Levels of skill (after Dreyfus and Dreyfus 1980; S. Lester, http://www.sld.demon.co.uk/dreyfus.pdf, 2005)

for intervention exists. And because traffic conditions change rapidly, there is a time limit on the window of opportunity for the user to participate in the decision.

In summary, at Level 5 it might be desirable to enable the user to participate in the decision-making loop and intervene at the level of tactical control, while leaving operational control to the system. If the user is not involved, this may lead to inefficiency, rigidity and sluggishness. Further research should tell whether users actually want this functionality and how the interface that enables the user to interact with the system should be designed. One possibility is to allow users to influence the decision making through settings that they may select in advance, such as Assertive/Relaxed. However, it is not certain that all situations can be handled by such general schemes, so that there might be a need to develop Human Machine Interfaces that allow users to participate in the decision loop in individual situations. The need for and the precise details of such HMIs need to be investigated through further research.

16.5.5 Skill Loss

One of the concerns that has been expressed with automated driving systems is the loss of driving skill by the driver. In particular, with levels L2 and L3, where the system takes care of operational (L2) and operational and tactical manoeuvres (L3) most of the time and the driver takes over the driving task only if the system runs into its limits, the opportunities for the driver to build and maintain driving skill will be substantially reduced. The fear is that, for L2 and L3, lack of experience will reduce the driver's ability to adequately take over control at moments when needed: The system can handle the routine situations, but the situations where the driver must take over control are, by their very nature, the more complex situations for which driving experience is needed.

Since there is no research about loss of driving skill in connection with automated driving, we can only analyse the claims carefully and bring insights from the existing literature, to formulate expectations about how serious the problem will be.

In the first place, as noted above, L3 automated driving systems will take care of the driving task mostly in routine situations. Even if the driver would be driving him-/herself in such situations, opportunities for further skill development and maintenance would be limited because of their routine character. In other words, the loss of skill in the case of L3 automation may be rather limited.

In the second place, there is no a priori reason to assume that what is complex for the system is also complex for the driver. In other words, situations that are difficult for the system and where the system runs into its limitations may not be equally difficult for the driver to handle, so that the task demands for the driver may be relatively modest and the driver may be very well capable to handle the situation, even if his driving skills have degraded. A detailed analysis of complex situations should point out whether and which situations that are difficult for the system are also difficult for drivers.

In the third place, in Sect. 16.3.1 we have made a distinction between expected and unexpected take-over situations, and we have argued that, in case of unexpected take-over situations, drivers are not well able to regain control in time. It is likely that this is the case regardless of loss of driving skill. Again, a detailed analysis should point out whether this reasoning is valid. With respect to expected take-over situations, the work of Wilde (1982) and Fuller (2005) (see Sect. 7.2) may provide clues to the effects of skill loss. They suggest that drivers may fit their behaviour to the task demands. If the driver feels the task is too complex, s/he may adjust by driving slower or taking more time before entering an intersection, in order to reduce the task demands. This will apply equally for experienced and unexperienced drivers.

In the fourth place, a more detailed analysis of driving skill may help to understand how driving skill will be affected precisely by lack of experience. We will first look at the situation of an experienced driver whose skill may be degraded by lack of experience when driving mostly in automated mode. Using the Situation Awareness model (see Sect. 5.1), the component driving tasks consist of perceiving and understanding the traffic situation, predicting how the situation evolves, deciding and acting. We assume that driving skill influences all aspects of the driving task, in the sense that, in the case of experienced drivers, mental processes for all component tasks have been largely automated and can be performed routinely. For instance, it has been shown that experienced drivers scan the situation in different ways than unexperienced drivers. Even if the speed and accuracy with which these tasks can be performed are affected by lack of experience, it may be assumed, as was stated above, that the driver adjusts his behaviour to lower the task demands, so that skill loss may not really be an issue. Of course, there is a need for research to validate these assumptions and find out how driving skill is affected by disuse—how are the different aspects of the driving task (perception, understanding, anticipation, deciding, acting in complex situations under temporal constraints) affected by disuse—and whether such regulatory mechanisms apply.

If we look at the situation of the unexperienced driver, i.e., a driver who passed his/her driving exam only recently and did not yet become an experienced driver, it appears safe to assume that his/her ability to take over control is lower than that of the experienced driver above. But again, we may assume that this driver will adjust his behaviour to lower the task demands.

In sum, upon closer inspection one might contend that the problem of skill loss is not so serious at it appears at first sight. We should also consider, however, that skill loss and lack of experience may influence the driver's feeling of being capable to handle the situation. In other words, it may lead to more driving anxiety (see Sect. 7.2). So, even if the degradation of cognitive skills has only minor effects due to regulatory mechanisms, drivers may become more anxious about having to take over control, and anxiety by itself may result in inadequate actions. It is unclear whether the regulatory mechanisms are suited to deal with this consequence. Again, further research is needed, also targeting design solutions.

As the systems become more competent, however, their ability to deal with non-routine situations will also improve, and this will then indeed reduce the opportunity for the driver to maintain and develop driving skill. For the current generation of drivers, who have engaged in manual driving most of their adult life and have already developed driving skill, this may not be really an issue, but, as argued above, further research on this topic is needed.

For future generations, who will likely have automated driving systems at their availability from the moment they acquire their driving license, the situation may be different, and one may question whether they will develop the skills to handle rare complex non-routine situations under temporal constraints when needed. Different solutions can be envisaged for this issue, each with its own disadvantages. For instance, one might require unexperienced drivers to drive manually until they reach a criterion level, and during this period they could be supported by assistance systems to ensure safety for themselves and other road users. Or one might require unexperienced drivers to spend time in driving simulators to learn to deal with non-routine situations until they reach a certain criterion level.

However, such solutions assume that there will remain a need (and a desire) for future generations to acquire driving skills. Once automated driving technology has become sufficiently capable to deal with most traffic situations, future generations may no longer be interested in driving themselves, and manual driving may become a leisure activity for hobbyists, so that the problem of skill loss dissolves.

16.5.6 Ethics

Automated driving systems are artificial agents that must coordinate with other road users. Since they operate in a highly dynamic environment and need to take decisions in fractions of second, automation failures or misjudgements potentially have fatal consequences. Thus, driving automation has also ethical consequences. The ethical aspects of driving automation were introduced first by Lin (2013) and Goodall (2014). Lin discussed the so-called trolley problem from ethical philosophy, shown in Fig. 16.11 Left. The trolley problem describes the situation where a trolley

Fig. 16.11 Left: The trolley problem. *Source* https://cvl-eng.ca/trolley-problem/. Unknown copyright holder. Right: The trolley problem translated to the traffic domain. *Source* Visualphotos.com. Unknown copyright holder

approaches a point where five people are lying on the track. An observer has the possibility of turning the lever, making the trolley switch tracks. On the other track, there is one person lying. If the observer does not intervene, five people will be killed. If the observer intervenes, only one person will be killed. The question is what the observer should do. Translated to the traffic context, the trolley problem describes a situation where an autonomous vehicle is heading towards a group of pedestrians crossing the street and cannot avoid crashing into the group, unless making an evasive manoeuvre that will probably result in crashing against an obstacle, which may kill the occupant(s) of the vehicle (Fig. 16.11 Right). Again, the question is what the vehicle should do, or, more specifically, how the decision-making algorithm should be designed.

While results of surveys usually show that people find that the vehicle should minimise the overall number of casualties and thus go for the evasive manoeuvre, therewith taking the risk of killing the occupant, manufacturers are unlikely to implement such algorithms. In the first place, the customers will probably be reluctant to buy a vehicle that might decide to take the risk and kill them. In the second place, manufacturers point to the fact that they are taking all kinds of passive and active safety measures to minimize the risk that either other road users or the occupant(s) will be killed. Thus, the trolley problem may well be an academic exercise rather than a problem that may become reality in the future. This does not mean that accidents will not happen. It needs to be noted, however, that the trolley problem involves a rational choice between two alternatives and does not follow from failures of the technology like sensor malfunctioning. In case of sensor malfunctioning, the vehicle may not even perceive the pedestrians and thus cause an accident with potentially fatal consequences. But in such cases, the decision algorithm will have little to say. Generally, it appears unlikely that algorithms can be built that make ethical decisions that do justice to the considerations that people apply when filling surveys. Instead, the algorithms and the hardware will probably be designed so that they try to minimise the damage, and the ethical outcomes of the decisions will be the by-product of the primary aims of these algorithms.

Two kinds of events may cause accidents. There may be unexpected events in the environment, e.g., a child running into the street from between parked cars, that the vehicle cannot avoid, even with all the on-board systems working well, because of mass inertia. Or there may be problems with the system itself, either because of sensor failure, sensor limitations or software problems. Both types of problems will exist, and fatal accidents cannot be fully excluded. From an ethical perspective, the question is what this means for automation. While people tend to be forgiving about human error and its consequences, they are not equally forgiving about the consequences of errors made by the system. Human error is a fact of nature that we can try to do something about by design, but cannot avoid altogether, while the technology was designed to avoid errors in the first place, and thus technology making mistakes is considered just bad design. However, the evidence so far suggests that accidents due to technical failure do not result in broad rejection of the technology. In the end, while "zero casualties" may not be achievable, at least the promise should be that the number of casualties is substantially less than without automation. If so, from a societal perspective the automation may be fully acceptable.

Other behaviours that have ethical implications have already been alluded to above (Sect. 16.4.4). As stated above, automated vehicles will probably exhibit a driving style that is not only defensive but also rigid, since it always complies with the traffic rules. In itself, a defensive driving style is good, because most likely it is safer that a risky driving style. However, it may be annoying to the occupants and to following vehicles. In addition, always complying with the rules may result in blunt behaviour (not relinquishing right of way if the situation would make this desirable), and it may lead to non-adaptive behaviour: sticking rigidly to the speed limit is non-adaptive if all other traffic violates the speed limit by 10 km/h. This is not to say that automated vehicles should be designed such that they violate the speed limit just because everyone else does. There is always a choice, and the point is that designers should think about which choice to make, and possibly provide the users with means to tune the system to their own preferences and make the choice themselves. Obviously, this raises the question of who is accountable if something happens. The general rule might be that users can tune the system to their own preferences, to the extent that the system can still perform adequately and maintain safety.

16.5.7 Usefulness

If autonomous vehicles succeed in considerably reducing the number of (fatal) accidents, from a societal perspective there will be no doubt about their usefulness. However, autonomous vehicles need to be bought by individual people, and it is unclear yet whether they will buy a fully automated car because of safety considerations. After all, most people have experienced few or no accidents, and therefore feel driving is safe already. Instead, they may make a buying decision based on the usefulness for their own situation and trade the costs of buying a vehicle with automation technology against the personal benefits, in terms of how often they can actually use the technology and what it brings to them. If for instance Level 3 automation only works on the highway, and people rarely ever drive on the highway but mostly in urban environments, the technology has little use for them. Or if the technology also works in urban environments, but people only use the vehicle for short trips, they may consider the fact that the system allows them to engage in non-driving-related activities not very convincing. One might point out that people already nowadays spend time on interacting with smartphones to engage in non-driving-related activities, but the question remains whether this would lead people to consider the technology useful for them.

Alternatively, if the automation technology becomes mandatory because of the societal benefits, or if it is implemented in vehicles that are available through car-sharing programs, there is little choice. In that case, people will have a vehicle with automation technology at their availability, even if they do not consider it useful

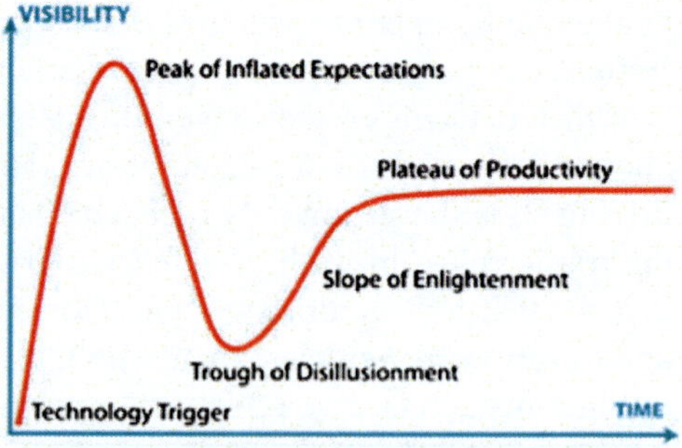

Fig. 16.12 The Gartner hype cycle™. Courtesy Gartner®

for their personal everyday life. The next question becomes then whether they will actually use it, provided that—for a long time to come—they will have a choice whether to use it or not. The topics that have been discussed in the previous sections will play an important role in determining their choice. In particular, the usability and the behaviour of the system in view of their goals and expectations will play a major role.

Currently, there is much to do in the media about driving automation, and expectations are high. It has been suggested that, in such cases, the Gartner® hype cycle™ (Gartner 2021; Fig. 16.12) may apply, where, after initially high expectations, people get disappointed about the technology if the actual results turn out not to meet the high expectations, as is suggested by recent (2022) reports about a Beta Full Self Driving (FSD) system available in the market. This will influence people's willingness to buy and use the technology. Clear and honest communication appears to be key to expectation management.

16.6 Interaction with Other Road Users

As automated vehicles move around in the space inhabited by other vehicles, bicycles, pedestrians etc., their relationship with other road users also needs to be studied. This presents a particular set of challenges, especially in the case of greater automation, where there is no driver to control or monitor the vehicle and be responsible for deciding how to interact with other road users. Much of the work to address these challenges has focused on the interaction between self-driving vehicles and pedestrians.

Let us first address the question of whether it is safe for pedestrians to cross the road in front of a self-driving vehicle. This is not just an imaginary question: Recent reports (2022) from users and experts about a Beta Full Self-Driving (FSD) system available in the market indicate that the FSD system has been reported to have problems dealing with pedestrians crossing the street, conducting "rolling stops" at pedestrian crossings instead of coming to a complete stop and having problems recognizing pedestrian walk signs, to such an extent that software updates are needed to disable some of the system's functionality.

In case of manually driven vehicles, there is communication between the driver and the pedestrian, generally using two types of information exchange, as shown in Fig. 16.13. Both parties can communicate implicit information, such as the direction and speed of movement and the distance. With regard to the vehicle, the implicit information also includes the make and model of the car and the characteristics of the driver, such as age and gender, state of attention and viewing direction. With regard to the pedestrian, the implicit information includes age and gender, state of attention and direction of viewing. Information that is explicitly exchanged between the driver and the pedestrian includes the car horn, flashing lights and gestures. Furthermore, Fig. 16.13 indicates that as the distance between the vehicle and the pedestrian becomes smaller, the type of information used in the decision changes (Dey et al. 2020). Of course, there may be other conditions, such as pedestrian crossings, both with and without pedestrian walk signs, surrounding traffic, and small islands in the middle of the street that may affect the decision, but the basic issue remains that the driver and pedestrian must coordinate the exchange of information between them, and the actions taken.

It is generally accepted that the exchange of explicit information (especially eye contact) plays an important role in the coordination between drivers and pedestrians. If so, the introduction of self-driving cars on the road raises the question of whether compensatory means should be designed to compensate for the absence of this exchange of explicit information in driverless situations.

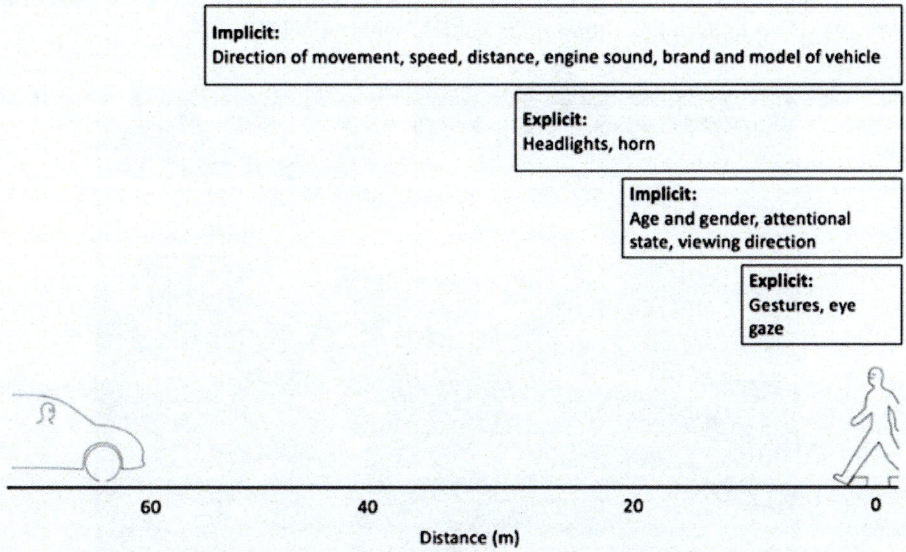

Fig. 16.13 Pedestrian-vehicle communication

Several in-depth studies of pedestrian-vehicle interaction have shown, however, that most decisions and coordination are based on implicit cues, i.e., the behaviour of the vehicle and the pedestrian (e.g., Dey and Terken 2017). Only at short distances can drivers and pedestrians interact explicitly by making eye contact, and mainly to solve 'deadlock' problems—i.e., both parties are waiting for the other to act. In other cases, the information that is implicitly communicated plays a more important role, as perceived directly or inferred from the orientation of the head: the pedestrian wants to know if the driver is aware of his/her presence, and the driver wants to know if the pedestrian is aware of the approaching vehicle. Nevertheless, even if explicit information only plays a minor role in decision-making and coordination, coordination between pedestrians and autonomous vehicles may still benefit from explicit information being sent from the vehicle to the pedestrian through a human–machine interface (eHMI) on the outside of the vehicle.

Several eHMI concepts have already been proposed. Some provide invitations or instructions to pedestrians with text displays on the windows of the car, such as "Walk/Don't walk", or project virtual pedestrian crossings on the road. Others use more abstract visual signals to provide information about the vehicle's intentions or emotions, such as a smiling car (see Fig. 16.14). Yet others use anthropomorphic signals, designing the front face of the car to look like a human face, where the two headlights in front look like human eyes that can be turned. User evaluations suggest that these concepts have little impact on the decision of pedestrians to cross the road in front of the approaching vehicle. The decision seems to depend more on the actual behaviour of the approaching vehicle, e.g., noticing that the vehicle is slowing down indicates that the vehicle is intending to yield to pedestrians. On the other hand, most evaluations suggest that people would have more confidence in the existence of an external HMI than in an autonomous vehicle without eHMI.

Fig. 16.14 Smiling car, signalling to pedestrians that it's safe for them to cross the street. *Source* https://semcon.com/news-media/media-library / © Semcon. Reprinted with permission

Another question is whether self-driving cars should signal to their surroundings that they are driving autonomously. Some companies have chosen to clearly mark the status of their vehicles by means of clearly visible sensors on the roof, while others have deliberately made self-driving cars indistinguishable from other cars so that other road users will see them as ordinary cars.

One of the considerations in whether to explicitly label self-driving cars or not is the issue of "bullying". It has been suggested that people might begin to take advantage of self-driving vehicles once they know that the vehicles are designed to maximise safety and therefore have a defensive driving style. Drivers of manually driven vehicles may decide not to grant right of way at intersections, because they know that self-driving vehicles will easily give up the right of way. As for pedestrians, even if they do not have the right of way, they may walk into the street to cross, because they know the vehicle will stop. It is not clear whether such bullying behaviour will actually happen and what factors will determine the behaviour. It will depend on how people adapt to the new technology, the results of which are difficult to obtain from research and simulator experiments. Only when new technologies are widely introduced into society, will it be possible to investigate how people adapt to them and whether bullying will occur.

Finally, explicit labelling of vehicles as self-driving may be beneficial in areas other than pedestrian-vehicle interaction. It has already been suggested that the number of rear-end collisions between leading automated vehicles and following manually driven vehicles has increased due to the way autonomous vehicles brake. Vehicles clearly marked as self-driving may warn other vehicles that they should be prepared for such unexpected behaviour to occur.

Further Reading Useful sources for further reading on the human factors of automated driving systems are Kun (2018) and Riener et al. (2021).

References

Dey D, Terken J (2017) Pedestrian interaction with vehicles: roles of explicit and implicit communication. In: Proceedings of the 9th international conference on automotive user interfaces and interactive vehicular applications, pp 109–113

Dey D, Holländer K, Berger M, Eggen B, Martens M, Pfleging B, Terken J (2020) Distance-dependent eHMIs for the Interaction between Automated Vehicles and Pedestrians. In: 12th international conference on automotive user interfaces and interactive vehicular applications, pp 192–204

Dreyfus SE, Dreyfus HL (1980) A five-stage model of the mental activities involved in directed skill acquisition. California Univ Berkeley Operations Research Center

Feldhütter A, Segler C, Bengler K (2017) Does shifting between conditionally and partially automated driving lead to a loss of mode awareness? I:n International conference on applied human factors and ergonomics, pp 730–741. Springer, Cham

Fuller R (2005) Towards a general theory of driver behaviour. Accid Anal Prev 37(3):461–472

Gartner® (2021) Understanding Gartner's Hype Cycles, Philip Dawson, Alexander Linden, Keith Guttridge, Katell Thielemann, Nick Jones, Jackie Fenn, 2 July 2021

Gold C, Damböck D, Lorenz L, Bengler K (2013) "Take over!" How long does it take to get the driver back into the loop? In: Proceedings of the human factors and ergonomics society annual meeting, vol 57(1), pp 1938–1942. Sage Publications, Sage CA: Los Angeles: CA

Goodall NJ (2014) Machine ethics and automated vehicles. In: Meyer G, Beiker S (eds) Road vehicle automation. Springer, pp 93–102

Hancock PA, Billings DR, Schaefer KE, Chen JYC, de Visser E, Parasuraman R (2011) A meta-analysis of factors affecting trust in human-robot interaction. Hum Factors 53:517–727

Helldin T, Falkman G, Riveiro M, Davidsson S (2013, October) Presenting system uncertainty in automotive UIs for supporting trust calibration in autonomous driving. In: Proceedings of the 5th international conference on automotive user interfaces and interactive vehicular applications, pp 210–217

Hoff KA, Bashir M (2015) Trust in automation: integrating empirical evidence on factors that influence trust. Hum Factors 57(3):407–434

https://www.gartner.com/en/documents/4003232

Kuehn M, Vogelpohl T, Vollrath M (2017) Takeover times in highly automated driving (level 3). In: 25th International technical conference on the enhanced safety of vehicles (ESV) national highway traffic safety administration

Kun AL (2018) Human-machine interaction for vehicles: review and outlook. Foundations and Trends® in Human–Computer Interaction 11(4):201–293

Lee J, Abe G, Sato K, Itoh M (2021) Developing human-machine trust: impacts of prior instruction and automation failure on driver trust in partially automated vehicles. Transport Res f: Traffic Psychol Behav 81:384–395

Lee JD, See KA (2004) Trust in automation: designing for appropriate reliance. Human Factors 46(1):50–80. Retrieved from http://www.ncbi.nlm.nih.gov/pubmed/15151155

Lin R, Ma L, Zhang W (2018) An interview study exploring Tesla drivers' behavioural adaptation. Appl Ergon 72:37–47

Lin R, Liu N, Ma L, Zhang T, Zhang W (2019) Exploring the self-regulation of secondary task engagement in the context of partially automated driving: a pilot study. Transport Res f: Traffic Psychol Behav 64:147–160

Lin P (2013) The ethics of autonomous cars. The Atlantic. http://www.theatlantic.com/technology/archive/2013/10/the-ethics-of-autonomous-cars/280360/. Accessed 24.03. 2014

Löcken A, Heuten W, Boll S (2015) Supporting lane change decisions with ambient light. In: Proceedings of the 7th international conference on automotive user interfaces and interactive vehicular applications, pp 204–211

Madhavan P, Wiegmann D (2007) Similarities and differences between human-human and human-automation trust: an integrateive review. Theor Issue Ergon Sci 8:270–301

Merat N, Jamson AH, Lai FCH, Daly M, Carsten OMJ (2014) Transition to manual: driver behaviour when resuming control from a highly automated vehicle. Transp Res Part F: Traffic Psychol Behav 27(PB):274–282. https://doi.org/10.1016/j.trf.2014.09.005

Michon JA (1985) A critical view of driver behavior models: what do we know, what should we do? In: Human behavior and traffic safety, pp 485–524. Springer, Boston, MA

Mulder M, Abbink DA, Van Paassen MM, Mulder M (2010) Design of a haptic gas pedal for active car-following support. IEEE Trans Intell Transp Syst 12(1):268–279

Parasuraman R, Molloy R, Singh IL (1993) performance consequences of automation-induced "complacency." Int J Aviat Psychol 3:1–23

Riener A, Jeon M, Alvarez I (2021) User experience design in the era of automated driving. Springer Studies in Computational Intelligence, 623

Rogers EM (1962) Diffusion of innovations, 1st edn. Free Press of Glencoe, New York

Rovira E, McGarry K, Parasuraman R (2007) Effects of imperfect automation on decision making in a simulated command and control task. Hum Factors 49:76–87

Ruijten PA, Terken J, Chandramouli SN (2018) Enhancing trust in autonomous vehicles through intelligent user interfaces that mimic human behavior. Multimodal Technol Interact 2(4):62–72

SAE (2021) Taxonomy and definitions for terms related to driving automation systems for on-road motor vehicles. J3016_202104. https://www.sae.org/standards/content/j3016_202104/

Terken J, Pfleging B (2020) Toward shared control between automated vehicles and users. Automot Innov 3(1):53–61

Wickelgren WA (1977) Speed-accuracy tradeoff and information processing dynamics. Acta Physiol (oxf) 41(1):67–85

Wiener EL, Curry RE (1980) Flight deck automation: promise and problems. Ergonomics 23:995–1012

Wilde GJ (1982) The theory of risk homeostasis: implications for safety and health. Risk Anal 2(4):209–225

Young MS, Stanton NA, Harris D (2007) Driving automation: learning from aviation about design philosophies. Int J Veh Des 45(3):323. https://doi.org/10.1504/IJVD.2007.014908

Yusof NM, Karjanto J, Terken J, Delbressine F, Hassan MZ, Rauterberg M (2016) The exploration of autonomous vehicle driving styles: preferred longitudinal, lateral, and vertical accelerations. In: Proceedings of the 8th International Conference on Automotive User Interfaces and Interactive Vehicular Applications, pp 245–252

Zeeb K, Buchner A, Schrauf M (2015) What determines the take-over time? An integrated model approach of driver take-over after automated driving. Accid Anal Prev 78:212–221. https://doi.org/10.1016/j.aap.2015.02.023

Zhang B, de Winter J, Varotto S, Happee R, Martens M (2019) Determinants of take-over time from automated driving: A meta-analysis of 129 studies. Transp Res Part F: Traffic Psychol Behav 64:285=307